The Idea of Principles in Early Modern Thought

This fascinating collection provides case studies allowing the reader to appreciate how many and how varied are the ways in which the concept of a principle has been deployed and to what effect in the early modern period.
—Margaret Atherton, University of Wisconsin-Milwaukee, USA

This collection presents the first sustained examination of the nature and status of the idea of principles in early modern thought. Principles are almost ubiquitous in the seventeenth and eighteenth centuries: the term appears in famous book titles, such as Newton's *Principia*; the notion plays a central role in the thought of many leading philosophers, such as Leibniz's Principle of Sufficient Reason; and many of the great discoveries of the period, such as the Law of Gravitational Attraction, were described as principles.

Ranging from mathematics and law to chemistry, from natural and moral philosophy to natural theology, and covering some of the leading thinkers of the period, this volume presents ten compelling new essays that illustrate the centrality and importance of the idea of principles in early modern thought. It contains chapters by leading scholars in the field, including the Leibniz scholar Daniel Garber and the historian of chemistry William R. Newman, as well as exciting, emerging scholars, such as the Newton scholar Kirsten Walsh and a leading expert on experimental philosophy, Alberto Vanzo. *The Idea of Principles in Early Modern Thought: Interdiscplinary Perspectives* charts the terrain of one of the period's central concepts for the first time, and opens up new lines for further research.

Peter R. Anstey is Professor of Philosophy at the University of Sydney. He specializes in early modern philosophy with a particular focus on the philosophy of John Locke, experimental philosophy, and the philosophy of principles. He is the author of *John Locke and Natural Philosophy* (2011) and editor of *The Oxford Handbook of British Philosophy in the Seventeenth Century* (2013).

Routledge Studies in Seventeenth-Century Philosophy

For a full list of titles in this series, please visit www.routledge.com

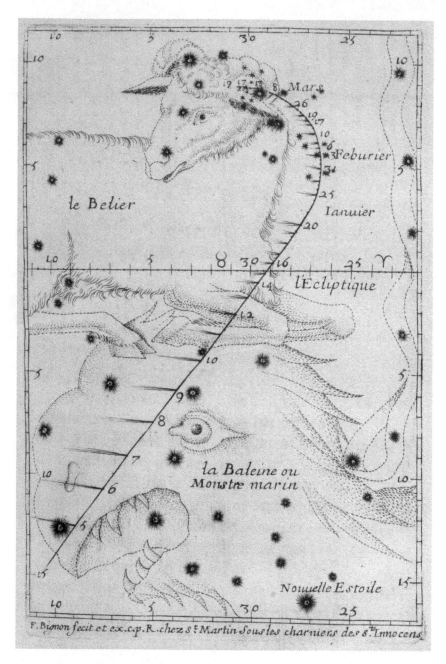

Frontispiece "le Belier—la Baleine ou Monstre marin," Observations on the passage of a comet made in December 1664 and January, February, and March 1665, in Pierre Petit, *Dissertation sur la nature des comètes,* Paris, 1665, p. 319

The Idea of Principles in Early Modern Thought

Interdisciplinary Perspectives

Edited by Peter R. Anstey

Routledge
Taylor & Francis Group

LONDON AND NEW YORK

First published 2017
by Routledge

2 Park Square, Milton Park, Abingdon, Oxfordshire OX14 4RN
52 Vanderbilt Avenue, New York, NY 10017

Routledge is an imprint of the Taylor & Francis Group, an informa business

First issued in paperback 2019

Library of Congress Cataloging-in-Publication Data
Names: Anstey, Peter R., 1962– editor.
Title: The idea of principles in early modern thought : interdisciplinary
    perspectives / edited by Peter R. Anstey.
Description: 1 [edition]. | New York : Routledge, 2017. | Series:
    Routledge studies in seventeenth-century philosophy ; 16 | Includes
    bibliographical references and index.
Identifiers: LCCN 2016053822 | ISBN 9781138211582 (hardback : alk.
    paper)
Subjects: LCSH: Philosophy, Modern—17th century.
Classification: LCC B801 .I34 2017 | DDC 190—dc23
LC record available at https://lccn.loc.gov/2016053822

ISBN: 978-1-138-21158-2 (hbk)
ISBN: 978-0-367-88425-3 (pbk)

Typeset in Sabon
by Apex CoVantage, LLC

Contents

Figures

Tables

Abbreviations

Bacon

SEH   *The Works of Francis Bacon*, 14 vols, eds. J. Spedding, R.L. Ellis, and D.D. Heath, London, 1861–1879. Cited as SEH, volume number, and page number.

Descartes

AT   *Œuvres de Descartes*, 11 vols, eds. C. Adam and P. Tannery, Paris: Vrin, 1996. Cited as AT, volume number, and page number.

Grotius

DJBP   *The Rights of War and Peace*, 3 vols, ed. R. Tuck, Indianapolis: Liberty Fund, 2005. English translation of *De jure belli ac pacis libri tres, in quibus jus Naturae & Gentium, item juris publici praecipua explicantur*, editio nova, Amsterdam: Apud Ioannem Blaev, 1667.

Guglielmini

Reflections   *Riflessioni filosofiche dedotte dalle figure de' sali: Espresse in un discorso recitato nell'Accademia Filosofica Esperimentale di Monsign. Arcidiacono Marsigli la sera delli 21. Marzo 1688*, Bologna, 1688.

Hobbes

Leviathan   *Leviathan,* ed. J.C.A. Gaskin, Oxford: Oxford University Press, 1996; 1st edn 1651.

Hume

Enquiry   *An Enquiry Concerning the Principles of Morals,* ed. T. L. Beauchamp, Oxford: Clarendon Press, 1998; 1st edn 1751.

Treatise   *A Treatise of Human Nature,* Vol. 1, eds. D. F. Norton and M. J. Norton, Oxford: Clarendon Press, 2007; 1st edn 1739–1740.

Leibniz

A   (1923—) *Sämtliche Schriften und Briefe,* Deutsche Akademie der Wissenschaften zu Berlin (eds.), Berlin: Akademie-Verlag. Cited as A, followed by series, volume, and page number.

AG   (1989) *Philosophical Essays,* eds. and trans. R. Ariew and D. Garber, Indianapolis: Hackett.

G   (1875–1890) *Die philosophischen Schriften,* ed. C. I. Gerhardt, 7 vols, Berlin: Weidmann.

GM   (1845–1863) *Leibnizens mathematische schriften,* ed. C. I. Gerhardt, 7 vols, Berlin: A. Asher.

LH   Leibniz Handschriften (manuscripts), Niedersächsische Landesbibliothek.

Spinoza

Opera   *Spinoza Opera,* 4 vols, ed. C. Gebhardt, Heidelberg: Carl Winter, 1925. Cited as *Opera,* volume number, page number, and, in places, line number.

TTP   *Tractatus theologico-politicus* in *Spinoza Opera,* vol. III, ed. C. Gebhardt, Heidelberg: Carl Winter, 1925; 1st edn 1670.

DPP   *Renati Des Cartes principiorum philosophiae Pars I & II, more geometrico demonstratae,* in *Spinoza Opera,* vol. I, ed. C. Gebhardt, Heidelberg: Carl Winter, 1925; 1st edn 1663.

Acknowledgments

All the chapters in this volume were presented at the colloquium on "Principles in Early Modern Thought" at the University of Sydney, 27–29 August 2014. The colloquium and the research for and production of this volume were funded by my future fellowship on "The Nature and Status of Principles in Early Modern Philosophy," Australian Research Council, grant number FT120100282. I should like to thank the Australian Research Council and the University of Sydney for their support, as well as the readers for the press for their constructive comments.

Introduction

Peter R. Anstey

Principles are almost ubiquitous in early modern thought. Consider these book titles from leading philosophers: Descartes, *Principia philosophiae*, 1644; Spinoza, *Renati Des Cartes principiorum philosophiae*, 1663; Newton, *Philosophiae naturalis principia mathematica*, 1687; Berkeley, *A Treatise Concerning the Principles of Human Knowledge*, 1710; Leibniz, *Principes de la nature et de la grâce fondé en raison*, 1714; Hume, *An Enquiry Concerning the Principles of Morals*, 1751; and Reid, *An Inquiry into the Human Mind on the Principles of Common Sense*, 1764. Simple title searches of books printed in England in Early English Books Online and Eighteenth-Century Collections Online confirm that it was not just the philosophers who were writing about principles. Among English language books printed from 1600 to 1700, well over 1,100 books have the term *principles* in their title. The number goes up significantly in the eighteenth century with about 3,500 titles using *principles*. And this does not count continental works in Latin, French, Italian, Dutch, or German. Nor does it include the enormous number of books that discuss or appeal to principles that lack the word in their title.

Principles were important in the early modern period, not just in philosophy but in many disciplines. There are books on principles of religion, politics, chymistry, navigation, music, geometry, arithmetic, husbandry, medicine, surgery, astronomy, astrology, writing, measurement, cosmology, fortifications, morality, law, grammar, surveying, rhetoric, painting, architecture, and the list goes on. So common is the notion of a principle, so often are they appealed to, so embedded are they in the basic conceptual resources of early modern thinkers, that very few modern day scholars have ever taken the time to look at them in a detailed and systematic way. This collection aims to provide a preliminary sounding of the subject by offering a selective overview that explores the nature and role of principles across a range of disciplines and individual thinkers.

The Centrality of Principles

Principles may have been appealed to very widely in the early modern period, but are they important? In some writers the notion is ill defined, in others principles serve merely a rhetorical purpose. Yet the notion of principles is central to a cluster of widely held doctrines and theories in the early modern period, and both the importance and the centrality of principles derive from this. For example, the most widely held account of knowledge acquisition of the period had principles at its heart. This view derived from Aristotle's *Prior Analytics* and *Posterior Analytics*. In summary, it held that any systematic body of knowledge derives ultimately from a set of determinate principles. Each body of knowledge or science has its own set of principles, and over and above these there are principles common to all the sciences. Once the principles of a science are established, one applies a method of demonstration, such as Aristotle's syllogistic, in order to generate the science from the principles. Thus, just as the truths of Euclidean geometry are derived from a finite set of postulates, so the sciences of natural philosophy or ethics can be derived from their own principles. This theory of knowledge acquisition applied to any discipline that aspired to be a science. To be sure, each of its constituents was contested in the seventeenth century: Locke, for example, contested the claim that demonstration requires the syllogistic.[1] And the theory underwent important changes and developments. Yet the basic contours of this approach to knowledge predominate at least until the end of the eighteenth century.[2]

In addition to the acquisition of knowledge, principles played a central role in early modern theories of explanation. Thus, in natural philosophy mechanical philosophers were committed to the principle that all natural phenomena can be explained by analogy with the functioning of machines. Again, the primary contours of early modern theories of matter, and their intersection with experiment, are best understood in terms of rival accounts of principles. This is definitively established in William R. Newman's chapter on the three different traditions of principles within early modern chymistry (see Chapter 3) and beautifully illustrated in Alberto Vanzo's chapter on the corpuscular theory of the Italian natural philosopher Domenico Guglielmini (Chapter 5).

Principles also play a central role in early modern moral philosophy, whether they be normative principles, such as the Golden Rule, or principles of human nature, such as humankind's natural propensity for sociability. Thus, for example, in his chapter, Kiyoshi Shimokawa shows how Hume opposed Grotius's founding of justice on a principle of humankind's disposition for sociability. Instead, Hume based his (somewhat narrow) conception of justice on the principle of the human passion of self-interest (see Chapter 10). As for the law, J. C. Campbell shows in his fascinating survey of the development of the law of equity from the pre-Reformation era until the late seventeenth century (Chapter 2), that principles had a central role to play here too. A series of contingent historical events, together with conceptual developments within the practice of the law, led, in the mid-seventeenth century, to the establishment of a foundational role for a set of principles of equity.

Furthermore, principles have a central and very prominent place in the views of many individual thinkers of the period, including leading philosophers, natural philosophers, jurists, and theologians. The most well-known and widely discussed example in philosophy is G. W. Leibniz. His Principle of Sufficient Reason and Principle of Plenitude are just two of a whole host of principles that were some of the constants in a constantly changing philosophy. And, as Daniel Garber shows in his chapter, from very early on the Principle of the Equality of Cause and Effect was at the centre of Leibniz's physics (see Chapter 8). Interestingly, Garber argues that part of the polemical context in which Leibniz's view of this principle was forged was his response to the necessitarianism of Spinoza. The question of Spinoza's conception of principles is taken up by Michael LeBuffe in his "The Principles of Spinoza's Philosophy" (see Chapter 6).

Another for whom principles are absolutely central is Jean Le Rond d'Alembert in the mid-eighteenth century. Take, for example, his article in the *Encyclopédie* on "Elements of the Sciences." This article uses the term *principle* (*principe*) forty-one times, which is indicative of the central role that principles play in his conception of the sciences. There, d'Alembert puts forward an architectonic version of the neo-Aristotelian theory of knowledge acquisition that gives principles pride of place. A few extracts will give the general tenor of his view:

> if we were able to observe without interruption the invisible chain that links all the objects of our knowledge, the *elements* of all sciences could be reduced to *one unique principle*, whose consequences would be the *elements* of each particular science.
>
> But what we must especially try to develop is the metaphysics of the propositions.
>
> Metaphysics, which has guided or should have guided its inventors, is nothing more than the clear and precise exposition of the general and philosophical truths on which the principles of the science were founded. Everything that is true, especially in those sciences that are based on pure reasoning, always has clear and perceptible principles, and consequently can be made clearly accessible to everyone. Indeed, how could the consequences be clear and certain if the principles themselves were obscure?
>
> (D'Alembert and La Chapelle 2011, underlining added)

A third philosopher for whom the notion of principle is absolutely central is the Scotsman Thomas Reid. Indeed, so important are principles to Reid that he develops a theory of principles.[3] And this brings us to a second feature of the nature and role of principles in the early modern period, namely, that a number of theories of principles with varying sophistication were developed.

Theories of Principles

Theories of principles were developed by early modern philosophers in order to account for their nature and uses. Many of these theories were heavily dependent on Aristotle's view of principles as found in his account of the acquisition of knowledge summarised earlier. Aristotle has (at least) three theses about principles. First, principles are propositions that must satisfy certain conditions: they are universal, necessary, and primitive (i.e. indemonstrable) propositions.[4] Second, principles are discovered by experience with the use of *nous*.[5] And third, each discipline has its own principles. Aristotle tells us that

> it is the business of experience to give the principles which belong to each subject. I mean for example that astronomical experience supplies the principles of astronomical science . . . Similarly with any other art or science.
>
> (*Prior Analytics*, 46a18–21, Aristotle 1984, 1: 73)[6]

In the early modern period this Aristotelian theory set the terms of reference for most theories of principles. However, there were many variations, and it is worth giving a thumbnail sketch of some of the more interesting theories in order to see how the approaches varied and changed over time. To that end, we will look briefly at the views of Isaac Barrow, Étienne Bonnot de Condillac, and Thomas Reid.

Isaac Barrow's theory of principles is the most elaborate English theory from the seventeenth century. It appears in his *Lectiones*, where he devotes two whole lectures to the theory of principles.[7] First, he discusses the nature of principles and then their different species. Along the way he critiques the theories of Aristotle, Proclus, Ramus, and the commentator on Archimedes, David Rivault (Barrow 1734: 118–128). Of particular interest in his own account are his typology of principles and the fine-grained distinctions that he makes between hypotheses, definitions, and axioms.

Barrow begins by claiming that "All the Truth, the Validity, and Evidence of every Science adheres to these inseparable Roots, and depends upon these unshaken Foundations" (Barrow 1734: 102), that is, principles. Principles are those propositions which are not themselves demonstrated from other propositions and which are the foundation of all other reasoning: "no Science can be erected but upon some Foundations. Such Propositions therefore are termed *Principles*" (Barrow 1734: 103). This is not to say that they are indemonstrable, but only that they are agreed upon. Taking the example of geometrical axioms, Barrow claims that they can be demonstrated from "some higher and more universal Science, as *Metaphysics*." And if there are no simpler axioms, then they can be formed from the definitions of their terms.[8] Thus, the principles of each science need not be self-evident or indemonstrable,[9] for while they may be fundamental to a particular science, they will be demonstrable outside of that science.

There is one particular principle, however, that Barrow regards as unique, namely, the Principle of Non-contradiction: "If this be not granted, all Reason, all Disquisition or Disputation about Truth is entirely vain and insignificant" (Barrow 1734: 107). Barrow goes so far as to claim that this principle is innate: "this Notion at least does seem to be immediately connate with our Minds, and implanted in us by God, together with the Faculty of Reasoning to which it is intrinsecally [sic.] annexed" (Barrow 1734: 107). John Locke famously argued that there are no innate principles in the first book of his *An Essay concerning Human Understanding*, but it is unlikely that he had Barrow in his sights. For, Barrow goes on to develop an account of hypotheses and definitions that are derived from experience and claims that there is no need for us to posit innate ideas, at least in the speculative sciences such as natural philosophy.[10]

Barrow also argues, given the structure of demonstrative reasoning based on principles, "it seems to follow, that every *Demonstration*, to make it effectually such, does in some sort suppose the Existence of God" (Barrow 1734: 109). This is merely an aside in Barrow's exposition of his theory of principles; however, it is illustrative of the close tie between theories of principles and natural theology in England in the period. In my own chapter in the collection (Chapter 9), I explore the meaning and use of the expression "principles of natural religion" in England, arguing that by the early eighteenth century, the newly discovered principles of Newtonian natural philosophy were being used as the principles that undergird the principles of natural religion themselves. The key players in this application of principles within natural religion were George Cheyne, William Whiston, and Samuel Clarke.

Returning to Barrow, he claims that there are three types of principles. First, there are hypotheses which "are Propositions assuming or affirming some evidently possible Mode, Action, or Motion of a Thing" (Barrow 1734: 128). Their foundation is sensory experience or experiment.[11] He gives numerous examples, such as a point may be carried from place A to place B. Second, there are definitions which are propositions "wherein a Name is imposed or ascribed from some possible Supposition of a Thing clearly resulting; which Supposition, being expressed in the Proposition, determines and circumscribes that Name" (Barrow 1734: 129). He gives the example of "The plane Figure which is produced from the Rotation of a Right Line may be called a *Circle*." Third, there are axioms, which are theorems in some science that are deduced from the definitions and hypotheses of a superior science but which are assumed without a proof in the subordinate science.[12] Barrow conceived of principles very much in the Aristotelian tradition as well as that of the Greek mathematicians such as Euclid and

Proclus. Nevertheless, he did use some examples from optics and was at-
tuned to the confirmatory role of experiment, even if there is no evidence of
him embracing experimental philosophy.

In this Barrow was not alone, for James Franklin shows (Chapter 1)
how "astonishingly successful" those early modern mathematicians were
who attempted to establish principles for the nature and behaviour of the
natural world without recourse to sustained observation and experiment.
Far from being a consequence of the new experimental philosophy, Franklin argues, the achievements of early modern mathematicians in discover-
ing many new principles is a legacy of the Aristotelian-Euclidean heritage.
One of the keys to this achievement was the application of *a priori* symmetry arguments in statics, hydrostatics, and the newly emerging field of
probability.

Nevertheless, there is abundant evidence that observation and experiment
did have an impact on the nature and status of principles in the period. In
her fascinating case study of cometary observations in France in 1664–1665
(Chapter 4), Sophie Roux shows how the various interpretations of these
observations ultimately turned on principles, principles which were used
to demarcate different positions in the debate. As for experiment, and, in
particular, the emergence of experimental philosophy, this is evident in the
theory of principles found in Condillac's *Traité des sistèmes* of 1749. While
Condillac's treatise is ostensibly on the nature of systems of knowledge, his
conception of a system is predicated on a theory about principles. The work
begins with a definition of a system: "[a] system is nothing other than the
arrangement of different parts of an art or science in an order in which they
all lend each other support and in which the last ones are explained by the
first ones" (Condillac 1982: 1). Notice here that the emphasis is on explanation rather than on demonstration.

He then goes on to define principles: "[p]arts that explain other parts
are called principles, and the fewer principles a system has the more perfect
it is." There are echoes here of the view of d'Alembert, discussed earlier,
and indeed, some scholars have claimed that in this passage Condillac is
paraphrasing an earlier similar statement of d'Alembert.[13] Be that as it may,
the salient feature of Condillac's discussion is his tripartite division of principles. The first class is abstract maxims, such as "Nothing cannot be the
cause of anything." These, according to Condillac, are the most popular
but also the most problematic. While he does not reject abstract ideas or
notions, he claims, following Locke, that systems built on abstract maxims
do not lead to new knowledge.

The second class is suppositions or hypotheses. These are also of little use
as principles. In fact, they can be positively dangerous, though Condillac
does not dismiss hypotheses outright as having no role in the acquisition
of knowledge. For, they can assist us to find the true principles on which a
system should be based.

The third class is that on which we should found our systems: "[t]he only proper scientific principles are established facts." "True systems, the only ones that merit that name, are based on principles of this last kind" (Condillac 1982: 3). Condillac gives the example of universal gravity as one of these principles and, while he is not entirely clear on the point, he seems to imply that the inverse square law of gravitational attraction discovered by Isaac Newton is the determinate principle that he has in mind. For Condillac, the new approach to natural philosophy exemplifies the correct method. He claims some natural philosophers "stick simply to collecting phenomena because they recognize that they should take in the effects of nature and discover their interdependence before formulating explanatory principles" (Condillac 1982: 10). He goes on to argue that analogous principles can be found that enable us to construct systems for politics and for art.

Condillac's theory of principles is set within his account of systems of knowledge. The model is not Euclidean geometry or the syllogistic, but the methods employed by experimental philosophers such as Newton. An entirely different approach is found in the theory of the Scots philosopher Thomas Reid. Reid's *Essays on the Intellectual Powers of Man* is one of the longest and most ambitious works on principles in English from the end of our period. Reid gives a long history of the notion and a critique of the views of the likes of Descartes and, especially, Locke on principles, and he advances his own theory. He devotes separate chapters to necessary principles and contingent principles, and in the latter he posits a principle of the uniformity of nature that provides him with his own account of the justification of induction.[14]

According to Reid the common sense is that faculty by which we come to grasp self-evident principles. His theory of principles is set out in a series of propositions. First, "all knowledge got by reasoning must be built upon first principles" (Reid 2002: 454). Second, "some first principles yield conclusions that are certain, others such as are probable, in various degrees" (Reid 2002: 455). Third, we need to determine what the first principles of knowledge actually are. Fourth, and finally, Reid acknowledges that there is sometimes disagreement as to what the first principles are, and he proposes remedies for this. He then goes on to list what he takes to be determinate contingent and necessary principles derived by common sense. Among the contingent principles are some that pertain to the mind: "the existence of everything of which I am conscious" and "[t]he thoughts of which I am conscious, are the thoughts of a being which I call *myself*, my *mind*, my *person*" (Reid 2002: 470, 472). And it is in this context that Reid lists his principle of uniformity of nature, namely, "in the phænomena of nature, what is to be, will probably be like to what has been in similar circumstances" (Reid 2002: 489). As for necessary principles, Reid lists logical axioms, mathematical axioms, and, interestingly, axioms of taste, and first principles of morals.[15]

Classification of Principles

Reid's theory of principles is very much centred on a propositional conception of principles. Yet the term *principle* had a variety of meanings beyond simply foundational propositions. A good entrée the semantic range of "principle" is John Harris's entry under "Principle" in his *Lexicon Technicum* of 1708. He lists six different meanings of "principle," describing it as

> a Word very commonly and variously used; sometimes it signifies the same as a Maxim, an Axiom, or a good Practical Rule of Action . . . Sometimes it signifies a Thing Self-evident and as it were Naturally known . . . Sometimes it hath the same sense with Rudiments or Elements; as when we say, the *Principles* of *Geometry* . . . And in Chymistry particularly, 'tis taken for first Constituent and Component Particles of all Bodies out of which they are made.
>
> (Harris 1708, sig. 5U2r)

Then after discussing various of the theories of chymical principles, Harris turns to a more fundamental conception: "That in general may be called a Principle which is the first cause of any Things *Existence*, or *Production*, or of its becoming *Known* to us." He then gives the example of the Aristotelian principles of earth, air, fire, and water; the Epicurean principles of magnitude, figure, and weight; Robert Boyle's view that after the creation "the Mechanical Principles, Matter, Motion, and Rest, are Principles sufficient to solve all the Phaenomena of Nature" (Harris 1708: sig. 5U2v); and, finally, the Cartesian principles, namely, the three forms of matter.

Interestingly, Harris omits to mention laws of nature and explanatory principles, such as the Principle of Sufficient Reason. Clearly the first four types of principle are propositional in nature and are related to the Aristotelian theory of knowledge acquisition. They are the sorts of principles that feature in a demonstrative science or *scientia*. The others, by contrast, are entirely different insofar as they are things in the world. Furthermore, as Kirsten Walsh argues in her chapter, none of the types of principle mentioned in Harris's article are equivalent to the way in which Isaac Newton used the notion (Chapter 7). Surprisingly, Walsh's chapter is the first-ever sustained treatment of Newton's concept of principle.

It is clear, therefore, that there is no single or fundamental notion of what a principle is. If we are to classify principles, it does seem that we need a pluralist approach: there is more than one concept of principle, and it may well be that the different types only loosely resemble each other. We might want to call it a family resemblance, or prototype concept.

This, in turn, brings us to the question of providing a classification of principles. There are various ways into this problem. One place to start might be with actors' categories; after all, when taken together, these constitute a simple historical classification system. Thus, we have principles of reason, principles of religion, principles of matter, principles of chymistry, principles of hydrostatics, principles of optics, moral principles, and so on. Another approach might be to examine the roles that principles play in early modern thought. Some provide epistemic foundations for systems of knowledge, others are explanatory, and some have rhetorical uses.

Alternatively, we might also classify by mode of epistemic access to principles. Before Locke's critique of the view in Book One of his *Essay* many claimed that principles were innate. For Reid they are self-evident, whereas for Locke and Condillac they are acquired by experience. Still others claimed that some principles are derived from more fundamental ones. Moreover, for some philosophers, such as Pascal, the mode of epistemic access to principles is tied to their faculty psychology (see Chapter 9).

Notwithstanding the value of each of these ways of classifying principles, I have taken a different approach. Having surveyed a host of determinate principles from the early modern period, it became apparent that there is a natural fault line among them. This is between propositional and ontological principles. Using this I have developed the following classification shown in Figure I.1.

The tree diagram in Figure I.1 is best expounded with some concrete examples. Among the propositional principles, the category of logical principles aims to capture a class of principles that were commonly deployed in early modern logic textbooks. While the precise divisions and nomenclature vary from one author to another, the general divisions are common enough, and the whole approach is set within the Aristotelian theory of knowledge acquisition as outlined earlier. Taking Thomas Blundeville's *The Arte of Logick*, 1617, as a typical example, we find first a definition of principles and then a clearly articulated division. Principles are defined as "true Propositions, having credit of themselves, and need no other proofe" (Blundeville 1617: 163). As for divisions, the basic dichotomy is between speculative and practical principles, which apply in different domains of knowledge. Thus, natural philosophy and mathematics are based on speculative principles, such as "The whole is more than his part" (Blundeville 1617: 163). Practical principles are the foundation of morals. They include "God is to be honored and obeyed" and "Justice is to be embraced" (Blundeville 1617: 164).

Over and above the speculative/practical dichotomy, there is a division between general and proper principles. General principles apply to many sciences, such as "if equall be taken from equall, equall doe remain." Proper principles are those that belong to one science only, such as the geometrical principle that "a line to be a length without breadth" (Blundeville 1617: 164). Proper principles are then further divided into dignities, maxims, and positions. Positions are divided into definitions, suppositions, postulata, and suppositions assumed. There is no need to elaborate on these finer distinctions, though it does help to see them set out diagrammatically (see Figure I.2), for this captures the sophistication of this category of propositional principle within Figure I.1 shown earlier.

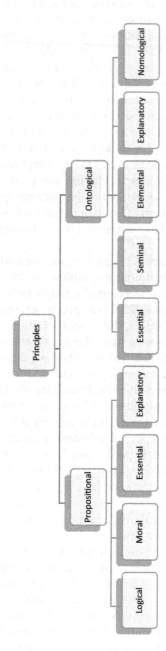

Figure I.1 Classification of early modern principles

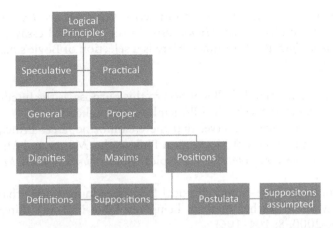

Figure I.2 Thomas Blundeville's classification of principles in *The Arte of Logick*, 1617

To this category of logical principles we might add what some philosophers called principles of reason, such as the Principle of Non-contradiction, so highly esteemed by Barrow. Leibniz regarded this principle and the Principle of Sufficient Reason as the two great principles upon which all our reasoning is founded.[16] Let us turn then to the Principle of Sufficient Reason. In *Monadology* §32 Leibniz claims, "There can be found no fact that is true or existent, or any true proposition, without there being a sufficient reason for its being so and not otherwise" (Leibniz 1969: 646). Clearly he regards this as a principle of reason. However, careful study of his appeals to this principle suggests that it is better classified as an explanatory propositional principle.

Turning to moral principles, on Blundeville's classification of logical principles these are practical, as distinct from speculative, principles. However, there are a number of reasons for treating moral principles separately from the scholastic distinction between speculative and practical principles. First, some important treatments of moral principles in the early modern period make no reference at all to the speculative/practical dichotomy. Second, and more importantly, the speculative/practical dichotomy itself underwent significant changes in the period, such that natural philosophy came in many quarters to be regarded as a practical science—experimental natural philosophy—and some philosophers, such as John Locke, came to regard morality as being analogous to the archetypal speculative science, mathematics.[17] In effect, for some philosophers, natural and moral philosophy switched sides across the speculative–practical divide.[18]

Among the ontological principles is the category of explanatory principles. Two good examples of principles in this category are matter and motion in Boyle's mechanical philosophy. That Boyle considered them as principles is most evident in his short work titled *Of the Excellency and Grounds of the Mechanical Hypothesis*. In this extended essay Boyle uses the term *principle* fifty-four times. Here is a selection of Boyle's main points in the essay:

I. The *first* thing that I shall mention to this purpose, is the Intelligibleness or Clearness of Mechanical Principles and Explications.

II. In the next place I observe, that there cannot be *fewer* Principles than the two grand ones of Mechanical Philosophy, *Matter* and *Motion*.

III. Nor can we conceive any Principles more *primary*, than *Matter* and *Motion*.

IV. Neither can there be any Physical Principles more *simple* than Matter and Motion; neither of them being resoluble into any things. (Boyle 1999–2000, 8: 104–106)

It might be thought that calling matter and motion explanatory is to conflate the *explanans* with the explanation itself: that to say that motion itself is a principle is shorthand for the claim that propositions about motion are explanatory. However, to take this line is to overlook an interesting feature of the role of principles in speculative theories such as the corpuscular hypothesis which provide purportedly reductive explanations of natural phenomena, as Boyle intimates in point IV. In Boyle's corpuscular philosophy, matter and motion are principles of being—"Physical Principles"—insofar as all phenomena just are variations of matter and motion, and the latter are not reducible to anything else. And, moreover, all explanations of natural phenomena ultimately appeal only to matter and motion. Matter and motion on this account might, therefore, be better regarded as hybrid principles with both an ontological and a propositional side. This is certainly the case for another of Boyle's principles, the seminal principles that are responsible for the replication of form. They are theoretical entities whose existence is assumed in order to explain the formation of crystals and the generation of plants and animals. For seminal principles, the main argument for their existence is their explanatory role.[19]

Finally, we may ask to what extent the classification of principles in early modern thought forms a taxonomy, that is, to what extent is there a hierarchy of principles. There is no doubt that some early modern thinkers proposed a hierarchical ordering of principles. Indeed, in some cases such a hierarchy was an important constituent of their theory of principles. Leibniz, for example, considered the two most fundamental principles to be the Law of Non-contradiction and the Principle of Sufficient Reason. Condillac and d'Alembert both claimed that there was one super principle from which all the sciences can be derived. They were also committed to a view that might be described as Principle Minimalism, namely, that the fewer principles on which a science is founded, the more fertile those principles are. Their objective in their own respective fields of inquiry was to discover the one master principle from which the sciences could be derived.[20]

Furthermore, the question of the hierarchical ordering of principles is intimately tied to the ranking of the sciences. For some, like Barrow in the 1660s and Samuel Formey of the revived Berlin Academy in the 1740s, the principles of metaphysics were of a higher order than those of any other science.[21] Yet it is fair to say that most thinkers who embraced the notion of principles, whether propositional or ontological, had no explicit theory of principles and were primarily concerned with those determinate principles that pertained to their disciplines of choice. It seems, therefore, that a flat, rather than a hierarchically ordered and overly reticulated classification of principles, such as the one proposed here, has greater utility in navigating the terrain of early modern principles.

Whatever the merits of this classificatory scheme, it is clear that the task of mapping the various types and roles of principles in the early modern period is complex and far-reaching. This volume does not aspire to provide a comprehensive treatment of the topic but, rather, offers a sampling of studies across a range of disciplines that it is hoped will be a stimulus to further research.

Notes

1 As Locke famously put it, "God has not been so sparing to Men to make them barely two-legged Creatures, and left it to *Aristotle* to make them Rational," *Essay* IV. xvii. 4, Locke 1975: 671.
2 For further discussion, see de Jong and Betti 2010.
3 See Reid 2002: 452–526.
4 *Posterior Analytics*, 73a24–27, Aristotle 1984, 1: 118.
5 *Posterior Analytics*, 100b5–14, Aristotle 1984, 1: 166.
6 For a full exposition of Aristotle's theory, see McKirahan 1992.
7 The lectures were written in 1664 but only published in 1683. An English translation appeared in 1734. See Barrow 1683 and 1734.
8 Barrow 1734: 105.
9 Barrow 1734: 106.
10 Barrow 1734: 115.
11 Barrow 1734: 133.
12 Barrow 1734: 130.
13 See Richard Schwab in d'Alembert 1995: 22, note 30 and W. R. Albury in Condillac 1980: 10.
14 See Anstey 1995.
15 Reid 2002: 492, 494.
16 See *Monadology* §§31 and 32, Leibniz 1969: 646.
17 *Essay* III. vi 16, Locke 1975: 516.
18 For the shift of natural philosophy from a speculative to a practical science, see Anstey and Vanzo 2012.
19 For further discussion, see Anstey 2002.
20 See Anstey forthcoming.
21 See Barrow 1734: 105 and Formey 1746: sig.)()(2r.

Bibliography

Anstey, P. R. (1995) "Thomas Reid and the justification of induction," *History of Philosophy Quarterly*, 12: 77–93.

———. (2002) "Boyle on seminal principles," *Studies in History and Philosophy of Biological and Biomedical Sciences*, 34: 597–630.

———. (forthcoming) "Principled Enlightenment" In eds. M. Lloyd, M. Sharpe and G. Boucher, *Rethinking the Enlightenment*. Lanham: Lexington Books.

Anstey, P. R. and Vanzo, A. (2012) "The origins of early modern experimental philosophy," *Intellectual History Review*, 22: 499–518.

Aristotle. (1984) *The Complete Works of Aristotle*, 2 vols, ed. J. Barnes, Princeton: Princeton University Press.

Barrow, I. (1683) *Lectiones Habitae in Scholis Publicis Academiae Cantabrigiensis*, London.

———. (1734) *The Usefulness of Mathematical Learning Explained and Demonstrated: Being Mathematical Lectures Read in the Publick Schools at the University of Cambridge*, London.

Berkeley, G. (1710) *A Treatise Concerning the Principles of Human Knowledge*, London.

Blundeville, T. (1617) *The Arte of Logick*, London.

Boyle, R. (1999–2000) *The Works of Robert Boyle*, 14 vols, eds. M. Hunter and E. B. Davis, London: Pickering and Chatto.

Condillac, É. Bonnot de. (1749) *Traité des sistèmes*, The Hague.

———. (1980) *La Logique: Logic*, trans. W. R. Albury, New York: Abaris Books.

———. (1982) *Philosophical Writings of Etienne Bonnot, Abbe de Condillac*, trans. F. Philip, Hillsdale: Lawrence Erlbaum Associates.

D'Alembert, J. Le Rond. (1995) *Preliminary Discourse to the Encyclopedia of Diderot*, trans. R. N. Schwab, Chicago: University of Chicago Press.

D'Alembert, J. Le Rond and La Chapelle, J. de. (2011) "Elements of the sciences" in eds. D. Diderot and J. Le R. d'Alembert 2011. http://hdl.handle.net/2027/spo.did2222.0001.133 (Accessed: 6 October 2016).

de Jong, W. R. and Betti, A. (2010) "The classical model of science: A millennia-old model of scientific rationality," *Synthese*, 174: 185–203.

Descartes, R. (1644) *Principia philosophiae*, Paris.

Diderot, D. and d'Alembert J. Le Rond, eds. (1751–1772) *Encyclopédie ou Dictionnaire raisonné des sciences, des arts et des métiers*, 17 vols, Paris.

———. (1751–1772) *Encyclopédie, ou dictionnaire raisonné des sciences, des arts et des métiers, etc.*, eds. Denis Diderot and Jean le Rond d'Alembert, Chicago: University of Chicago, ARTFL Encyclopédie Project (Spring 2016 Edition), Robert Morrissey and Glenn Roe (eds.), http://encyclopedie.uchicago.edu/.

———. (2011) *The Encyclopedia of Diderot & d'Alembert Collaborative Translation Project*, Ann Arbor: Michigan Publishing, University of Michigan Library. http://hdl.handle.net/2027/spo.did2222.0001.133 (Accessed: 6 October 2016)

Formey, S. (1746) "Preface" to *Histoire de l'académie royale des sciences et des belles lettres de Berlin*, Berlin.

Harris, J. (1708) *Lexicon Technicum*, vol. 1, London.

Hume, D. (1998) *An Enquiry Concerning the Principles of Morals*, ed. T. L. Beauchamp, Oxford: Clarendon Press; 1st edn 1751.

Leibniz, G. W. (1714) "Principes de la nature et de la grâce fondé en raison," *L'Europe savante*, 6: 101–123.

———. (1969) *Philosophical Papers and Letters*, 2nd edn, ed. L. E. Loemker, Dordrecht: D. Reidel.

Locke, J. (1975) *An Essay concerning Human Understanding*, ed. P. H. Nidditch, Oxford: Clarendon Press; 1st edn 1690.

McKirahan, R. D. Jr. (1992) *Principles and Proofs: Aristotle's Theory of Demonstrative Science*, Princeton: Princeton University Press.

Newton, Sir I. (1687) *Philosophiae naturalis principia mathematica*, London.

Reid, T. (1997) *An Inquiry into the Human Mind on the Principles of Common Sense*, ed. D. R. Brookes, Edinburgh: Edinburgh University Press; 1st edn 1764.

Reid, T. (2002) *Essays on the Intellectual Powers of Man: A Critical Edition*, ed. D. R. Brookes, Edinburgh: Edinburgh University Press; 1st edn 1785.

Spinoza, B. (1663) *Renati Des Cartes principiorum philosophiae Pars I & II, more geometrico demonstratae*, Amsterdam.

1 Early Modern Mathematical Principles and Symmetry Arguments

James Franklin

Mathematics is the home ground of principles. Since Euclid, mathematics has been the model of a body of knowledge organized as a deductive structure based on self-evident axioms. The prestige of that model was highest in early modern times, lying after the vast extension of the realm of mathematics in the Scientific Revolution but before the discovery of non-Euclidean geometries and the foundational crises of the late nineteenth century. When the Jesuit missionaries chose Euclid as the first book to be translated into Chinese (Engelfriet 1998) and when Spinoza offered to exhibit ethics as a system demonstrated *more geometrico* from definitions and simple self-evident axioms, they paid tribute to the place of the Euclidean model at the summit of intellectual achievement.

Seventeenth-century thinkers who applied mathematics to physics—Stevin, Galileo, Descartes, Pascal, Huygens, Barrow, Newton, and Leibniz, to name only the most prominent—were captivated by a model of applied mathematics, apparently realized in Euclid's *Optics* and Archimedes's mechanics, according to which pure thought could establish principles for empirical reality. Their ambition stood at the opposite extreme from the Baconian style of "experimental philosophy," which recommended generalization from carefully collected facts (Domski 2013).

Their hopes were not fully realized because of such awkward facts as the contingency of the constant of gravitational attraction. But their vast optimism concerning the power of mathematical reasoning and the possibilities of understanding reality through pure symmetry arguments proved astonishingly successful and created modern physics.

The Background: Aristotle's *Posterior Analytics* and Euclid's *Elements*

The model of mathematics as a deductive structure of propositions proved from self-evident truths was backed by the two greatest possible authorities, Aristotle and Euclid, and was in any case open to inspection if one followed the proofs in Euclid oneself. Virtually all thinkers accepted it without question.

The essentials of the model were laid down in Aristotle's *Posterior Analytics*. A true or fully developed science should demonstrate its truths by syllogistic deduction from self-evident first principles, explaining why the truths of the science must be as they are. The first principles should be simple enough to be evident to the pure light of reason, or *nous* (Latin *intellectus*), a divinely granted faculty of the soul capable of grasping necessities.

If to modern ears that seems a promise too good to be true, for early modern thinkers it was confirmed by the existence and obvious success of Euclidean geometry, a science that seemed to conform exactly to Aristotle's model (McKirahan 1992: Ch. 12). Euclid's *Elements* begins with twenty-three definitions, such as

1.   A *point* is that which has no part . . .
15. A *circle* is a plane figure contained by one line such that all the straight lines falling upon it from one point among those lying within the figure equal one another . . .

These are meant to be not arbitrary stipulations but contentful truths about real geometrical objects. Though Euclid omits any philosophical commentary, modern writers rightly emphasize that to appreciate the definition of a circle, for example, one needs an act of insight to understand that something uniformly round does, in fact, consist of all the points equidistant from a centre (Lonergan 1997: Ch. 1). There follow five postulates or axioms, such as

1.   To draw a straight line from any point to any point . . .
5.   That, if a straight line falling on two straight lines makes the interior angles on the same side less than two right angles, the two straight lines, if produced indefinitely, meet on that side on which are the angles less than the two right angles.

The way in which the Fifth Postulate is posed is such as to make it appear as self-evident as possible, despite the worries, expressed since ancient times, as to whether anything involving lines "produced indefinitely" could be truly self-evident. Indeed, behind the scenes there were considerable doubts among the best mathematicians as to whether Euclid's logic was watertight (De Risi 2016), but from the perspective of outsiders the Euclidean structure was monolithic and impregnable.

There follow a large number of propositions or theorems derived logically from the postulates, beginning with

I.1 To construct an equilateral triangle on a given finite straight line.

These truths were taken to be necessary truths about the real space we live in, even though Aristotle, to some extent, adopts the Platonist idea that the geometer's lines and circles are assumed perfect and hence are idealizations of the real shapes of rulers and wheels. Any doubts about the principles of geometry and about its direct applicability to the physical world were suppressed in the seventeenth century. Even as contra-suggestible a mind as Hobbes, initially doubtful about Euclid, was convinced by the proofs. Virtually the only doubter was the Chevalier de Méré, a "man of the world" without intellectual standing. When he wrote to Pascal with some puzzles about probability which he took to show that mathematics was self-contradictory, Pascal wrote to Fermat that de Méré's incompetence in mathematics was clear from his belief that space was atomic (contrary to the infinite divisibility of space in Euclid) (Pascal 1964–1970, 2: 1142; embarrassingly, it turned out that whether space is continuous or atomic is not provable; see Franklin 1994).

To appreciate the early modern understanding of mathematical principles, it is essential to put to one side certain more recent philosophical views which make it appear implausible. Right or wrong, certain contemporary received ideas about mathematics and its relation to reality impede a clear view of early modern assumptions. We discuss briefly the two main ones, which both assume a divorce between mathematics and reality that was not part of, or even one of the possibilities considered by, the early modern view of mathematics.

First, the legacy of four hundred years of experimental science is an assumption that substantial scientific knowledge is contingent and must be established by experiment; hence, the role of mathematics is as a "theoretical juice extractor," a collection of methods to codify the generalizations arising from experiment and to enable predictions from them. Mathematics itself is not seen as directly about reality. As Einstein put it, "[a]s far as the propositions of mathematics refer to reality, they are not certain; and as far as they are certain, they do not refer to reality" (Einstein 1954: 233; argued against in Franklin 2014: Ch. 5).

The second and related contemporary obstacle to understanding early modern views on mathematics is the oscillation in post-Fregean philosophy of mathematics between nominalism and Platonism. Nominalist philosophies that take mathematics to be not about anything but a language of science, or a collection of tautologies, or manipulation of formal symbols, or an investigation of what follows from arbitrary axioms, stand against Platonism, which regards mathematics as about "abstract objects" such as sets and numbers (or idealized geometrical objects like perfect circles). (The dichotomy explained and criticized in Franklin 2014: Ch. 7; a textbook on mathematical proof from an Aristotelian point of view is Franklin and Daoud 2011.)

Nothing could be further from the early modern conception of mathematics, which was neither nominalist nor Platonist but Aristotelian (or "moderate") realist—mathematics was said to be the "science of quantity," with quantity understood as a property of extended physical things (e.g., Barrow

1734: 10–15; further in Barrow 1860 [1664]: lectiones 4–8; Gillette 2009; *Encyclopaedia Britannica* 1771: article 'Mathematics', vol. 3: 30–31; Mancosu 1996: 16, 35–37, 56, 88; Jesseph 1993: Ch. 1. Gassendi and Hobbes do have nominalist tendencies: Sepkoski 2005, Pycior 1987; Dear 1995 puts a constructivist interpretation on the topic; eighteenth-century continuations in Franklin 2006). Quantity, one of the basic Aristotelian categories, comes in *discrete* (studied by arithmetic) and *continuous* (studied by geometry). (For the bridging of this gap in the sixteenth and seventeenth centuries, see Malet 2006; Neal 2002a.) As Barrow writes,

> [i]t is plain the *Mathematics* is conversant about two Things especially, viz *Quantity* strictly taken, and *Quotity*; or if you please, *Magnitude* and *Multitude*. By others they are called *Continued* and *Discontinued Quantity* . . .
>
> (Barrow 1734: 10)

There are also *subordinate* or *mixed* sciences, such as music (subordinate to arithmetic) and optics and astronomy (subordinate to geometry). Barrow says,

> But because both *Magnitude* and *Multitude* may be considered in a double respect; *viz.* either as they are mentally separated, or abstracted from all Matter, material Circumstances, and Accidents; . . . or as they inhere in some particular Subject, and are found conjoined with certain other physical Qualities, Actions and Circumstances: Hence arises the Division of *Mathematics* into *Pure* or *Abstract*, and *Mixed* or *Concrete*.

Being concrete does not impede the certainty of the mixed mathematical sciences. They are again conceived as sciences of real but necessary aspects of the world. The search for self-evident principles for the subordinate sciences is the most interesting aspect of the topic and we will return to it.

The Aristotelian-Euclidean background to seventeenth-century mathematics means that mathematics does not fit at all into the old picture of the Scientific Revolution as a revolt against scholastic obfuscation in favour of empiricism and experimentalism. That model, even if oversimplified, does make some sense as applied to the natural sciences, especially chemistry, biology and the more empirical parts of physics. Mathematics, including applied mathematics, is otherwise. As Derek Whiteside put it after his extensive investigation of early British mathematics,

> I have neglected a prevailing fashion which sees mathematics as a mere handmaiden to the sciences, and the 17th century scientific achievement as a revolution in which scientific thought was freed from the largely sterile dominance of scholastic authority under a universal guiding principle of the primacy of theory induced from observed instances in phenomena.
>
> (Whiteside 1961: 180)

The story of mathematics is more like an extension of scholasticism than a retreat from it. Where the old scholastics had been excessively modest about the possibilities of reducing the contingent physical world to quantitative order and demonstration, the mathematicians showed that it could be done by doing it. Early modern applied mathematics is the pursuit of the scholastic vision by other means.

Mathematical Developments in the Seventeenth Century

Before the Scientific Revolution was a revolution in *science*, it was a mathematical revolution. Copernicus's astronomy was purely geometrical, an attempt to rearrange the geometry of the heavens without any considerations of the causes of celestial motions. The same is largely true of the successful parts of Kepler's astronomy, though he did add some speculations about magnetic attractions as a possible cause of orbits. John Dee's preface to the first English Euclid, of 1570, describes a culture of some thirty useful mathematical sciences like perspective, navigation, astrology, and statics (Euclid [1570]: Billingsley's Preface). The important mathematical achievements around 1600 of Viète, Stevin, and Harriot in fields like algebra, statics, navigation, and logarithms preceded the first new experimental results of Galileo and Bacon (Franklin 2000).

Then the seventeenth century itself saw a flowering of mathematical genius—much more so than in the two adjacent centuries. Galileo, Descartes, Fermat, Pascal, Huygens, Newton, and Leibniz were mathematicians of the first rank; Kepler, Cavalieri, Barrow, Wallis, Jacob Bernoulli, and others, important contributors. All were trained in Euclid's model of how to do mathematics. All had, so to speak, a Euclid baton in their knapsack; they harboured the ambition of becoming the new Euclid of their fields. In many but not all cases, as we will see, that involved explicitly laying down definitions and axioms for new fields and deriving theorems from them.

What they hoped for is expressed in a confident, even impudent, exchange about theory and experiment from Galileo's *Dialogue Concerning the Two Chief World Systems*:

> SIMPLICIO: So you have not made a hundred tests, or even one?
> SALVIATI [for Galileo]: Without experiment, I am sure that the effect will happen as I tell you, because it must happen that way. (2nd day, Galileo 1953: 145)

The major mathematical developments in the seventeenth century included the following:

- Number theory, algebra, and Cartesian geometry
- The science of the continuous: from Cavalieri's infinitesimals and Descartes's tangents to the calculus of Newton and Leibniz

- Kepler's and Newton's laws in astronomy
- Galileo's law of free fall and Newton's laws of motion and gravity in mechanics
- Laws of proportion in physics
- Mathematical probability theory.

The seventeenth-century plan was to organize these as far as possible as Euclidean structures founded on self-evident principles. But the principles were not to be found in Euclid. We will examine a number of these areas of mathematics and consider the attempts to provide them with principles. While success was not total, especially in exhibiting principles that were fully self-evident, the overall success of the project was remarkable.

Before proceeding, a word on terminology: writers of the time used more or less interchangeably the Latin "principles/principia," sometimes "postulates" and "laws/leges," and the Greek "axioms/axiomata." Thus, Newton has "Principia" in the title of his most celebrated book but then "axiomata sive leges motus." There is, however, a tension between the traditional language of "principles" and "axioms," suggesting absolute necessity, and the new language of "laws," meant to suggest obedience to the commands of God (Oakley 1961). If God has to command something, presumably it could be otherwise and hence cannot have the necessity that gives rise to self-evidence. A voluntarism with regard to divine action is a departure from Aristotelian orthodoxy. The issue was not directly joined (unlike in the Ockhamist voluntarism of earlier centuries), but in thinkers like Barrow and Newton, one must be aware of a tension that is suppressed (Malet 1997).

Number Theory: Free of Principles?

It is something of a mystery why the central part of pure mathematics, number theory, has rarely been seen to need principles or axioms in the way that geometry does. Euclid does number theory but mentions no axioms. Undoubtedly one reason is that the basic truths of arithmetic just look obvious, lacking, for example, the subtleties of continuous ratios and infinite divisibility that afflict geometry. That is the point of view even in the unusually detailed account of discrete quantity by Wallis in his *Mathesis universalis*. The basic truths about natural numbers can just be written down and are obvious (Wallis 1657: Ch. 10; Whiteside 1961: 186–188). It is arguable that number theory was seen as so general that it needed no axioms, while geometry's need for subject-specific axioms was comparatively a defect (De Risi 2016, §1.2.4).

In the late nineteenth century, Peano did successfully axiomatize arithmetic, revealing the crucial role of the principle of mathematical induction. Almost all truths about numbers need to be proved using this principle (which says that if a property is true of the number 1, and its being true of any number n implies its truth for $n + 1$, then it is true of all numbers).

While no such principle was thought of before Peano as an axiom of number theory, the use of mathematical induction as a method for rigorously demonstrating arithmetical truths did become established in the seventeenth century, after isolated earlier uses of lesser degrees of explicitness. Pascal uses the method fully explicitly in his *Treatise on the Arithmetical Triangle*, written around 1654. He writes of a complex proportion among consecutive numbers in any row ("base") of Pascal's triangle (as we now call it):

> The first, which is self-evident, that this proportion is found in the second base . . . The second, that if this proportion is found in any base, it will necessarily be found in the following base. Whence it is apparent that it is necessarily in all the bases. For it is in the second base by the first lemma; therefore by the second lemma it is in the third base, therefore in the fourth, and to infinity.
>
> (Pascal 1665: twelfth consequence)

Pascal plainly understands a very general template that is applicable to the proof of any arithmetic statement that can be expressed as a proposition about an arbitrary number n.

Fermat's Method of Infinite Descent, which is logically equivalent to the principle of mathematical induction, was also understood by Fermat as a very general technique that relies on the structure of the natural numbers to prove arithmetical statements (Bussey 1918). While it is not usual to call a method a principle, the reliance of the method on the essential linear structure of the numbers makes it close to an axiom.

While there was debate about the status of algebra, the normal perspective on it was as just a method rather than as a body of knowledge in its own right which might require principles (for debates on algebraic reasoning with symbols, see Pycior 1987).

Infinitesimals and Calculus

In the area of mathematics that deals with the continuous—involving limits, rates of change, tangents to curves, areas and volumes of curved figures—the seventeenth century had inherited the cumbersome "method of exhaustion" of Eudoxus and Archimedes. It permitted rigorous answers to be derived in a limited range of questions about areas and volumes, but was a barrier to progress because it ruled out as unrigorous methods that seemed intuitively fruitful and justified. Four centuries of difficult mathematics since then have shown that the field is a quagmire where unrigorous methods can easily lead to disaster, so the Greeks had a point in demanding absolute rigour. But in the meantime, it was natural to invest hope in intuitively attractive new "shortcut" methods and to look for new principles which would undergird them.

The essential idea of these methods was to use indivisibles or infinitesimals, that is, quantities that were larger than zero but could be considered less than any positive number. After a significant start by Kepler in calculating the volumes of wine barrels by thinking of them as made up of infinitesimal slices (Kepler 1615; Struik 1969: 192–197; Galileo's beginning on infinitesimals in Bascelli 2014), the first of these methods that claimed rigour, and the first to show both the suggestiveness, calculatory power, and logical slipperiness of all such attempts, was Cavalieri's Principle (full account in Andersen 1985; on the century's appreciation of the logical problems with infinitesimals, see Jesseph 1989; full account in Jullien 2015).

Cavalieri's Principle applied to the classic problem of finding areas or volumes of curved figures.

In the volume case, as in Figure 1.1, the principle states that if two solids lie between two parallel planes (top and bottom in Figure 1.1) and if every plane parallel to those two intersects the two solids in slices of equal area, then the solids have the same volume. One naturally thinks of the solid as

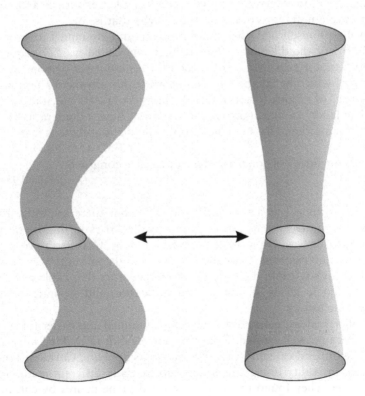

Figure 1.1 A modern explanation of the idea of Cavalieri's Principle

made up of the infinitely many infinitely thin slices, which slide horizontally to make one figure from the other. However, that creates a logical problem, since one has the dilemma of choosing between

(a)   having the volume actually made up of infinitely many zero-volume slices, in which case one has non-rigorous and dangerous ideas of "zero times infinity"; who knows how to multiply zero by infinity to obtain the "right" answer?
        or
(b)   having the volume made up of finitely many slices, which would each have a non-zero volume but when slid horizontally would not exactly fit the other volume (the edges of the slices cannot be exactly right to fit both curved volumes).

Option (a), which appears cleaner and more natural, is also more danger-ous, as it can easily lead to wrong answers. For example, if one argued that in Figure 1.1 the surface areas of both figures were made up of the circum-ferences of the cross-sectional circles, and hence were equal, that would be wrong. The more curvy figure on the left has a larger surface area. It is not easy to explain in terms of infinitesimals why that is so.

Cavalieri himself understood the problem and attempted to straddle the two horns of the dilemma. What is normally called his "principle" is, in his work, not a principle in the sense of a beginning of demonstration but a theorem which he attempts to demonstrate. He advances a notion of "all the planes" of a figure (i.e. its infinitely many cross sections), and argues that "all the planes" of one figure can, in the mass, have a finite ratio to "all the planes" of another. His fundamental postulate about these is that

"All the lines" of congruent figures . . . are congruent.

(Andersen 1985: 316)

That has appeal as a symmetry principle, but there are difficulties ap-plying it to more general cases by, as Cavalieri tries to do, dissecting and rearranging figures. On a modern understanding, Cavalieri's project is im-possible, as no rigorous foundation of limits and continuity was discovered until Cauchy's work in the early nineteenth century that explained the con-cept of limit in terms of multiple quantification and separated safe from unsafe conclusions.

Cavalieri's ideas, reduced to a slightly simplified and easier form by Tor-ricelli, became well known. Roberval and Pascal, in making advances to-ward the integral calculus, saw themselves as applying Cavalieri's methods but were in reality doing something substantially different at the founda-tional level. They found the areas of curved plane figures by cutting them into finitely many strips and taking a genuine limit as the number of strips increases (in the manner of the modern Riemann integral). That is unlike

Cavalieri, whose principles avoid any transition from the finite to the infinite (Andersen 1985: §X).

In developing the methods of the differential and integral calculus, Newton and Leibniz were not unaware of their lack of a rigorous foundation for limits. They reacted by such means as issuing magisterial lists of rules and keeping to algorithms for differentiating and integrating functions rather than trying to lay down principles from which the results would follow. Both Leibniz and Newton found themselves manipulating "infinitesimals," quantities that were regarded as non-zero when convenient and zero when convenient—indeed, the calculus came to be commonly called "infinitesimal calculus." The obvious problem of regarding a quantity as both zero and non-zero—which was to attract Berkeley's entirely justified sarcasm that infinitesimals were "the ghosts of departed quantities"—provoked various attempted responses from the founders of calculus, including Wallis's attempt to regard infinitesimals as really zero, Johann Bernoulli's attempt to establish them as really positive, Leibniz's attempt to regard them as fictions, and Newton's hope of bypassing them by depending on intuitions of continuous movement to understand "fluxions" (Brown 2012: Ch. 6; Jesseph 1998).

A serious attempt to expound the calculus as a structure of theorems following from principles occurs in the first textbook on differential calculus, L'Hôpital's *Analyse des Infiniment Petits, pour l'intelligence des lignes courbes* of 1696 (actually written in large part by Johann Bernoulli). It displays definitions and postulates in a Euclidean style, such as

> Definition II: The infinitely small part whereby a variable quantity is continually increased or decreased, is called the differential of that quantity.
> Postulate I (literally "demande ou supposition"): Grant that two quantities, whose difference is an infinitely small quantity, may be taken (or used) indifferently for each other . . .
> Postulate II: Grant that a curved line may be considered as the assemblage of an infinite number of infinitely small right lines . . .
> (L'Hôpital 1696: 2–3; discussion in Mancosu 1996: 151–152)

However, merely declaring as a principle that quantities that differ by an infinitesimal are equal does not render that statement meaningful.

Applied Mathematics/Subordinate Sciences/ Mixed Mathematics

The most interesting aspect of the topic of early modern mathematical principles concerns their appearance in (what we call) applied mathematics. As described in the introduction, the division of mathematics into pure and applied is an anachronism as applied to the seventeenth century. To think of mathematics as essentially a "pure" discipline whose results are then "applied" to a physical reality which it models imperfectly is a Platonist conception of mathematics, inapplicable to the Aristotelian conception that was dominant until at least the eighteenth century (Franklin 2014: Ch. 14).

The ancient division of mathematics, largely accepted in the seventeenth century, was into the science of discrete quantity (arithmetic) and the science of continuous quantity (geometry). In addition there were sciences "subordinate" to these, namely music or harmony (subordinate to arithmetic: really the study of discrete ratios of notes such as the octave and the fifth) and optics (subordinate to geometry), and astronomy (geometry in motion). The division of the mathematical sciences into the Quadrivium (arithmetic, geometry, music, and astronomy) was old enough to be attributed vaguely to "the Pythagoreans," (e.g., Proclus 1970: 29–30) while Aristotle said in the *Posterior Analytics* that optics is subordinate to geometry in the same way as harmonics is subordinate to arithmetic (Aristotle, *Posterior Analytics* 75b16 and 78b37; McKirahan 1978; Lennox 1986).

That way of organizing the mathematical sciences suggests that the subordinate sciences inherit their principles from arithmetic and geometry, with the implication that those principles should be just as certain in the subordinate sciences as they are in the higher ones. That is a view opposite to the usual modern one (asserted by Einstein earlier) that there must be a model–reality gap that prevents any certainty of pure mathematics from transferring to the applications of mathematics.

As the successes of mathematics built up in the seventeenth century, high hopes were entertained of extending the certainty of mathematics to a wide range of areas of what we now call physics. (For early seventeenth-century mixed mathematics, see Brown 1991; Lennox 1986; Malet and Cozzoli 2010). Some of those areas are more promising for that approach than others. Astronomy, once the jewel in the crown of the mathematization of nature, was proving hard going, what with the debate about alternative "hypotheses" and with a resolution impeded for theological reasons. Mechanics (dynamics) was a prize not yet ready to fall. (Details on attempts in Bertoloni Meli 2006; Capecchi 2017.) The two sciences that *prima facie* fit best the model of a subordinate science simply taking whole its principles from its higher science while retaining their certainty were the very old pair, optics and statics. As Barrow rightly said, in the course of extending mathematics to applications beyond the actual world—to worlds which are self-consistent and which God might create though he has chosen not to—statics, like optics, is a specially central and certain part of mixed mathematics:

> that Part of Mechanics treating of the Center of Gravity, and that Branch of Optics vulgarly called Perspective, are not unfitly numbered among the Parts of Geometry, because they scarce require any thing which is not granted and proved in that Science, nor use any other Principles or Reasonings than what are strictly Geometrical.
>
> (Barrow 1734: 27; Malet 1997)

Seventeenth-Century Optics: A Wave of "Principles"

Optics is especially certain, it seems. To all appearances, it is just geometry. The seventeenth century was very familiar with the model of Euclid's *Optics*, which deals with the basic geometry of vision. It ignores all the physical and psychological aspects of vision, and starts with postulates that relate vision to straight-line geometry, such as

1. That rectilinear rays proceeding from the eye diverge indefinitely
2. That the figure contained by a set of visual rays is a cone of which the vertex is at the eye and the base at the surface of the objects seen
3. That those things are seen upon which visuals rays fall and those things are not seen upon which visual rays do not fall
4. That things seen under a larger angle appear larger, those under a smaller angle appear smaller, and those under equal angles appear equal

and so on (Lindberg 1976: 12; a complete translation in Euclid 1945). (It does not have separate definitions, presumably on the grounds that the definitions of *line* and other terms in the *Elements* are sufficient.)

How are these postulates established? It seems that one can just look at experience and analyse it. We see what is before us in a straight line, if nothing closer in that direction obscures it. It is true that the "indefinitely" in postulate 1 is a big claim, given limited perceptual experience, but since the indefinitely distant stars are visible, it has a solid claim to be obvious from experience.

In any case, if the postulates in some way extend beyond experience, devotees of the "experimental philosophy" of the seventeenth century were in no position to complain, as experiments are particular and their results too are generalized by some semi-magical process to other times and places.

Euclid avoids discussing reflection and refraction, whose principles seem harder to establish *a priori*. How could the basic principle of equality of the angles of incidence and reflection be certain prior to observation? The difficult Snell's formula for the angle of refraction seems even less *a priori*. However, the principles of Euclid are wide ranging, being essentially sufficient for the science of perspective (as in painting), which was recognized in Renaissance times as one of the most useful areas of mathematics. Euclid's *Optics* provided a model of a substantial applied mathematical science with self-evident principles true of the real world.

But the great advance of optical theory in the seventeenth century— from Kepler's understanding that the eye focuses on the retina (Gal and Chen-Morris 2010; Lindberg 1976) to the achievements of Newton's *Opticks*—had another major contribution to make to the development of mathematical principles. Snell's (or Descartes's) law of refraction, Fermat's principle of least action, and Huygens's principle of wave propagation certainly explained optical phenomena in terms of very general mathematical principles. But they did not appear to be self-evident principles. As in

Newton's principles for explaining planetary motions, but more perspicuously in the simpler and more geometrical field of optics, the principles need some degree of empirical support, and it is the combination of their inherent power, generality, and simplicity with their ability to exhibit the phenomena as logical consequences, that gives them their credibility.

Why exactly it is a logical virtue in a theory to explain phenomena when it is not based on self-evident principles was unclear at the time, and is still contested (although logical Bayesian philosophy of science has a credible in-principle explanation, it is still unknown, for example, how to quantify the strength of "arguments to the best explanation").

The mystery was deepened by the strange fact that the apparently contingent and less-than-intuitive law of reflection and Snell's law of refraction could both be explained in terms of two principles that were entirely different from each other, even if probably compatible.

Fermat's Principle of Least Time (or "Least Action") explains the equality of angles of a reflected ray of light as in Figure 1.2.

Similar but more difficult reasoning can be used to derive Snell's law of refraction from Fermat's Principle.

However, there was something unsatisfactory about regarding Fermat's Principle as a *principle*. Not only did it lack self-evidence, it appeared to require that light travel with finite speed, in order that the least distance from A to B should translate into least time. If light travelled instantaneously, why should the shortest path be preferred? The finite speed of light was suspected, but neither provable *a priori* nor able to be established by experimental evidence. Even worse, the form of Fermat's Principle seemed wrong for a free-standing physical principle, as it seemed to suggest that Nature somehow "knew" ahead of time which was the shortest path and directed

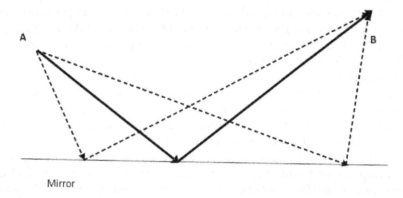

Figure 1.2 Fermat's Principle to explain reflection: Among all paths from A to B via the mirror, the solid path, with equal angles at the mirror, has the shortest length

light along it. As Clerselier objected to Fermat, "the principle which you take as foundation of your demonstration, viz. that nature acts always but the shortest and simplest route, is but a moral and not physical principle, which is not and cannot be the cause of any physical effect" (Mahoney 1994: 401).

Huygens's Principle explained, or purported to explain, the same phenomena (reflection and refraction) by a different supposition: that each point in propagating light acts as a source of a circular (or in 3D, spherical) wave, and the resultant light paths are the envelope of the waves (Dijksterhuis 2005; Shapiro 1989; Animation at http://www.physics.ucdavis.edu/Classes/Physics9B_Animations/ReflRefr.html). The circular waves themselves may or may not be purely fictional.

Huygens, unlike Newton, was prepared to bite the bullet and conclude that hypotheses of this sort, from which phenomena follow naturally but which cannot be independently established, are probable—with "a degree of probability which very often is scarcely less than complete proof" (Franklin 2001: 369; Huygens 1912: Preface, vi–vii) but nevertheless not complete proof. His choice was not a popular one.

Statics

After Euclid's *Optics*, the best-known science where it seemed that real-world certainties could easily be derived from self-evident geometrical principles was statics, especially the central part concerning the law of the balance. Pierre Duhem's original work on the late medieval influences on early modern physics concerned statics (Duhem 1905–1906) because that science is very simple—much more so than dynamics—and is easily appreciated by those looking for mathematical principles in real-world examples. To the medieval contributions was added the powerful and clearly expressed work of Archimedes in his *Equilibrium of Planes*.

The law of the balance states that two weights on a beam supported by a fulcrum balance each other if the ratios of the weights is the inverse of the ratios of their distances from the fulcrum. That is, lighter weights far away can balance heavy weights closer in. Archimedes does not regard this as a physical law needing experimental support but offers to demonstrate it from absolutely certain first principles. In the simplest example, where one weight is twice the other, his proof works like this:

Archimedes first extends the (weightless) beam in each direction so that the fulcrum is now in the middle (the weights do not change position).

The weights are now imagined as malleable, like clay, and are gently patted down to rest with a uniform thickness on the beam, as in Figure 1.3c.

We see that in the final state, weight is uniformly distributed, with equal amounts each side of the fulcrum. So by symmetry, the weights must balance. Therefore, the original pair of weights must balance. Seventeenth-century

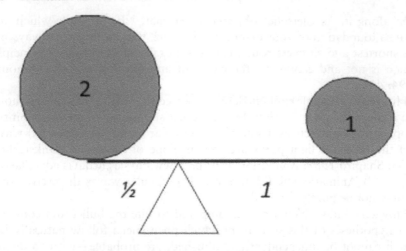

Figure 1.3a A weight of 2 close to the fulcrum balances a weight of 1 twice as far away

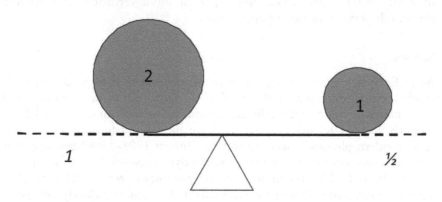

Figure 1.3b The beam is extended so that the fulcrum is in the middle

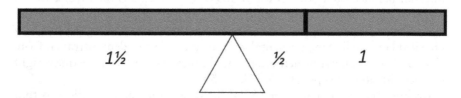

Figure 1.3c The weights are patted down to rest uniformly on the beam

thinkers were impressed, as well they might be (Vampoulis 2010). The demonstration of what can be done by extending a pure symmetry is very powerful.

A well-known similar example, where pure symmetry considerations lead to a surprising and apparently empirical result, is Stevin's "wreath of spheres." Stevin printed the diagram as the title page of his 1586 *Elements of the Art of Weighing* a work which imitates and extends Archimedes's deductive structure (Stevin 1586, discussed in Devreese and Vanden Berghe 2008: 136–139).

The text which decorates it says, approximately, "Wonderful, but not incomprehensible." That is correct. It is clear that the circle of balls does not tend to rotate either clockwise or anticlockwise (that would be perpetual motion, and in any case there is no reason to prefer clockwise to anticlockwise rotation). But the balls hanging below the horizontal line are in equilibrium. One could cut them in the middle and allow them to hang down, or even remove them altogether, without disturbing the balls resting on the two inclined planes. These upper balls, then, are at rest, in balance: the many on the lightly inclined plane balancing the few on the steeper plane. The numbers of balls on each side of the apex are in inverse proportion to the sine

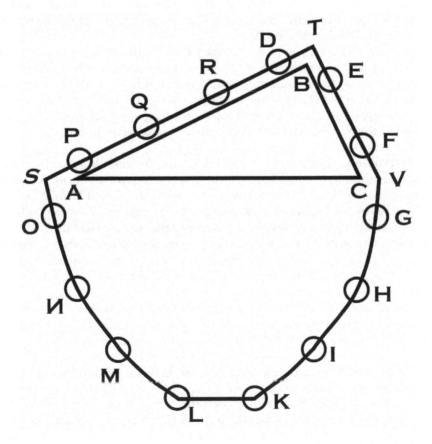

Figure 1.4 Stevin's "wreath of spheres" showing how weights on inclined planes balance each other according to the angle of inclination

Source: http://en.wikipedia.org/wiki/Simon_Stevin#mediaviewer/File:StevinEquilibrium.svg)

of the angles at which the planes are inclined. One has derived, therefore, the law of the inclined plane, or, equivalently, the resolution of forces into components. It is significant that this, probably the only significant discovery of the early Scientific Revolution concerning forces, is expressed as pure geometry.

Galileo: Uniform Acceleration of Fall

Kinematics, the science of motion in itself without regard to its causes, is somewhat harder than statics to reduce to purely mathematical principles, but it can to some extent be done. The necessary connections between acceleration, speed, and distance travelled are laid out in the simplest case by the medieval Merton Mean Speed Theorem, repeated by Galileo (Boyer 1959: 82–85) and cleared up in general by Newton's and Leibniz's calculus. Feats such as Newton's derivation of Kepler's elliptical orbits from a central law of acceleration are purely mathematical and need only principles of calculus (though as we saw earlier, the state of that subject left something to be desired at the foundational level).

It is otherwise with dynamics, which deals with the causes of motion such as gravity and other forces. It cannot be established *a priori* that the fall of heavy bodies is uniformly accelerated, much less what the constant of acceleration is. (That is why the contingency of the value of G and other constants of nature plays a crucial role in contemporary debates on the Anthropic Principle.) Galileo's observations were necessary to find the law of fall of heavy bodies.

Nevertheless, Galileo was at the same time responsible for one of the more remarkable demonstrations of the power of *a priori* mathematical reasoning in dynamics. When first considering what law should be followed by falling heavy bodies, once it is accepted that they go faster as they fall, he wondered about how to distinguish between the two simplest theories: the perhaps most natural one, that speed is proportional to distance travelled from the start, and the equally simple but perhaps less natural one, that speed is proportional to time from the start (i.e. the body is uniformly accelerated, which is the correct answer).

Galileo realized, and was able to demonstrate, that the first theory needs no observations to refute it. It is absolutely impossible that acceleration should be proportional to the distance travelled. Galileo argues thus:

> When speeds have the same ratio as the spaces passed or to be passed, those spaces come to be passed in equal times; if therefore the speeds with which the falling body passed the space of four braccia were the doubles of the speeds with which it passed the first two braccia, as one space is double the other space, then the times of those passages are equal; but for the same moveable to pass the four braccia and the two in the same time cannot take place except in instantaneous motion.
>
> (Galileo 1974: 160; Norton and Roberts 2012)

time taken

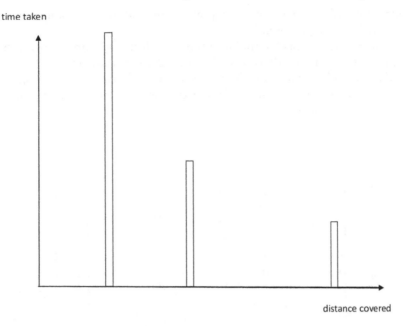

distance covered

Figure 1.5 Galileo's proof that speed cannot be proportional to distance

That reasoning is less than totally clear. But it suggests a diagram, surely present in some form in Galileo's mind, which clarifies it.

Figure 1.5 should be read from right to left. If the time taken to cover a small distance is represented by the column of time above it (right column), then at half that distance, it needs twice the time, at a quarter the distance, four times, and so on. As distance approaches zero, it needs indefinitely more time to cover the same interval. So the motion can never get started.

From the falsity of the theory of the proportionality of speed to distance there does not follow, of course, the truth of the (true) alternative theory of the proportionality to time. But it leaves that theory as the natural, simple alternative, guiding the effort of empirical confirmation.

A proper understanding of the matter required some principle to connect the cause of motion (a force) to the way the motion itself develops (the acceleration). Galileo did not attempt any such thing—indeed, he avoided thinking about forces in connection with motion at all, although he was familiar with them in statics. That was left to Newton. Before looking at Newton's mathematical principles, we consider the other area of physics in which early seventeenth-century mathematical science had signal success, the physics of fluids.

Pressure/Hydrostatics

A priori symmetry arguments proved very successful in hydrostatics, the science of pressure in water. Stevin was again in the lead. There is something extraordinarily simple about how pressure in fluids works, sometimes called the "hydrostatic paradox."

As in Figure 1.6, one's initial intuition that different-shaped vessels should support water differently is false. Pressure depends only on depth.

Why should such a simple law be true? As with the "wreath of spheres," Stevin offered a mathematical demonstration from first principles in place of empirical evidence.

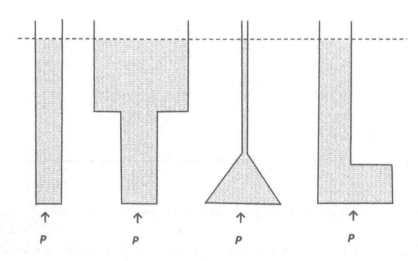

Figure 1.6 Modern illustration of the hydrostatic paradox: Pressure at the bottom is equal, irrespective of the shape of the vessel above

Stevin argues that in a rectangular tank ABCD as in Figure 1.7a, with imaginary vertical divisions GE and HF, there is no tendency of water to move across the vertical boundaries in either direction; hence, on each part of the bottom, such as EF, there rests the weight of the water directly above it. Next he considers a more complex figure of water resting on EF (Figure 1.7b).

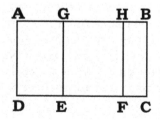

Figure 1.7a Stevin's diagram of water at rest on different vertical bottoms

Source: Gaukroger and Schuster 2002: Figure 1, reprinted with permission from Elsevier Science.

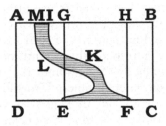

Figure 1.7b Stevin's diagram of a tube of water of complex shape resting on bottom EF

Source: Gaukroger and Schuster 2002: Figure 2, reprinted with permission from Elsevier Science.

He supposes that the parts of the water *not* in the shaded complex-shaped tube are replaced by a solid of the same density as water. That makes no difference to the forces in the water. So the pressure on the bottom EF, now caused solely by water in the complex shape above it, must be the same as before, that is, the weight of the column GHFE directly above EF. Therefore, the pressure on the bottom depends only on the height of the column, not on its shape (Gaukroger and Schuster 2002: 542–545).

Stevin's work impressed the next writers in the field, such as Descartes and Pascal. Pascal put forward a demonstration of the proportionality of pressure to depth that is more local and calculus-like than Stevin's global symmetry argument. He also made sense of the brute empirical fact that there is a limit of 10.3 metres to the height to which water can be raised by a vacuum pump (again, irrespective of the shape of the column of water). Pascal explained it as an effect of air pressure—air being also a fluid with weight and the differences between the pressures of air and water being due solely to the differences in their densities.

Practical Mathematics at Sea

Mathematics, as known in early modern times—and indeed today—was not just a theoretical study but in certain aspects a very practical and lucrative business. A snapshot of the state of mathematics during the "Mathematical Revolution," but at the very beginning of the Scientific Revolution, as usually calculated, can be found in Billingsley's Euclid, the first English translation, of 1570. It has a preface by John Dee, "specifying the chief Mathematicall Sciences, what they are, and wherunto commodious." He arranges in a tree the two principal sciences, arithmetic and geometry, and some thirty derivative sciences. (A reproduction of the diagram is available at http://www.math.ubc.ca/~cass/euclid/dee/dee.html.) There are "vulgar" arithmetic and "vulgar" geometry, the latter divided into eleven sciences, concerned with surveying, in one, two, and three dimensions, on both land and sea. There follow nineteen miscellaneous sciences, most with a strongly geometrical aspect (hence, these sciences are principled ones, not empirical ones like alchemy). Of these, perspective, astronomy, music, cosmography, astrology, statics, and navigation are clear enough, but the remaining eleven have names now unrecognisable. "Anthropographie" is about the proportions in the human body; "Zographie" is something like the modern theory of rendering in computer graphics; "Trochilike" studies circular motions, simple and compound; "Hydragogie" "demonstrateth the possible leading of Water by Natures Law, and by artificiall helpe, from any head" and "Pneumatithmie" "demonstrateth by close hollow Geometrical figures (Regular and Irregular) the straunge properties (in motion or stay) of the Water, Ayre, Smoke and Fire." Some of these are plainly more commodious unto nascent capitalism than are others, but the total picture is of a suite of mathematical, mostly geometrical, sciences, in common and successful use, with investors queuing (Euclid 1570, Preface; survey of early modern practical mathematics in Bennett 1986). Two of the mathematical sciences that certainly did attract serious money, especially in England, where it was already appreciated that naval superiority was the key to national survival, were navigation (Eden 1561; Dee 1577—though more propaganda than technical navigation; Shirley 1985; Neal 2002b) and the hydrostatics and hydrodynamics of hull design. Navigation is straightforwardly geometry on the sphere (plus the difficult task of finding where on the sphere one is located). Hydrodynamics is much harder. (Today fluid modelling and hull design are computer-intensive and lack simple principles.)

Hull design is heavily geometrical but with many complexities. For one thing, it has to take into account both hydrostatics (to make the ship float high enough out of the water to ensure the gun ports do not take on water when the ship rolls but stable enough so that the ship does not roll over) and hydrodynamics (so it cuts through the water without excessive drag). Hydrodynamic and hydrostatic considerations are to some extent in conflict since one can gain speed at the expense of stability.

As was usual and natural, books such as Manoel Fernandes's *Livro de Traças de Carpintaria* (1616) were highly geometrical, but the text consists of recipes more than principles. There were attempts to apply to naval architecture the principles of the pseudo-Aristotelian *Mechanical Problems*, but success was limited (Ferreiro 2010). Deane's *Doctrine of Naval Architecture* (1670) is heavily mathematical and concentrates on hydrostatics. Problems were still experienced with buoyancy, as shown when the *Royal Katherine* at her launch in 1664 proved to have her gun ports only three feet out of the water before guns and provisions were loaded (Deane 1981 [1670]: 16). Deane offers recipes for construction and formulas to calculate how far out of the water the resulting ship will be.

Ship design was typical of the more applied mathematical disciplines, in that the advantages of deriving results from basic mathematical principles were well appreciated, but the complexities of the subject matter made it very difficult to achieve more than modest success. It was necessary to fall back on and cautiously adapt recipes that had proved successful for practitioners in the past.

The Scientific Revolution's Laws of Proportion

We will treat only lightly the high period of the Scientific Revolution when a large suite of scientific laws were discovered. These laws were mathematical in one sense, in that they ascribed to nature simple formulas—indeed, formulas in general of simple proportion. They were not *purely* mathematical, in the sense that (to the disappointment of some) they are not derivable solely from mathematical principles. They need some—however small—input of empirical and observationally derived fact.

A list of the Scientific Revolution's laws of proportion, with approximate dates, includes the following:

- Kepler's Second Law: The area swept out by a radius from the sun to a planet is proportional to the time taken (1609)
- Snell's Law: When light is refracted at a surface, the sine of the angle of refraction is proportional to the sine of the angle of incidence (1602, 1621, 1637)
- Galileo's Law of Uniform Acceleration: The speed of a heavy body falling from rest is proportional to the time from dropping (1638)
- Pascal's Law: The pressure in an incompressible fluid is proportional to depth (1647)
- Hooke's Law: The extension of a spring is proportional to the force exerted to stretch it (1660)
- Boyle's Law: For a fixed quantity of air at constant temperature, pressure is inversely proportional to volume (1662)
- Newton's proposition on the prism: there is some kind of proportionality between refrangibility and colour of light (1672)
- Newton's Second Law of Motion: The acceleration of a body is proportional to the total force acting on it (1687)
- Newton's Law of Gravity: The force of gravity exerted by one body on another is proportional to the masses of each and inversely proportional to the square of the distance between them (1687)
- Newton's Law of Cooling: The rate of temperature loss from a body is proportional to the difference in temperature between the body and its surroundings (1701).

Debate still proceeds on the relationship between empirical evidence and these very general proportionalities. It is the inherent logical difficulty of putting together the Euclidean model with the establishment of basic laws by experiment that gave rise to the endless problems over Newton's claim to have not "feigned hypotheses"—already in his first paper on the prism, which plainly rests on experiments, he insists that everything is "not an

Hypothesis but the most rigid consequence" (discussion in Walsh 2012). He kept to that view, creating a conundrum whose irresolvability provokes debate that shows no sign of ending. Let us look briefly at the most celebrated case, Newton's laws of motion.

Newton's Mathematical Principles

Newton's *Mathematical Principles of Natural Philosophy* opens in the Euclidean style, with definitions and axioms. The definitions include

1. Quantity of matter is a measure of matter that arises from its density and volume jointly.
4. Impressed force is the action exerted on a body to change its state either of resting or of moving uniformly straight forward.

The axioms or laws (*axiomata sive leges*) of motion are

1. Every body perseveres in its state of being at rest or of moving uniformly straight forward, except insofar as it is compelled to change its state by forces impressed.
2. A change in motion is proportional to the motive force impressed and takes place along the straight line in which that force is impressed.
3. To any action there is always an opposite and equal reaction . . .

Deduction then proceeds:

> Corollary 1: A body acted on by two forces acting jointly describes the diagonal of a parallelogram . . .
> > (Newton 1999: 403–405, 416–417; clarifications of Newton's use of the geometrical model in Dunlop 2012)

These principles have a number of strange features, both from the point of view of what has gone before and from the purely logical perspective. First, if it is asked whether these are *mathematical* principles, in the self-evident sense of Euclid's geometry, or principles that need to be established by observation, then *prima facie* they are neither. They appear natural yet contingent, so not totally self-evident. Yet the observations that would establish them are hard to pin down, mainly because of the difficult logical and observational status of Newton's primitive, "action" or "force."

The notion of "action" is primitive and its observational meaning unclear. That is what has led to modern physicists' persistent worrying whether the law $F = ma$ (the modern formulation of Newton's Second Law) is merely a definition of force and lacks real content (e.g., Wilczek 2004). Plainly that was not Newton's view, as he thinks of action as a real cause "exerted on a body," nor was it the view of philosophers like Reid, who pointed to the possibility of the direct perception of force as pressure on one's body. But Newton does not make clear how force may be measured other than through observing the acceleration it causes. He hardly helps clarify matters with his gnomic Scholium: "The principles I have set forth are accepted by mathematicians and confirmed by experiments of many kinds" (Newton 1999: 424).

That is doubly false. They are *Newton*'s laws because they had not been received by earlier mathematicians, nor does Newton make clear how they may be confirmed directly by abundance of experiment. He merely refers to some phenomena explained by them, such as Galileo's law of fall and the period of pendulums. Only the success of the global project of explaining planetary motion in terms of the principles gives them, in some overall sense, confirmation. The view of his contemporaries is understandable that Newton simply communed with nature and understood her principles (except, that is, for a few such as Huygens, who remained completely unconvinced by Newton's principle of gravitation; Guicciardini 1999: Ch. 5).

Probability

Probability is surely among the hardest of quantities to establish principles for. It was far from clear even that it was a quantity, in the sense of something to which a precise number could be assigned: surely the realm of uncertainty and chance of its nature resists pinning down with precision? And if we consider just the phenomena of games with dice and coins, the subject of Fermat and Pascal's 1654 correspondence that began the mathematical theory of probability, it is quite unclear what the quantity is to which a number should be assigned—the modern answer, the long-run relative frequency of outcomes, was not considered in Fermat and Pascal's work nor in any other of that century.

Thus, the initial problem considered, very unnaturally from a later point of view, was a moral one: that of the just division of the stake in an interrupted game of chance. The quantity to be calculated is a moral right. Initially, there was a complete lack of principles or calculations, and no apparent place to start in finding principles. Only Pascal realised that one might establish a realm of mathematical certainties in a new and unlikely field, the fluctuating realm of uncertainty. He wrote,

> the matter has hitherto wandered in uncertainty; but now what has been rebellious to experiment has not been able to escape the dominion of reason. For we have reduced it to art so securely, through Geometry, that, participating in [Geometry's] certainty, it now goes forth daringly, and, by thus uniting the demonstrations of mathematics to the uncertainty of chance, and reconciling what seem contraries, it can take its name from both sides, and rightly claim the astonishing title: the Geometry of chance (*aleae Geometria*).
>
> (Franklin 2001: 312; Pascal 1654)

Where might the principles be found? Pascal reaches for that old staple that had been so successful for Archimedes and Stevin, among so many others, in introducing *a priori* reasoning into applied mathematics, symmetry:

> The first principle for discovering how one should make the division is this:
>
> If one of the players is in a situation such that, whatever happens, a certain sum will belong to him in case of either a loss or a win, without chance being able to take it away from him, then he should make no division of it, but take it all, as assured . . .
>
> The second is this. If two players are in a situation such that, if one wins, a certain sum belongs to him, and if he loses, it belongs to the other; and if the game is of pure chance and there are as many chances for one as for the other, and consequently no more reason why one should win rather than the other, then if they wish to separate without playing, and take what legitimately belongs to each, the division is that they divide the sum that is at stake in halves, and take one each.
>
> (Franklin 2001: 311; Pascal 1665)

Soon Huygens, writing the first book on probability and establishing it as a coherent body of theory, also laid down a symmetry principle as the basis for all calculations:

> Although in games determined solely by chance the outcomes are uncertain, there is always a fixed value for how much one has for winning over losing . . . I take it as a foundation that in a game the chance (*sortem seu expectation/kansse*) that one has towards something is to be estimated as such that, if one had it, one could procure the same chance in an equitable game (*aequo conditione certans*). For example, if someone hides from me 3 shillings in one hand and 7 in the other, and gives me the choice of taking either hand, I say this is worth the same to me as if I were given 5 shillings. For if I have 5 shillings, I can again arrive at having an equal chance (*aequam expectationem*) of getting 3 or 7 shillings, and that by an equitable (*aequo*) game.
>
> (Franklin 2001: 314; Huygens 1657)

The ability of *a priori* symmetry arguments to establish absolutely certain principles even in such an unlikely subject matter as pure chance was the most perfect demonstration of their power.

According to views descending from the early Royal Society's propaganda, what is now known as the Scientific Revolution was essentially entirely new, English, experimental, inductive, and Baconian (in the sense of building up theories from masses of carefully collected facts). Although some developments from the seventeenth century do fit that schema, the story of mathematical principles suggests a far more complicated picture of the nature of the Scientific Revolution. Continental (mostly), top-down and theoretical, early modern mathematical science was the product of the minds of individual geniuses penetrating necessities behind the flux of appearances, much as Aristotle had claimed science should be.

Bibliography

Andersen, K. (1985) "Cavalieri's method of indivisibles," *Archive for History of Exact Science*, 31: 291–367.
Anstey, P. R., ed. (2013) *The Oxford Handbook of British Philosophy in the Seventeenth Century*, Oxford: Oxford University Press.
Barrow, I. (1860 [1664]) "Lectiones Mathematicae" in ed. W. Whewell 1860.
———. (1734) *The Usefulness of Mathematical Learning Explained and Demonstrated*, London.
Bascelli, T. (2014) "Galileo's *quanti*: Understanding infinitesimal magnitudes," *Archive for History of Exact Sciences*, 68: 121–136.
Bennett, J. A. (1986) "The mechanics' philosophy and the Mechanical Philosophy," *History of Science*, 24: 1–28.
Bertoloni Meli, D. (2006) *Thinking with Objects: The Transformation of Mechanics in the Seventeenth Century*, Baltimore: Johns Hopkins University Press.
Boyer, C. B. (1959) *The History of the Calculus and Its Conceptual Development*, New York: Dover.
Brown, G. I. (1991) "The evolution of the term 'mixed mathematics'," *Journal of the History of Ideas*, 52: 81–102.
Brown, R. C. (2012) *The Tangled Origins of the Leibnizian Calculus*, Singapore: World Scientific.
Bussey, W. H. (1918) "Fermat's method of infinite descent," *American Mathematical Monthly*, 25: 333–337.
Capecchi, D. (2016) "A historical reconstruction of mechanics as a mathematical physical science," *Mathematics and Mechanics of Solids*, 21: 1095–1115.
Deane, A. (1981) *Deane's Doctrine of Naval Architecture, 1670*, ed. B. Lavery, London: Conway Maritime Press.
Dear, P. (1995) *Discipline and Experience: The Mathematical Way in the Scientific Revolution*, Chicago: University of Chicago Press.
Dee, J. (1577) *The Perfect Arte of Navigation*, London.
De Risi, V. (2016) "The development of Euclidean axiomatics: The systems of principles and the foundations of mathematics in editions of the *Elements* in the Early Modern Age," *Archive for History of Exact Sciences*, 70: 591–676.
Devreese, J. T. and Vanden Berghe, G. (2008) *"Magic is No Magic": The Wonderful World of Simon Stevin*, Southampton: WIT Press.
Dijksterhuis, F. J. (2005) *Lenses and Waves: Christiaan Huygens and the Mathematical Science of Optics in the Seventeenth Century*, New York: Springer.
Domski, M. (2013) "Observation and mathematics" in ed. P. R. Anstey 2013, pp. 144–168.
Duhem, P. (1905–1906) *Les origines de la statique*, Paris: Hermann, trans. as *The Origin of Statics*, Dordrecht: Kluwer, 1991.
Dunlop, K. (2012) "The mathematical form of measurement and the argument for Proposition I in Newton's *Principia*," *Synthese*, 186: 191–229.

Eden, R. (1561) *The Arte of Navigation*, London, trans. of Martín Cortés' *Breve compendio de la sphera y de la arte de navegar* (1551).

Einstein, A. (1954) *Ideas and Opinions*. New York: Random House.

Encyclopaedia Britannica (1771) 1st edn, Edinburgh.

Engelfriet, P. M. (1998) *Euclid in China: The Genesis of the First Chinese Translation of Euclid's* Elements *Books I–VI (Jihe yuanben; Beijing, 1607) and Its Reception up to 1723*, Leiden: Brill.

Euclid. (1570) *The Elements of Geometrie*, trans. H. Billingsley, London. Reprinted in *The Elements of Geometry*, Ann Arbor: University of Michigan Press, 1967.

———. (1945) "The *Optics* of Euclid," trans. H. E. Burton, *Journal of the Optical Society of America*, 35: 357–372.

Fernandes, M. (1989) *Livro de Traças de Carpintaria*, Lisbon: Academia de Marinha; 1st edn 1616.

Ferreiro, L. D. (2010) "The Aristotelian heritage in early naval architecture, from the Venice arsenal to the French Navy, 1500–1700," *Theoria*, 68: 227–241.

Franklin, J. (1994) "Achievements and fallacies in Hume's account of infinite divisibility," *Hume Studies*, 20: 85–101.

———. (2000) "Diagrammatic reasoning and modelling in the imagination: The secret weapons of the Scientific Revolution" in eds. G. Freeland and A. Corones 2000, pp. 53–115.

———. (2001) *The Science of Conjecture: Evidence and Probability Before Pascal*, Baltimore: Johns Hopkins University Press.

———. (2006) "Artifice and the natural world: Mathematics, logic, technology" in ed. K. Haakonssen 2006, vol. 2, pp. 817–853.

———. (2014) *An Aristotelian Realist Philosophy of Mathematics: Mathematics as the Science of Quantity and Structure*, Basingstoke: Palgrave Macmillan.

Franklin, J. and Daoud, A. (2011) *Proof in Mathematics: An Introduction*, Sydney: Kew Books.

Freeland, G. and Corones, A., eds. (2000) *1543 and All That: Image and Word, Change and Continuity in the Proto-Scientific Revolution*, Dordrecht: Kluwer.

Freudenthal, H. (1980) "Huygens' foundations of probability," *Historia Mathematica*, 7: 113–117.

Gal, O. and Chen-Morris, R. (2010) "Baroque optics and the disappearance of the observer: From Kepler's *Optics* to Descartes' doubt," *Journal of the History of Ideas*, 71: 191–217.

Galileo. (1953) *Dialogue Concerning the Two Chief World Systems*, ed. S. Drake, Berkeley: University of California Press.

———. (1974) *Two New Sciences*, trans. S. Drake, Madison: University of Wisconsin Press.

Gaukroger, S. W. and Schuster, J. A. (2002) "The hydrostatic paradox and the origins of Cartesian dynamics," *Studies in History and Philosophy of Science*, 33: 535–572.

Gillette, G. (2009) *The Philosophical Mathematics of Isaac Barrow (1630–1677): Conserving the Ancient Greek Geometry of the Euclidean School*, Lewiston: Edwin Mellen Press.

Guicciardini, N. (1999) *Reading the Principia: The Debate on Newton's Mathematical Methods for Natural Philosophy from 1687 to 1736*, Cambridge: Cambridge University Press.

Haakonssen, K., ed. (2006) *The Cambridge History of Eighteenth-Century Philosophy*, 2 vols, Cambridge: Cambridge University Press.

Huygens, C. (1657) *De ratiociniis in ludo aleae/Van reeckening in spelen van geluck* (*On Reckoning in Games of Chance*), Leiden. Reprinted in Huygens 1888–1950, 14: 52–95; partial English translation in H. Freudenthal 1980.

———. (1888–1950) *Oeuvres complètes de Christiaan Huygens*, 22 vols, The Hague: Nijhoff.

———. (1912) *Treatise on Light*, trans. S. P. Thompson, London: Palgrave Macmillan.

Jesseph, D. M. (1989) "Philosophical theory and mathematical practice in the seventeenth century," *Studies in History and Philosophy of Science*, 20: 215–244.

———. (1993) *Berkeley's Philosophy of Mathematics*, Chicago: University of Chicago Press.

———. (1998) "Leibniz on the foundations of the calculus: The question of the reality of infinitesimal magnitudes," *Perspectives on Science*, 6: 6–40.

Jullien, V., ed. (2015), *Seventeenth-Century Indivisibles Revisited*, Cham: Springer.

Kepler, J. (1615) *Nova stereometria doliorum vinariorum*, Linz.

L'Hôpital, G. de. (1696) *Analyse des infiniment petits, pour l'intelligence des lignes courbes*, Paris.

Lennox, J. G. (1986) "Aristotle, Galileo and 'mixed sciences'" in ed. W. A. Wallace 1986, pp. 29–51.

Lindberg, D. C. (1976) *Theories of Vision from al-Kindi to Kepler*, Chicago: University of Chicago Press.

Lonergan, B. (1997) *Insight: A Study of Human Understanding*, 5th edn, Toronto: Toronto University Press.

Maclaurin J., ed. (2012) *Rationis Defensor: Essays in Honour of Colin Cheyne*, Dordrecht: Springer.

Mahoney, M. S. (1994) *The Mathematical Career of Pierre de Fermat, 1601–1665*, 2nd edn, Princeton: Princeton University Press.

Malet, A. (1997) "Isaac Barrow on the mathematization of nature: Theological voluntarism and the rise of geometrical optics," *Journal of the History of Ideas*, 58: 265–287.

———. (2006) "Renaissance notions of magnitude and number," *Historia Mathematica*, 33: 63–81.

Malet, A. and Cozzoli, D. (2010) "Mersenne and mixed mathematics," *Perspectives on Science*, 18: 1–8.

Mancosu, P. (1996) *Philosophy of Mathematics and Mathematical Practice in the Seventeenth Century*, New York: Oxford University Press.

McKirahan, R. D. (1978) "Aristotle's subordinate sciences," *British Journal for the History of Science*, 11: 197–220.

———. (1992) *Principles and Proofs: Aristotle's Theory of Demonstrative Science*, Princeton: Princeton University Press.

Neal, K. (2002a) *From Discrete to Continuous: The Broadening of Number Concepts in Early Modern England*, Dordrecht: Kluwer.

———. (2002b) "Mathematics and empire, navigation and exploration: Henry Briggs and the Northwest Passage voyages of 1631," *Isis*, 93: 435–453.

Newton, Sir I. (1999) *The Principia: Mathematical Principles of Natural Philosophy*, eds. I. B. Cohen and A. M. Whitman, Berkeley: University of California Press.

Norton J. D. and Roberts, B. W. (2012) "Galileo's refutation of the speed-distance law of fall rehabilitated," *Centaurus*, 54: 148–164.

Oakley, F. (1961) "Christian theology and the Newtonian science: The rise of the concept of the laws of nature," *Church History*, 30: 433–457.

Paipetis, S. A. and Ceccarelli, M., eds. (2010) *The Genius of Archimedes: 23 Centuries of Influence on Mathematics, Science and Engineering*, Dordrecht: Springer.

Pascal, B. (1654) "Address *Celeberrimae Matheseos Academiae Parisiensi*" in ed. Pascal 1964–1992, vol. 2, pp. 1034–1035.

———. (1665) "Traité du triangle arithmetique, third tract," in Pascal 1964–1992, vol. 2, p. 1308.

———. (1964–1992) *Oeuvres complètes*, 4 vols, ed. J. Mesnard, Paris: Desclée de Brouwer.

Proclus. (1970) *A Commentary on the First Book of Euclid's Elements*, trans. G. R. Morrow, Princeton: Princeton University Press.

Pycior, H. M. (1987) "Mathematics and philosophy: Wallis, Hobbes, Barrow, and Berkeley," *Journal of the History of Ideas*, 48: 265–286.

Sepkoski, D. (2005) "Nominalism and constructivism in seventeenth-century mathematical philosophy," *Historia Mathematica*, 32: 33–59.

Shapiro, A. E. (1989) "Huygens' *Traité de la lumière* and Newton's *Opticks*: Pursuing and eschewing hypotheses," *Notes and Records of the Royal Society*, 43: 223–247.

Shirley, J. W. (1985) "Science and navigation in renaissance England" in eds. J. W. Shirley and F. D. Hoeniger 1985, pp. 74–93.

Shirley, J. W. and Hoeniger, F. D., eds. (1985) *Science and the Arts in the Renaissance*, Washington: Folger Shakespeare Library.

Stevin, S. (1586) *De Beghinselen der Weeghconst*, Leiden.

Struik, D. (1969) *A Source Book in Mathematics, 1200–1800*, Cambridge, MA: Harvard University Press.

Vampoulis, E. (2010) "Archimedes in seventeenth century philosophy" in eds. S. A. Paipetis and M. Ceccarelli, pp. 331–343.

Wallace, W. A., ed. (1986) *Reinterpreting Galileo*, Washington, DC: Catholic University of America Press.

Wallis, J. (1657) *Mathesis universalis*, in *Operum mathematicorum pars prima*, Oxford.

Walsh, K. (2012) "Did Newton feign the corpuscular hypothesis?" in ed. J. Maclaurin 2012, pp. 97–110.

Whewell, W., ed. (1860) *The Mathematical Works of Isaac Barrow*, Cambridge: Cambridge University Press.

Whiteside, D. T. (1961) "Patterns of mathematical thought in the later seventeenth century," *Archive for History of Exact Sciences*, 1: 179–388.

Wilczek, F. (2004) "Whence the force of $F = ma$? I: Culture shock," *Physics Today*, 57, October: 11–12.

2 The Development of Principles in Equity in the Seventeenth Century*

J. C. Campbell

Legal history is an important part of the history of ideas—a part that historians of ideas overlook too often. For, the law provides many of the institutions that enable a society to operate. It is one of the main means by which a minimum content is given to ideas about how one person should treat another and to ideas about how a society should work—a minimum content because it identifies those obligations that can be enforced, rather than being merely matters of exhortation. It has the advantage over many theories of ethics and political philosophy that it descends to particulars and can provide specific answers to questions about what is to be done in a particular situation.

English legal history is the product not only of events in the court system but also of events in English history, indeed sometimes European history, that happened outside the court system. Many contingent factors have shaped the strand of English law known as equity. The more important ones include the huge religious dislocation of the Reformation, and the rise of scientific knowledge and the doubts it brought about how certain knowledge could ever be obtained. Other influences came from changes, most acutely expressed in events leading to the English Civil War, in theories about how the king and the church related to each other, the source of law, and the proper roles of the monarch and the parliament.

This chapter documents how, in the course of the seventeenth century, the Court of Chancery underwent a paradigm shift. It moved from deciding cases using relatively unformed criteria of conscience, to deciding them by reference to articulated principles and rules. For centuries after the chancellor first began to decide disputes as the king's delegate, he based his jurisdiction on the king's obligation and prerogative right to administer justice. But after the violent political struggles of the seventeenth century had confined the royal prerogative, it was only if the prerogative was exercised in accordance with particular rules and principles that it provided an acceptable basis for the court's operation. Originally Chancery had used the requirements of conscience, understood to be objectively knowable, as its standard of judgment. But during the late sixteenth century and the seventeenth century, ideas from the Reformation and the scientific revolution had the effect that the concept of conscience could no longer provide an objective standard of judgement. It was ascertainable rules and principles that provided the objective standard of judgement that Chancery came to use instead.

But a rule or principle of equity operated differently to the rules and principles of the sciences. It arose through a casuistical process and could provide a standard of decision for a factual situation different to the one concerning which it had been formulated in only a provisional way. However, other equitable rules and principles, and the procedures by which equity litigation was conducted, set limits to the extent to which the rule or principle could be departed from.

Lightning Sketch of Role of Court of Chancery in English Law

About one hundred years after the Norman Conquest, Henry II set about establishing some royal courts[1] that were open to any of his subjects. Part of the justification for doing so was that the King claimed to be the fount of all justice. The coronation oaths of kings traditionally included an oath to dispense justice to all.[2] This obligation of the king had as its converse that the king had a prerogative power to create courts and to decide how they would operate.

There had previously been local courts, which treated local customs as having the force of law. Those customs could differ from one local court to another. The law that the royal courts applied was, in theory, a custom that was common to the whole of England. Hence, it was known as the common law. English customary law was different to the civil law, based on Roman law, which came to be applied in many parts of Europe. However, the royal courts did not provide a remedy for every type of legitimate complaint of wrongdoing. Every action in the royal courts had to be initiated by a writ, issued from the office of a royal official akin to the king's secretary, the Lord Chancellor.[3] There was a limited number of types of writs available, for a limited number of types of complaint.

At first the royal courts granted a wide range of remedies.[4] However, by the early fourteenth century the royal courts had restricted themselves so that they provided only a limited range of remedies. Broadly, they could order the defendant to pay monetary compensation (damages), or they could order that the sheriff put the plaintiff into possession of a particular parcel of land. The limited types of complaints that the royal courts would hear, and the limited remedies provided for those types of complaints that they did hear, led to people petitioning the king that the justice offered in his courts was inadequate and that he should render them justice himself. As the fount of justice, the king had the power to fix defects in the way in which his own system of justice had operated in the common law courts.

Moreover, because of his coronation oath he had a sacred obligation to do so.[5] The king delegated the hearing of those petitions to the chancellor.[6]

The chancellor dealt with the petitions in a systematic way, and by the fifteenth century it came to be recognized that the chancellor was holding his own court in doing so.[7] The chancellor's court was called the Court of Chancery. In time the chancellor came to have the power to issue orders to fix defects in how the common law courts had operated in particular cases or to satisfy a complaint about a type of wrongdoing that the common law did not recognize. The chancellor came to have the power to do this himself, rather than merely advising the king what the king should do.[8]

From the Norman Conquest until the reign of Henry VIII the chancellors were nearly always clerics. Sometimes a cleric appointed as chancellor was also trained in the common law, but usually not. Their views about proper behaviour were moulded by the moral teaching of the church. By the time the chancellor was holding his own court, the church's moral teaching was significantly influenced by St Thomas Aquinas. A central notion in Thomist ethical teaching is that of conscience.[9] Conscience was the guiding principle that the chancellor used in dealing with petitions and the Court of Chancery was referred to as a court of conscience.

In keeping with the clerical training of chancellors, the primary objective of the Court of Chancery was cathartic—to save the soul of the defendant by not permitting him to engage in the mortal sin of acting contrary to conscience. It did this by sometimes ordering a defendant not to engage in an action that was contrary to conscience and sometimes by ordering him actually to carry out actions that conscience required. Sometimes conscience required a defendant to act in a way that was different from the way in which the common law would hold he was either entitled or at liberty to act.

The justification for the Chancery requiring action that differed from the common law's requirements is found in the common medieval view about the nature of law. It was that the law of God governed the universe, and hence his law and the law of nature and reason, which were nearly synonymous, predominated over the rules of any state. The human law could not be valid in contradiction to divine law.[10] Thus, it was not only permissible, but also necessary, for the chancellor (or the king on the advice of the chancellor) to interfere with the course of the law in a particular instance, even where the general rule was just, if according to conscience it would work against the law of God.[11]

The English Reformation in the 1530s sowed seeds that would later influence the development of Chancery. It ended the medieval theory of the relations between church and state, under which the church was supreme concerning spiritual matters and the king supreme concerning temporal matters. The Reformation made the king the supreme authority in both temporal and spiritual matters.[12]

Pre-Reformation Chancery and the Thomistic Understanding of Conscience

Up to near the end of the sixteenth century the way conscience was spoken of, concerning the Chancery, was not as an attribute of an individual person. Rather, an action was alleged, or found, to be "against conscience"—not against the conscience of the defendant, or of any other particular individual. The requirements of conscience were treated as being objectively ascertainable.

The objectivity of the requirements of conscience arose through the account the scholastics gave of moral obligation. It involved two separate concepts, *synderesis* and *conscientia*. *Synderesis* was seen as being a faculty or disposition[13] that was innate in people by which they perceived the moral law. The moral law had an external, objective existence, being laid down by God. To some extent it was part of the natural law; to some extent it was revealed in both the Old Testament and the New Testament. *Conscientia* "involves the application of the knowledge afforded by *synderesis* to actual, particular situations" (Klinck 2010: 33). When the moral law ascertained by *synderesis* was objective, and the facts of the particular situation under examination were known, the requirement of conscience in that situation was itself an objective matter. This usage in the late medieval law is illustrated by remarks of Fortescue CJ (ca. 1394–1476) in an Exchequer case. He said, "We are not to argue law in this case but conscience," and said the word *conscience*

> comes of *con* and *scioscis*. And so together they make "to know with God" to wit: to know the will of God as near as one reasonably can.
> (See Klinck 2010: 15–17)[14]

One of the few recorded pre-Reformation Chancery judgements contains Chancellor Morton's[15] response to an executor who relied on having no liability at common law for having released a debt that was owing to the testator:

> Every law should be in accordance with the law of God; and I know well that an executor who fraudulently misapplies the goods and does not make restitution, will be damned in hell, and to remedy this is, as I understand it, in accordance with conscience.
> (Holdsworth 1945: 222)[16]

That the Chancery required defendants to act to carry out their obligations of conscience was consistent with the church's teaching. For many sins the church required not just penitence and reformation but also taking action to undo or compensate for the wrong that had been done.[17] A priest vicariously exercised God's powers concerning conscience, so confession to a priest and performance of the penance the priest required gave the possibility of obtaining absolution. Consistent with this, a pre-Reformation chancellor (a priest, though with judicial authority) sought to ascertain what conscience required a particular litigant to do and ordered him to do it.

Types of Inadequacy of the Common Law Corrected by the Chancery

The Chancery corrected inadequacies of the common law concerning the substantive legal obligations that it recognized a person as owing, concerning the remedies available if an obligation had been breached and concerning the procedure through which litigation was conducted.

Substantive Law

The Chancery supplemented the common law by recognizing obligations unknown to the common law about the ways in which rights and duties that the common law recognized should be exercised. For example, if property had been transferred to X on the basis that X would use the property for the benefit of Y, the Chancery would require the property to be used in that way; if a mortgagee of land had a common law right keep the land because the mortgagor did not repay punctually the debt secured over the land, Chancery would sometimes require the mortgagee not to exercise his right to forfeit the land.

The Chancery also recognized obligations that did not relate to common law rights and duties. For example, if X had taken on the task of administering Y's business affairs, Chancery would require X to provide detailed accounts of what he had spent and received in administering those affairs, and to correct any shortcomings in his administration of Y's affairs. In each case the Chancery created these obligations because it saw the defendant as having an obligation of conscience to act in the way that the Chancery required. One of the effects of the pervasive influence of the concept of conscience in the Chancery was that the emphasis of the common law was on the plaintiff's right, but the emphasis of Chancery was on the defendant's duty.[18]

Remedies

Once rigidity had set in in the fourteenth century, the only remedy that the common law offered was damages, or an order restoring possession of land. The Chancery had a far more extensive range of orders. Unlike the common law, the Chancery's orders were moulded to fit the facts of the particular case and stated with considerable precision what the defendant must do to undo his particular breach of the requirements of conscience. Without being exhaustive, they included orders requiring specific performance of obligations that the defendant had undertaken,[19] injunctions directing a defendant to do or to not do some specific act, orders for the undoing of transactions that had been procured by means that were contrary to conscience like fraud or misrepresentation or undue influence and orders stripping a defendant of any profit he had made through unconscientious dealing with the plaintiff. The common law could offer a remedy only once a wrong had been done, but the Chancery could issue an injunction to prevent a wrong being done. The common law allowed fraud as a defence to an action, but only the Chancery could order the delivery up and cancellation of documents that had been entered as a result of fraud.[20]

Procedure

The common law required a plaintiff to prove his case by evidence that he mustered himself. All the evidence had to be given on the one occasion, at a trial before a jury, by witnesses giving oral testimony and by the tendering of documents and other physical objects. The parties were not permitted to give evidence. If the facts were known only to the plaintiff, the defendant or to other people who would not cooperate with the plaintiff, the allegations could not be proved.

The Chancery's procedure for conducting litigation derived from the canon law.[21] Unlike the common law not permitting the parties to give evidence, the Chancery *required* the defendant to attend at court and to answer questions on oath concerning the allegations and to disclose documents that related to them. Usually the plaintiff was required to swear to the correctness of his allegations. The Chancery could require witnesses who were not parties to attend court to answer questions on oath.[22] There was no jury. The evidence might be given on different occasions rather than at a single trial. If someone was too ill to attend the court,[23] or might die before the hearing could take place[24] the Chancery could order that evidence on oath in written form be taken from him. It could require someone to disclose the identity of a person he had dealt with so that the plaintiff could sue that person.[25]

Chancery Procedures and Conscience

A significant reason why the Chancery adopted these procedures arose from the way in which *synderesis* and *conscientia* were understood to interact. It was only if the chancellor was confident that he knew the facts concerning a case that he could apply the revelations of *synderesis* and so ascertain what conscience required in the case. If the chancellor was to know the facts correctly he could not leave it to the chance of whatever evidence the parties might put before him. He followed the example of the canon lawyers, of extracting confessions upon oath from the defendant, and requiring evidence on oath from the plaintiff and witnesses. The evidence was required to be given on oath because the litigant would understand that a lie on oath was sinful[26] and that sin could lead to eternal damnation.

Furthermore, even non-parties were required to provide information and documents that were relevant to a piece of litigation. This was because for a person to remain silent about a matter that he knew about, where his silence prejudiced another, was seen as being a mortal sin.[27]

Conscience dictated the form of the Chancery decrees: they were orders *in personam* directed to the defendant and telling him what he must, or must not, do. Conscience also affected the way in which the Chancery enforced its decrees. A person who did not obey an order of the Chancery was in contempt of court, and could be committed to prison until he had purged his contempt. His failure to obey the decree meant that he had not yet acted as conscience required, so strong means were justifiable to save his soul.

The Chancery never enquired whether a judgement that had been given at common law had wrongly applied the common law. Rather, it enquired whether, by reason of circumstances that were not taken into account in the judgment, it was against conscience to enforce it. Similarly, if judgement had not been given, the Chancery enquired whether there were matters beyond those which gave someone a right under the common law, which provided a reason why in conscience that right ought not be exercised. The Chancery could issue a *common injunction*, that is, an order to a litigant not to enforce a common law judgement that had been given, not to proceed with a common law case that he had started, or not to rely on some particular argument in the common law case.

Equity

The word *equity* has multiple shades of meaning. Even concerning the law, it has been used in varying ways. Without being exhaustive, it has meant treating like cases alike,[28] construing statutes in accordance with their purpose, an intrinsic aspect of the common law[29] whereby all laws are interpreted in a way that is just,[30] a filling of gaps in the law and a power of ameliorating harsh consequences of rigid laws. In the course of the sixteenth and seventeenth centuries *equity* came to be used in England to refer to the principles that were applied in the Chancery court[31] to supplement and correct the common law.

It is in the writing of Christopher St Germain (1460–1540) that the notion of equity being administered by the Chancery gains currency. In 1528 he published, in Latin, *Dialogus de fundamentis legum Anglie et de conscientia*. This text is more commonly known as *Doctor and Student* because it takes the form of a dialogue between a Doctor of Divinity and a student of common law. It was the first extended account of the role of equity in English law. An English translation appeared in 1530 or 1531 and was regularly reprinted into the seventeenth century. It continued to be relied on even after the seventeenth century.[32] St Germain continued to regard the role of conscience as fundamental, but he also gave currency to the notion that the Chancery was a court of equity. By the end of the seventeenth century conscience had by no means disappeared from the discourse about the Court of Chancery, but the notion that the Chancery administered equity was well established.

Equity's Change to Being Based More on Identifiable Principles in the Seventeenth Century

Well before the seventeenth century there had been recognized *types* of cases concerning which equity would grant relief. Sir Thomas More said, "[T]hree things are helped in conscience, fraud, accident and confidence" (Rolle 1668: 374).[33]

Furthermore, from the fifteenth century the Chancery had a recognized jurisdiction to relieve against mistake and against penalties and forfeitures. It had a recognized jurisdiction to enforce "uses" of land.[34] The Chancery's ground for interfering in all these areas was to prevent behaviour contrary to conscience, but the topics concerning which it could grant remedies operated as a primitive statement of the sorts of things that conscience could require—for example it operated to undo breaches of conscience concerning forfeitures of mortgaged land or operated to undo breaches of conscience concerning fraud. There was some predictability about the sorts of circumstances that would be seen as being contrary to conscience concerning those topics.

Holdsworth says that in the period after 1616,

> [e]quity tended to become less a principle or a set of principles which assisted, or supplemented, or even set aside the law in order that justice might be done in individual cases, and more a settled system of rules which supplemented the law in certain cases and in certain defined ways. We can see the beginnings of this change in the first half of the seventeenth century. But, during this period, its progress was hindered by the victory of the Parliament, because Parliament suspected the equity administered by a Chancellor in intimate relations with the King. Thus, although we can see the origins of some of our later equitable rules, they are, as yet, very rudimentary. We must wait till the latter half of the seventeenth century for marked progress in this process of transformation. It is not till then that the lineaments of our modern system of equity begin to emerge with any distinctness.
>
> (Holdsworth 1945: 217–218)

The change occurred concerning the substantive obligations that the Chancery enforced: the court continued to grant the remedies and use the procedures that it had previously developed. The judge who made the greatest contribution to this transformation of the Chancery in the seventeenth century was Lord Nottingham, chancellor from 1673 to 1682. He is rightly referred to as the father of modern equity.[35]

The transmutation of the Chancery into a court governed by identifiable principles and rules is illustrated by comparing accounts of the Chancery given in 1615 and around 1674. In 1615 the chief justice of the Court of Common Pleas said that the common law courts and the Chancery

> are fundamental Courts, as ancient as the kingdom itself, and known to the law, for all kingdoms in their constitution are with the power of justice both according to the rule and of law and equity, both which being in the King as sovereign, were after settled in several Courts, as the light being first made by God, was after settled in the great bodies of the sun and moon. But that part of equity being opposite to regular law, and in a manner an arbitrary disposition is still administered by the King himself and his Chancellor, in his name *ab initio*, as a special trust committed to the King, and not by him to be committed to any other. And it is true, that the one is bound to rules, the other absolute and unlimited, though out of discretion they entertain some forms, which they may justly leave in special cases.
>
> (*Martin v Marshall* (1615) Hobart 63; 80 ER at 212)[36]

This passage continues to recognize the king as sovereign, and the fount of justice, and that he has delegated his power to dispense justice to the courts. It also sees the Chancery as arbitrary and not bound by rules as the common law courts are—that the power of the Chancery is unlimited and that while there are some usual practices ("forms") in the Chancery, they are not binding, but rather are adhered to or departed from as a matter of discretion.

This is to be contrasted with Nottingham's account of Chancery jurisdiction, written in about 1674, near the beginning of his period of judicial office.[37] Nottingham first explained the origin of the power of the Chancery in terms that reflected centuries of convention:

> kings being sworn at their coronation to deliver their subjects *aequam et rectam justitiam*, the King must either reserve to himself or refer to others a power to supply or correct the rigor of positive law, which neither is nor can be a perfect rule in all cases.
>
> (*Prolegomena* Ch. III [25], in Yale 1965: 193)

However, he immediately went on to deny that the Chancery would correct all action that was contrary to conscience and to insist that the Chancery's decisions were based on rules:

> And now the matter is so settled that it is become a maxim in our books that the Chancery can only relieve in such cases where the party has no remedy at the Common Law ...
>
> And yet all cases that are without remedy at common law, are not relievable in equity; nor is the rule *nullus recedat a cancellaria sine*

remedio[38] so to be understood: for some cases are only to be considered between a man and his confessor. As for example, a man swears off a true debt by wager of law,[39] or the grand jury affirm a false verdict in an attaint,[40] no court of conscience can help these cases. Yet the party is bound to restitution *sub periculo animae*.[41] *Angusta est innocentia ad legem bonam esse*.[42] And God forbid, a man should use no better conscience than the Chancery can compel him, however the rule must always hold, that 'tis not fit for a court of equity to do everything that is fit to be done; for there is a twofold conscience, *viz conscientia politica et civilis, et conscientia naturalis et interna*. Many things are against inward and natural conscience, which cannot be reformed by the regular and political administration of equity: for if equity be tied to no rule, all other laws are dissolved, and everything becomes arbitrary. Some say, he is called *cancellarius, qui intra cancellos agit*.[43]

(*Prolegomena* Ch. III [26] and [27], in Yale 1965: 193–194)

In 1676 Nottingham echoed these thoughts in his decision in *Cook v Fountain*:[44]

If after all this a man will still suppose that there was a secret trust, security, or agreement between the parties to repurchase this rent, which no bill charges, no proof can make out, and the defendant denies upon oath, then it must be such a trust, security or agreement as is only between a man and his confessor. With such a conscience as is only *naturalis et interna* this Court has nothing to do; the conscience by which I am to proceed is merely *civilis et politica*, and tied to certain measures;[45] and it is infinitely better for the public that a trust, security or agreement, which is fully secret, should miscarry, than that men should lose their estates by the mere fancy and imagination of a chancellor.

((1676) 3 Swans 585 at 600, 36 ER 984 at 990)

Concerning the need for equity to be rule based, in 1678 he wrote,

But yet Justice is a severe thing and knows no compliance nor can bend itself to any man's conveniences, and equity itself would cease to be Justice if the rules and measures of it were not certain and known. For if conscience be not dispensed by the rules of science, it were better for the subject that there were no Chancery at all than that men's estates should depend upon the pleasure of a Court which took upon itself to be purely arbitrary.

(*Earl of Feversham v Watson*, LNCC 2, p. 637 at 639)

Nottingham applied the notion of the difference between private conscience and the principles enforced in his court in a case in which a son had given a note in which he promised to pay his father's debts:

If this note were voluntary and without consideration,[46] though it did bind the son in honour and private conscience, with which I had nothing to do, it could not bind him in legal and regular equity . . .

(*Honywood v Bennett* (1675), case 307 in Yale 1957: 214)

Causes of the Development of Principles and Rules in Seventeenth-Century Equity

Several quite heterogeneous causes of the Chancery's change to being governed by principles and rules can be identified. Some of the causes predated the seventeenth century, but it was only in the course of the seventeenth century that they came to interact with other causes to result in the Chancery acting in a much more principled fashion.

Chancery's Jurisdiction to Issue Common Injunctions Finally Established

The significance of Holdsworth's nominating 1616 as the start of the time during which the Chancery became more rule-bound[47] is that it was then that King James issued a proclamation that settled a long-running jurisdictional battle between the Chancery and the common law courts over whether the Chancery had power to issue an injunction requiring a defendant not to enforce a common law judgement that he had obtained.[48] James's decision was in favour of the Chancery having the power to make that sort of order.[49] It is an irony that James granted the Chancery superiority over the common law courts so that his royal prerogative to relieve against the harshness of the common law could be exercised, but by the time of Nottingham even if the prerogative continued to be acknowledged as the basis of the Chancery's power, that prerogative was required to be exercised in accordance with known principles and rules.

Common Lawyers Replace Clerics as Chancellors

Sir Thomas More (chancellor, 1529–1533) was an eminent common lawyer. His chancellorship marks the beginning of an almost unbroken line of lawyers, rather than clerics, being appointed as chancellors. The appointment of trained lawyers as chancellors meant that they were familiar with the common law and with its methods of reasoning. In the sixteenth and seventeenth centuries there was no rigid split between the Chancery and common law bars, so any barrister was likely to be familiar with both types of courts. Indeed, many of the seventeenth-century chancellors had been high public officials whose office had required them to appear in the common law courts, or had been judges of common law courts, before their appointment as chancellor.[50] It was fairly common for the chancellor to ask common law judges to hear a case with him.

Negatively, the discontinuance of appointment of clerics contributed to a decline in the influence of Catholic moral teaching and of the canon law. Conscience as a generative notion continued to be important, but, as I discuss in the following, there was a shift in the meaning of the word.

One of the characteristics of common law reasoning is that it seeks to confine statements of principle to what is necessary for the decision of the case at hand. This encourages the statement of principles in terms closely connected to the facts of the case. Sometimes the reasoning proceeds by consideration of precedents without explicit statement of a principle to be drawn from them, and the instant case is decided as an exercise in analogy. Sometimes a judge states a rule in terms that are sufficient to decide the instant case, but that cannot be applied as a truly universal rule—in other factual circumstances the rule might be qualified or treated as being subject to an exception. These characteristics came to be part of the way the Chancery cases were often decided.

Precedent

Up to and during most of the Tudor period the record-keeping relating to the business of the Chancery was very patchy. Most of the petitions that the chancellors dealt with have not been kept, and there are only very incomplete records of the reasoning by which decisions were arrived at. Orders of the Chancery were recorded in the Year Books from the time of Henry VI,[51] but they give little guide to the reasons for the order. This had the effect there was no basis on which a doctrine of precedent could have operated in the Chancery, even if having such a doctrine had seemed desirable.

There was no abrupt change, but from about the beginning of the reign of Elizabeth records of pleadings came to be better kept so that it became possible for judges and barristers to identify what had been the issues in earlier cases, as well as the result of the case, embodied in the court order. The availability of the records meant that it became possible for a doctrine of precedent to operate in the Chancery, and the training and mind-set of the common lawyers who had come to be the chancellors and barristers predisposed them to a doctrine of precedent. Gradually, one came to operate.[52]

Francis Bacon provided an account of the significance of previously decided cases: they provided the raw material from which a legal rule could be derived:

> It is a sound precept not to take the law from the rules, but to make the rule from the existing law. For the proof is not to be sought in the words of the rule, as if it were the text of law. The rule, like the magnetic needle, points at the law, but does not settle it.
>
> (quoted in Holdsworth 1945: 240)[53]

Here, by "rule" he means a statement by a judge that in cases of type X result, Y follows. To paraphrase Bacon, a judge deciding a case should not apply the rules stated in earlier cases as though they were binding but, rather, should take decided cases as evidence of the operation of a guiding legal principle ("the law") and ascertain that principle by an inductive process from the cases. From it the judge should then formulate a rule appropriate to decide the case at hand.

Nottingham also held this view, at the time he was a law officer:

> there are precedents that prove too much and that it will concern the Lords as well as the Commons to resort to the good and wholesome rule *Judicandum est legibus non exemplis*.[54] For precedents do not make law; they are only evidences of it, and no few evidences neither. He that thinks so [i.e. that precedents make law] must first conclude there was never an illogical thing done in former times and so every precedent must pass for demonstrated and a general rule may be built upon one or two examples. What is this but to argue *a particulari ad universales* and to put a fallacy upon all the laws and liberties of England?
>
> (Finch Papers, quoted in Yale 1957: xlix)

The acceptance of the importance of precedent made the outcome of particular cases more predictable. In a 1670 case where Lord Keeper Bridgman sat with three common law judges, Vaughan CJ,[55] a common lawyer, was voicing a view of equity that had already become *passé* when he said,

> I wonder to hear of citing of precedents in matters of equity. For if there be equity in a case, that equity is an universal truth, and there can be no precedent in it. So that in any precedent that can be produced, if it be the same with this case, the reason and equity is the same in itself. And if the precedent be not the same case with this, it is not to be cited, being not to that purpose.

Bridgman LK disagreed, saying,

> In them we may find the reasons of the equity to guide us; and besides the authority of those who made them is much to be regarded . . . It would be very strange and very ill if we should distrust and set aside what has been the course for a long series of times and ages.
>
> (*Fry v Porter* (1670) 1 Mod 300 at 307; 86 ER 898 at 902)[56]

Notice that he accepts the authority of like cases decided in the same way over time but says nothing about the authority of a single decision or a very recent decision.

Books Explaining the Chancery

After *Doctor and Student* other theoretical writing about the basis on which the Chancery operated appeared from the late fifteenth century. Robert

Snagg,[57] William West,[58] and Edward Hake[59] all wrote treatises on the Chancery. Richard Crompton wrote a work on the jurisdiction of the various royal courts, which included a section on the Chancery,[60] including an account of the orders recorded in the Year Books.

Law Reports

The use of precedents in the Chancery was assisted by the publication in the mid-seventeenth century of summaries of the results of decided cases and indexes to decided cases arranged by subject matter, with occasional terse notes concerning the reason why the court reached its conclusion.[61] Notwithstanding the mid-century publication dates for these summaries, before then "many manuscript copies circulated, especially of Cary."[62] There were no regular published reports of the reasons for judgement in Chancery cases until 1693.[63] Even so, before then the summaries and indexes of the decisions made the court records of the decisions locatable.

Effects of Seventeenth-Century Struggles over the Prerogative

When James I became king of England in 1603 he was already influenced by continental notions that in a state a particular body or person was sovereign, in the sense of having a right to command that was not bound by any civil law,[64] and that in England, it was the king who was sovereign. In Tudor times it had been recognized that the monarch had prerogative powers, but the question of how those prerogative powers related to the powers of the courts and of the Parliament had not arisen.[65] James came believing that the King was both "supreme ruler and supreme judge; that he was above the law, which he could make, mitigate, or suspend; and that he was answerable for his acts to God alone" (Holdsworth 193: 20).

This led to protracted disputes with both the common law courts led by Sir Edward Coke,[66] and the Parliament, about whether the royal prerogative was the supreme power in England. The judges took the view that the common law was the supreme law, and the judges the sole expounders of that law; the king took the view that the judges were officers of the Crown who decided cases unless the king decided that he should decide the case himself.[67] The Parliament took the view that it had privileges of its own that the king could not remove.

The dispute with Parliament led to King Charles I ruling without summoning Parliament from 1629 to 1640. The Long Parliament, from 1640 to 1658, passed legislation that limited various aspects of the prerogative.[68] The ongoing dispute about the extent of the prerogative became a fundamental cause of the English Civil War, which lasted, on and off, from 1642 to 1651. During the Commonwealth period there was no Lord Chancellor actually exercising judicial functions.[69] Instead, Parliament issued its own Great Seal of Parliament in November 1643. It was held by a succession of groups of men called commissioners, usually three at a time, until the Restoration in 1660. During these seventeen years the commissioners carried out the functions that the chancellor had formerly carried out.

Dislike of the Chancery during the Commonwealth resulted in pressure for it to become more like an ordinary court. Part of the basis for the attack on it was its slowness and expense, but as well it was attacked for its absolutism and arbitrariness and interference with ordinary legal rights.[70] When the basis of the Chancery's existence had been the king's prerogative power, it was a natural target for the men who had found the King guilty of treason and cut off his head in 1649.

Procedural reforms of the commissioners in 1649 dealt with some of the Parliamentarians' objections, but even so the Barebone Parliament of 1653 voted to abolish the Court of Chancery. However, that vote did not come to be embodied in legislation. The Chancery survived because by then it was recognized to "occupy an integral part in the machinery of the law" (Yale 1965: 8).[71]

Even though the Restoration in 1660 was in form a restoration of the monarchy as it had been at the time of the execution of Charles I, the political reality was that it was no longer possible for there to be personal rule by a monarch, in the way the Tudors and the first two Stuarts had ruled. It was no longer possible for the King to override, or to dispense with, the Parliament. Parliament had attained a power to mould even the way in which the prerogative was exercised:[72] "the experience of the nation under a Protectorate, which had constantly found itself under the necessity of violating the law, had increased the national desire to see the law really supreme."[73]

When the Restoration occurred, the Chancery had been functioning for seventeen years with judges whose authority did not derive from the king. The old theory that Chancery operated as a delegate of the king, who had a personal power and responsibility—and the Stuarts would add, a God-given right—to administer justice could not explain how the Chancery had continued during this time. Some other justification for its operation was needed.

Changes in the Concept of Conscience

While Thomas More's appointment as Chancellor preceded the Reformation,[74] continuing to appoint lay Chancellors after the Reformation was consistent with the Reformation. After the Reformation everyone had access to the scriptures, one of the ultimate sources of the moral law, so clerical chancellors were in no better position than lay chancellors to understand the requirements of conscience.[75]

At first the account of conscience, as dependent on God's law apprehended through *synderesis*, was not changed. The change was rather that every individual had both the means, and the responsibility of ascertaining God's law, and therefore of ascertaining what conscience required.

But there came to be a multiplicity of views about the content of God's law. By the seventeenth century the Reformation had led to there being no longer a single Catholic Church but a proliferation of religious opinions, many of them stressing the importance of the individual deciding what God required him or her to do. Notwithstanding Henry VIII's initial intention that the English Reformation should change the governance of the Church of England without changing its doctrine, the doctrine changed from previous Catholic doctrine to give greater emphasis to the importance of an individual reading and seeking to understand the Bible.[76] As well, there continued to be Catholic recusants, Puritans contended that the Church of England needed further purification, there were Calvinist-influenced Presbyterians, and there were many minor sects. When different interpreters of the scriptures came to different conclusions about what the law of God was, this had the potential to undermine the objectivity of conscience. It also had the potential to undermine the view that human laws were justifiable because they approximated to the law of God.

A more fundamental problem for the pre-Reformation account of conscience came from a shift in the epistemological *zeitgeist* that occurred from about the middle of the seventeenth century. It acknowledged the difficulties in having certain knowledge concerning many subject matters and recognized that in many subject matters only probable knowledge was attainable.[77] Even in natural philosophy certainty was not attainable, and other subject matters, including moral and civic ones, were in their nature even less able to provide certain knowledge.[78] It was a logical consequence of this that *synderesis*—traditionally seen as a means of gaining knowledge—must be on shaky ground. Practising lawyers did not fully appreciate or adopt this logical consequence, and in any event the law has never regarded logic as its sole inspiration,[79] so the old language of equity being based on conscience continued. But even so, it is hard to believe that lawyers would be totally immune to such an important shift in the climate of intellectual opinion in their country.

The major premise of a pre-Reformation argument about the requirements of conscience was the content of the moral law; its minor premise was what the facts of the instant situation were. It was not only different possible views of the content of the moral law that led to an undermining of the pre-Reformation view that conscience provided the basis for the intervention of the chancellor; the minor premise became unreliable too. One emphasis in Protestant casuistic writing was that, consistent with conscience being "knowing with another," the relevant "other" was God. But it was only the person himself, and God, who could know *all the facts* that related to what conscience required of that person. Thus, on the Protestant account it was only the person himself and God who could know what conscience required him to do. Any Protestant could say, "No person can tell me what conscience requires me to do. And that means that a judge cannot tell me what conscience requires me to do."

The potential for the objectivity of conscience to be undermined had become a reality in the early seventeenth century. Sir John Selden (1584–1654) wrote,

> [I]f we once come to leave that out-loose, as to pretend conscience against law, who knows what inconveniency may follow? For thus, suppose an anabaptist comes and takes my horse; I sue him, he tells me he did according to his conscience; his conscience tells him all things are common among the saints, what is mine is his; therefore you do ill to make such a law, if any man take another's horse he shall be hanged. What can I say to this man? He does according to his conscience. Why is not he as honest a man as he that pretends a ceremony, established by law, is against his conscience? Generally to pretend conscience against law is dangerous . . .
>
> (Selden 1892, Title 26: Conscience)

The language that came to be used concerning the Chancery was that, instead of it being an objective and impersonal conscience, it was the conscience of an individual person that the Chancery applied. However, there was vacillation about exactly whose conscience it was. Sir Christopher Hatton, chancellor from 1587 to 1591, spoke of it as being the conscience of the Queen.[80] Similarly, a letter of advice from Sir Francis Moore to Bishop John Williams in 1621 says that the Chancery has

> [a] power of jurisdiction according to Equity and Conscience which burden his Lordship is to take upon himself and undergo. But that Conscience is the King's committed to the Chancellor and if the Chancellor shall in his own private Conscience be of another opinion then he is persuaded the King his master would be, he is to judge according to the King's Conscience and not his own, which rules he means ever to hold.
>
> (Yale 1965: 78)[81]

Sometimes, though, the language used was that it was the conscience of the chancellor that should be applied,[82] and the conscience of the litigant that should be corrected. But regarding the relevant conscience as being that of the chancellor produced difficulties concerning the legitimacy of the chancellor's orders. It led to Sir John Selden's jibe:

> Equity is a roguish thing: for law we have a measure, [and] know what to trust to. Equity is according to the conscience of him, that is Chancellor; and as that is larger, or narrower, so is Equity. 'Tis all one, as if they should make the standard for the measure we call a Foot, a Chancellor's

foot. What an uncertain measure would this be? One Chancellor has a long foot; another a short foot; a third an indifferent foot. 'Tis the same thing in the Chancellor's conscience.

(Selden 1892, title 37: Equity)

While these remarks are not precisely dated, they necessarily precede the date of Selden's death in 1654. They illustrate clearly the dangers there would have been in conscience remaining the sole standard by which equity acted.

Before the English Reformation conscience had been a theoretically suitable standard for a law because it was understood to be objective, and applied to everyone regardless of their consent. But once conscience had its objectivity undermined, if equity was to be part of a body of law, rather than a manifestation of the individual will or power of the king or the chancellor, some other sort of objectivity was required. It came in the form of stating principles and rules, more specific than a bare invocation of conscience, which applied in deciding cases of particular types. This was done in a piecemeal fashion, case by case, and with greater concreteness concerning some topics than others. The aim of the principles and rules might well be to articulate what conscience was likely to require in a particular type of case, but they still had an existence of their own, and could count as law. It was these principles and rules that made up the "civil and political conscience" that the Chancery applied. Their historical origins in the church's notions of what conscience required gave them the distinctive ethical flavour that equity continues to have.

Casuistry

There was a mass of writing in seventeenth-century England, in contexts separate from the law, about how to reconcile the claims of conscience with other general principles of conduct. Mostly it was in religious and political works. Much of this writing involved *casuistry*. There was already a body of Catholic casuistic writing, originally written to provide guidance to a priest who had to decide what to require of a person who made confession to him. As well a body of Protestant casuistic writing grew up, that applied the more individualistic Protestant concepts of conscience.

At the time *casuistry* did not have the overtone it now has of sophistry, or excessively clever or cynically Jesuitical reasoning. The name derives from *casus*, Latin for a "case." It is a type of reasoning about what proper behaviour in a particular case requires, when the decision about what to do is arrived at by exploring the limits of a general proposition about how a person should behave. Typically it accepts that in a paradigm case a particular result should apply, but then considers whether that result should continue to apply in the particular factual situation concerning which a decision must be made. It does so by focusing on the detail of the particular case and on the circumstances that explain why the general proposition correctly states how one should usually behave, rather than on general theories or concepts.[83]

The method of the common law in the seventeenth century was casuistic,[84] taking the rules[85] articulated in previous decisions as given, then considering whether the detail of the facts of the present case was such that the rule should not be applied and, if necessary, articulating a modified rule that could be applied in the instant case. The Chancery followed the casuistical method of the common law. In so doing it started from the Chancery's imprecisely expressed historical notions of what conscience required, tested how and the extent to which they applied in particular situations, and arrived at more precisely stated principles of conduct.

By the end of the seventeenth century or beginning of the eighteenth, conscience had come to be regarded as such an internal and personal thing that casuistry itself stopped being part of the public debate.[86] But at least it was in accord with the casuistic theory on foot for much of the seventeenth century, and also with its assimilation of common law methods of reasoning, for the Chancery to develop its reasoning consistently with the casuistic approach. This had, and continues to have, a profound influence on the nature of the principles and rules that the law of equity implements.

Effects of the Chancery's Pleading Practices and Changes to Them

By the middle of the sixteenth century the Chancery had developed well-understood procedures for pleading—identifying the issues for decision in a case—which differed from those of the common law.[87] A plaintiff began a Chancery suit by a *bill*, which was required to "show an equity"—that is, set out facts that if true, and the only relevant facts, would justify the Chancery in granting him a remedy. For example, if a person wished to sue to enforce a right arising under a bond or other document under seal but that document had been lost, the common law gave no remedy because tendering of the document was essential.[88] However, the Chancery would allow a remedy, if it was established that the document had indeed been lost.[89] The facts that the bond had once existed, had not been paid, and had been lost gave the plaintiff the entitlement to approach the Chancery for relief—that is, constituted the plaintiff's equity. Nottingham recognized this equity when he wrote,

> [A] bill founded upon no other equity but oath made of the loss of the bond or deed is not to be admitted unless all who are concerned in law or equity be made parties.
>
> (Nottingham, *Manual*, Title III [10] at Yale 1965: 94)

Once a plaintiff had filed a bill the defendant then had several courses of action open to him. One was to file a *demurrer*, by which he could seek to have the case dismissed at that stage because the bill did not show an equity.

A bill might fail to show an equity because it did not allege all the facts that were necessary to make out a type of relief that the Chancery could give or because it claimed relief concerning a type of situation for which the Chancery lacked jurisdiction to give any relief at all.

Having an equity was necessary but not sufficient for a plaintiff to win— there was always the possibility that a set of facts that was recognized as giving rise to an equity when that set of facts stood alone, might not entitle the plaintiff to win if some additional facts were added. Thus, another course of action open to a defendant was to file a *plea*—a document alleging that there were facts beyond those alleged in the bill which, when taken together with those alleged in the bill, showed that the situation was one concerning which the Chancery would not grant the remedy that the plaintiff claimed.

One type of facts alleged by a defendant that, if made out, would show that the plaintiff was not entitled to a remedy, or not entitled to the remedy he claimed, were known as *defensive equities*.[90] Typically a defensive equity was stated in language of high generality, and that defensive equity could be applied to counter a wide variety of different equities. But it was also possible for a plea to allege extra facts that did not amount to a defensive equity of a general type but were more closely connected to the facts alleged in the bill.

One defensive equity was known as *unclean hands*—conduct of the plaintiff relating to the facts on which his claim was made that showed that he ought not be granted relief because of his own bad conduct. An eighteenth-century statement said that before unclean hands provided an equitable defence, "it must have an immediate and necessary relation to the equity sued for."[91] However, Nottingham had earlier recognized in substance this requirement for the defendant's unworthy conduct to be closely related to the facts that constituted the plaintiff's equity:

> Williams sued the Countess of Arundel to be relieved against the forfeiture of a lease for non-payment of rent. The Countess showed many injuries the plaintiff had done her against conscience, as felling of timber, selling lead, forging acquittances, and refusing to account on pretence of some releases, which were given in trust not to be used if his account could be falsified; and [she] offered to confirm his lease, if he would account. Yet decreed for the plaintiff, for no ill conscience of a plaintiff in collateral matters shall foreclose him of relief, but it should be *in eâdem*; else no sinner should be relieved in equity.
> (Nottingham, *Prolegomena*, Ch. XXV [7], in Yale 1965: 305)

Another defensive equity expressed in unspecific language and able to counter a wide variety of equities was *laches* (long delay on the part of the plaintiff in bringing the claim).[92]

In the course of the seventeenth century a procedure was introduced requiring that a demurrer be decided in open court rather than by a Master.[93]

Inevitably this practice fostered the growth of known rules because the limits of the principles concerning the matters that the court had authority to hear, and what facts were sufficient for the plaintiff to have an equity, were decided in court, at an early stage of proceedings. Furthermore, the judge's decision concerning the demurrer would be recorded in the court records. In that way what it decided about whether the bill disclosed an equity would become known to practitioners.

If a defendant filed a plea, frequently the court decided the case by determining the adequacy of the plea without having a full argument of the case.[94] If that happened, the only enquiry the court engaged in was whether the extra facts that the defendant alleged showed that the plaintiff's bill lacked an equity:

> When a plea is allowed and replied to, the defendant has nothing to do
> at the hearing but to prove his plea; and therefore if the plaintiff prove
> all the equity suggested in his bill, yet he shall not be suffered to read,
> because nothing is material but the truth or falsehood of the plea, so the
> defendant never examines to the equity suggested in the bill . . .
> (Nottingham, *Prolegomena*, Ch. XXIII [8], in Yale 1965: 298)

The practice of often deciding the adequacy of a plea before any full hearing had the effect of producing decisions on what combination of facts constituted a successful defensive equity, or a reason for some specific type of equity to fail.

The Nature of Equity's Principles and Rules

At the start of the seventeenth century, it was possible to identify, from previous decisions and from some of the writing about the operation of the court, principles expressed in quite open-textured and imprecise language that the Chancery required people to adhere to in their dealings in matters of property and business. These included not obtaining a benefit by fraud or accident, that a person should faithfully carry out tasks concerning the performance of which another had reposed confidence in him, that an obligation to which several people were subject should be shared fairly, and that the substance of a commercial arrangement should be respected rather than its legal form. These principles are expressed in language that contains an element of evaluation, either explicitly ("fairly," "faithfully") or implicit in its terminology (what counts as "fraud" or "accident"). They had all originated as general statements of what conscience required in certain types of circumstances.

The Chancery applied these very general principles, like requiring faithful performance of tasks concerning which confidence had been reposed, to different types of subject matter, for example, to the responsibilities of an executor and to the duties of a person who was administering the business affairs of another person. Uses of land[95] had largely faded from significance in the sixteenth century of as a consequence of some legislation of Henry VIII,[96] but in the course of the seventeenth century, the Chancery recognized that if property was conveyed to X (the trustee) on trust for Y (the beneficiary), there were obligations of conscience on X similar to those that the use had imposed. Once the trust was recognized, the general principle requiring faithful performance of a task undertaken was applied to the trustee.

There were other principles of a high level of generality that the Chancery applied in granting its own remedies. They have come to be referred to as the equitable maxims. For example, by reference to the maxim that "he who seeks equity must do equity," the Chancery would grant relief to a plaintiff only if the plaintiff carried out his own obligations of conscience concerning the subject matter in question.[97]

In the course of the century these principles were developed by successive decisions so that the principles themselves, and rules derived from them, became sufficient to provide a fair measure of predictability of the outcome of most disputes, without any need for the invocation of conscience.

It is possible to identify three different ways in which the Chancery developed these principles. One was by the Chancery making decisions about whether in a particular set of facts a general principle had been infringed. The multiplication of instances gave greater content to the evaluative terminology in the principle, by acting as a kind of ostensive definition of what counted as fraud or accident, of when a task of a particular kind had not been faithfully carried out, or of when a common obligation had not been fairly shared. However, the multiplication of instances did not deprive the original principle of its force, so that if, for example, some new method of obtaining a benefit by fraud arose, it was possible for the Chancery to give a remedy concerning it.

Similarly, the course of decisions provided a multiplication of instances of what type of factual situation was enough to give rise to a defensive equity. For example, the Chancery refused to assist a plaintiff who had obtained a security from the defendant while the defendant was drunk,[98] and the cases gave some specificity to how long a delay in enforcing a particular type of claim, in what circumstances, could give rise to a defence of *laches*.

A second way in which principles were developed is that the broad open-textured principles generated some more specific statements of the type of outcome that the law would require in a particular type of case. These more specific statements were of the type "When there is a factual situation of type A, result B will follow." These more specific statements are sometimes referred to as rules of equity. For example, in the course of the seventeenth century, equity developed specific rules about the rights and obligations of trustees and beneficiaries and the rights and obligations of mortgagors and mortgagees. One such rule is that a trustee is entitled to a full indemnity from the trust property for his costs of bringing litigation concerning the trust.[99] Another is that an executrix who used estate money to trade on her own account was liable for any loss of the money and also obliged to pay interest on it even if it was not lost.[100] These rules were statements of what one of the Chancery's general principles would require in a particular type of factual situation. For example, the general principle that a trustee must faithfully perform the office he had undertaken requires that he take no more care of the trust property than he would take of his own property; thus, when a trustee of gold was robbed of that gold and also of some of his own property he was not liable for the loss.[101]

A third way in which the principles were developed was by the recognition of exceptions to a rule, so that the rule did not apply in a certain type of situation. Thus one co-trustee was not liable for breaches of trust by his co-trustee—but there was an exception if he took part in them,[102] or by his own breach of duty made them possible.[103] A trustee was not obliged to account for more than he has actually received from the investment or utilization of the trust property—but there was an exception if there was a "very supine negligence," in which case he could be charged for what he would have received if he had not acted negligently.[104] The exceptions just mentioned can all be seen as arising from considering what qualifications to the rule were required to give effect to the more basic principle, that there should be faithful performance of a task concerning which trust had been reposed.

However, these principles and rules do not have their full significance revealed by the language in which the judges expressed them. Nottingham frequently stated, in terms that look absolute, propositions about the result that a court would arrive at in a particular set of circumstances—for example "a purchaser shall be charged with an incumbrance though it be defective in law, if he had an allowance for it in the price";[105] "A man pays a statute[106] or obligation without acquittance, and is sued again; he shall be relieved in Chancery";[107] "If one joint tenant takes all the profits, his companion may sue him in equity, for in conscience he has a right but to a moiety."[108] Such statements cannot be treated as though they are universally applicable. They must be understood bearing in mind that all the judges and lawyers realised that they were made in the context of a contest between opposing claims in litigation, that their purpose was to state a principle or rule that would resolve the issues that had been identified by the process of pleading in that particular case, and that the law of equity worked through the recognition of equities asserted by competing parties. Each of the apparently absolute propositions can be transposed as a standard for deciding a different case only if it is understood as being subject to an implied "unless some equity is shown why that result should not arise" and an implied "unless there are extra facts that show that the equity asserted does not arise." Thus, each of the principles and rules articulated by the judges provided a guide to the result that the court would come to in a particular type of factual situation, but that guidance was only ever provisional: it was dependent on the "unless" not arising.

Furthermore, the principles and rules themselves might sometimes prove inadequate to dictate what was required in a particular factual situation. It remained possible that two general principles that the Chancery had developed led to different results when applied to a particular factual situation. It remained possible that in a particular factual situation a rule that the Chancery had developed was in conflict with a different equitable principle to the one that had generated the rule. It remained possible that a principle of the Chancery led to a different result to a principle of the common law. In each of these sorts of situations it was necessary for the judge to decide which principle or rule should prevail.

In time, decisions about which of two general principles, or which of a particular rule and a general principle, would prevail when they were in conflict, themselves became precedents, capable of providing a new principle about which in which circumstances one of two competing claims should prevail. But the principles and rules themselves were always capable of being re-examined in the light of the very general statements of principle that had been derived from the Chancery's earlier reliance on conscience as its standard of judgement.

By the end of the seventeenth century, rules and principles already established were enough to provide a fair measure of certainty of the outcome of most litigious disputes in the Chancery. They could not always provide certainty because the Chancery's casuistic method inevitably left open some possibility of pre-existing principles and rules not dictating a single outcome concerning some factual situations. However, the extent to which this created uncertainty was not a serious obstacle to equity providing a workable and intelligible system for dispute resolution. This was because these conflicts between principles or rules did not arise often and, when they did the area of any dispute was limited to which of the principles and rules already recognized in decisions of the court, should prevail.

Notes

* A version of this chapter was given at the colloquium on Principles in Early Modern Thought conducted by the School of Philosophical and Historical Inquiry and the Sydney Centre for the Foundations of Science at Sydney University, 27–29 August 2014.
1 There were three such courts, the Court of Kings Bench, the Court of Common Pleas, and the Court of Exchequer.
2 The precise words of the oath changed from time to time, but the central ideas in it had a remarkable consistency for over a millennium. It is accurate enough to say, "The king bound himself by a threefold promise to preserve peace and protect the church, to maintain good laws and abolish bad, to dispense justice to all. This oath had been taken by English kings from the tenth century: it was taken by William the Conqueror and by his successors," Richardson 1941: 129. Condren 2006: 254 distills the central obligations as, "to act justly, and maintain law, custom, religion and the office itself." The earliest extant version of the oath, administered by Dunstan, Archbishop of Canterbury in the period

959–988, includes words translated as, "I promise and command justice and mercy in all judgments": http://www.earlyenglishlaws.ac.uk/laws/texts/sacr-cor/view/#edition,1/translation,1. The oath of Edward II included a promise, "to do equal and right justice and discretion in mercy and truth": *Statutes of the Realm*, 1: 168, cited in Baker 2002: 98. The coronation oath of Henry VIII included an affirmative answer to the question, "Will you so far as in you lies, cause justice to be rendered rightly, impartially and wisely, in compassion and in truth?.," Stephenson and Marcham 1937: 192. For the accession of William and Mary in 1688 Parliament's victory over the Stuarts concerning the prerogative led to the Coronation Oaths Act 1688 prescribing a different version of the second promise in the oath, to make clear that the monarch was bound to govern "according to the statutes in Parliament agreed upon, and the laws and customs of the same." However, the 1688 version continued to contain a promise to "cause law and justice in mercy to be executed in all your judgments." Also see Yale 1965: 193.

3 The Lord Chancellor held the Great Seal, a symbol of royal authority, which was used to authenticate the writs he issued. Holding the Great Seal became a symbol of exercising the functions of the Lord Chancellor.

4 The judges of the common law courts were also often clerics and were influenced by Church moral teaching. This affected the comparatively broad approach they adopted to available remedies in the first two or three hundred years of the existence of the common law courts. See Holdsworth 1945: 215; Potter 1948: 551–552.

5 See note 26 of this chapter for the significance in medieval England of an oath.

6 Holdsworth 1956: 403. Sometimes the chancellor's legal duties would be carried out by a temporary chancellor, called the Lord Keeper of the Great Seal. Because their legal functions were the same, throughout this chapter I refer to the *chancellor* as covering both the Lord Chancellor and the Lord Keeper, unless a distinction is explicitly made between a Lord Chancellor and a Lord Keeper.

7 Holdsworth 1956: 404–407.

8 The chancellor was assisted by twelve Masters and a large number of clerks: Holdsworth 1956: 417, 421–423.

9 An important and thorough account of the role that conscience played in the Court of Chancery, particularly in the sixteenth and seventeenth centuries, is Klinck 2010.

10 Potter 1948: 559. This was made explicit in St Germain's *Doctor and Student* (discussed in the text following note 32): "When the law eternal or the will of God is known to his creatures reasonable by the light of natural understanding, or by the light of natural reason, that is called the law of reason: and when it is showed of heavenly revelation . . . then it is called the law of God. And when it is showed unto him by order of the Prince, or any other secondaries governor, that has power to set a law upon its his subjects, then it is called a law of man, though originally it be made of God . . . For if any law made of men bind any person to anything that is against the said laws (the law of reason or the law of God) it is no law but a corruption and a manifest error": quoted Potter 1948: 559.

11 Potter 1948: 559.

12 The *Statute of Appeals 1533* declared the king to have power in both spheres and outlawed appeals to the pope in ecclesiastical matters. The *Act of Supremacy 1534* declared the king the head of the English church. See, generally, Holdsworth 1956: 588–592.

13 There was some debate about whether it was a faculty (i.e. a potentiality of the human mind to apprehend a particular type of thing, in the way the faculty of

70   *Campbell*

sight enables visible things to become known) or a disposition (i.e. a habit or in-clination). The difference does not matter in understanding the medieval notion of conscience for present purposes.

14 My translation. The case is reported under the heading "conscience" in Nicholas Statham's *Abridgement des libres annals*, published 1490, recounted in Klinck 2010: 15–17. The Exchequer was a court whose particular concern was disputes that related to the royal revenues. From at least a decade before the accession of Queen Elizabeth it applied both common law and equity: Bryson 1975: 3.

15 John Morton was chancellor from 1487 to 1500, at a time when he was also the archbishop of Canterbury.

16 Slightly different words, to the same effect, are cited from Y. B. Hil. 4 Hen. 7, f.4, p. 8 (1489), in Fifoot 1949: 326.

17 Klinck 2010: 27.

18 This expressed itself in the forms of the orders of the respective courts: "Upon a decree in Chancery the Chancellor directed the defendant to pay the money due: the Court of law adjudged the plaintiff entitled to recover a certain sum," Yale 1965: 20.

19 That is an order that the defendant actually do what he had undertaken to do.

20 Holdsworth 1956: 457.

21 Browne 1933: 22.

22 Originally the questioning of both the defendant and the non-party witnesses was oral, but by the reign of Elizabeth the questioning was done before examiners, with the questions and answers recorded in depositions. See Browne 1933: 24.

23 For example, *Wade v Gwye* (1558–1559) Cary 41, 21 ER 22; *Bagnold v Green* (1559–1560) Cary 48, 21 ER 26.

24 For example, *Barentine v Harbert* (1559) Cary 42, 21 ER 23.

25 Holdsworth 1945: 332.

26 Its sinfulness was shown by Numbers 30:1–2: "Moses said to the heads of the tribes of Israel: 'This is what the Lord commands: When a man makes a vow to the Lord or takes an oath to obligate himself by a pledge, he must not break his word but must do everything he said.'" It was also shown less directly by Exodus 20:7 and Numbers 32:20–24.

27 Whitman 2008: 94 and 104 attributes this view to Vincent of Beauvais, Hostien-sis, Guillaume Durand (in *Speculum iuris*, Venice, 1576), and to a work Ordo Judicarius "Scientiam."

28 Cicero, *Topica* 4 23: "Valeat aequitas, quae paribus in causis paria iura de-siderat." The glossator Martinus in twelfth-century Bologna paraphrased it as "Aequitas est rerum convenientia, quae paribus in causis paria jura desiderat," and Martinus's language was adopted by Bracton in the 1250s in *De Legibus et Conseutudinibus Angliae*, 2: 25.

29 Hake, cited in Klinck 2010: 100.

30 The Roman law concept of *aequitas*.

31 A variety of English courts have administered equitable jurisdiction—the Chan-cery, the Court of Requests at Whitehall, the equity side of the Exchequer, and minor courts like the courts of the Palatinates and the City of London. However, by far the most significant was the Chancery, and I confine my attention to that court.

32 St Germain's notion of equity was not confined to the role of the Chancery—he also has a concept of equity of a statute (something like its purpose), which is administered by the common law courts. One of his examples of this was that if there is a statute that forbids citizens to open the city gates before dawn, but the gates are opened to allow in some citizens fleeing an enemy, the statute is not breached. Another is that there is no breach of a statute forbidding the giving of

alms to sturdy beggars if the gift is of clothing to protect a beggar in imminent danger of freezing to death. At other times he speaks as though the common law courts as well as the Chancery dispensed equity. See Klinck 2010: 44–50.

33 "Accident" refers to a person losing a legal right through a misfortune not of his own making. "Confidence" is the type of confidence that a person displays in transferring land to X to the use of Y, or in giving X power to administer business affairs on his behalf.

34 See text at paragraph preceding note 18.

35 Strictly Nottingham was Lord Keeper from 1673 to 1675 and Lord Chancellor from 1675 to 1682. Yale 1965 says, "The nine brief years during which he held the Great Seal were decisive for the future of Equity." Some but not all of Nottingham's judgements were published in the early nineteenth century in Swanston's reports (reproduced in volume 36 of the English Reports). All his judgements were published in Yale 1957 and 1961.

36 Hobart was chief justice of the Court of Common Pleas from 1613 to 1625.

37 Yale 1965: 76.

38 "No-one should depart from the Chancery without a remedy."

39 Wager of law was a medieval procedure where the defendant in a criminal trial swore that he was innocent, or the defendant in a civil claim swore that the claim was not well founded, and other people swore that he was a credible person. It survived concerning actions for debt until the nineteenth century: Plucknett 1956: 115–116.

40 The grand jury was a body of men drawn from the locality where a crime had been committed who had the responsibility of deciding whether a person should be required to stand trial for the crime. They were required to swear that they would give a true verdict. They acted on the basis of notorious local knowledge, rather than evidence in the modern sense, and were a combination of compelled witness and compelled accuser: Whitman 2008: 136–137. A grand jury was also used in medieval times to decide civil claims on the basis of the local knowledge of the jurors. An attaint was, in general terms, the loss of all civil rights suffered by someone who was convicted of treason or a felony. It had a particular meaning concerning decisions by a grand jury in a civil trial, namely that a second jury was empanelled to decide whether the first jury had given an inaccurate verdict. It was also an inquiry into whether the members of the first jury had perjured themselves, by giving a verdict they knew to be false. If it was found that the members of the first jury had perjured themselves they suffered the usual punishment for an attaint: Plucknett 1956: 131–133.

41 "At the peril of his soul."

42 "What narrow innocence it is to be good only according to the law" (Seneca, *De Ira* II 27).

43 "The Chancellor, who acts within barriers/boundaries/limits."

44 (1676) 3 Swans 585 at 600, 36 ER 984 at 990. Nottingham was considering a contention that counsel for the plaintiff had advanced in the course of argument even though it had not been alleged in the bill. The contention was that a rent-charge (a type of interest in land) that the plaintiff's deceased father had given to the defendant was given on the basis that it was to be a security for the father later making a substantial gift to the defendant, and that when that gift was later made the rent-charge came to be held on trust for the father, because the purpose of transferring the rent-charge had then been accomplished.

45 By "certain measures" he means clear known standards.

46 "Voluntary" is used in the lawyers' sense of being done when there was no obligation to do it, "without consideration" in the sense of without a price or other benefit having been provided to the son in exchange for his executing the note.

47 See Holdsworth 1945: 217–218.

48 The issue had been simmering since the early sixteenth century but came to a head in the *Earl of Oxford's Case* in 1615–1616.

49 The text of his order is found at (1616) 1 Chan Rep at 48, 21 ER 588.

50 Ellesmere (chancellor, 1596–1617) had been solicitor-general and then attorney-general from 1581. Francis Bacon (chancellor, 1618–1621) had been attorney-general from 1613. Thomas Coventry (chancellor, 1625–1640) had been attorney-general from 1621. John Finch (chancellor, 1640–1641) had been chief justice of the Common Pleas from 1634. Sir Orlando Bridgman (chancellor, 1667–1672) had been chief baron of the Exchequer for several months in 1660, from which position he was appointed chief justice of the Common Pleas. Nottingham (chancellor, 1673–1682) had been first solicitor-general and then attorney-general from 1660. North (chancellor, 1682–1685) had been chief justice of Common Pleas from 1674.

51 Spence 1846: 416.

52 On one occasion Ellesmere found that a trust "was not so fully proved as the Lord Chancellor would make a full decree thereupon, so as it should be a precedent for other causes, yet so far forth proved, as it satisfied him as a private man; and therefore . . . he thought fit to write his letters to the defendant to conform himself to reason; and affirmed, that if you should find the defendant obstinate, then would he rule this cause specifically against the defendant, *sans la tires* consequence": *Mynn v Cobb* (1604) Cary 25, 21 ER 14.

53 *De augmentis scientiarum*, Bk VIII, Ch. 3, Aph 85, SEH 5 106.

54 A quotation from Justinian Codex. 7.45.1, roughly, "decide in accordance with the law, not previous cases."

55 Sir John Vaughan was chief justice of the Court of Common Pleas from 1668 to 1674.

56 Counsel were ordered to attend on the judges with precedents. These "precedents" were likely to have been records of the court, particularly those in the books of decrees and orders, rather than published reports of cases. See Yale 1957: xliv.

57 Died 1605. His work was a reading to the Middle Temple in 1581, part of which was reduced to writing in 1587 and presented to Sir Christopher Hatton on his becoming chancellor. It was published decades later as Snagg 1654.

58 West 1594. It was republished in 1637. See Spence 1846: 416.

59 In Hake 1953. Hake lived circa 1545 to circa 1604. His treatise was written in the reign of Elizabeth, and presented to James I in 1604. That he used the Greek word for *equity* in the title is indicative of the extent to which Aristotle's treatment of equity had become known in English legal circles at the time.

60 Crompton 1594.

61 William Tothill (1560–1620), who had been one of the Six Clerks, composed an index of the Chancery's cases, divided into topics that were arranged alphabetically. Often it gave just the name of the case and the year it was decided, though for about half of the entries there is also a one- or two-sentence summary of the principle involved in the case and for very occasional entries a longer note concerning the case. It was published in 1649. Brief summaries of the Chancery's decisions in the period 1557 to 1604 were collected by Sir George Cary. They were printed in 1650. *Choyce Cases in Chancery* consisted of a 103-page treatise on the practice of the Court of Chancery, and 75 pages of case summaries, covering cases from 1576 to 1605. The case summaries had marginal notes that identified the topic of the case. It was printed in 1652.

62 Yale 1957: xlii.

63 The first thing approaching proper reports was *Reports and Cases taken and adjudged in the Court of Chancery in the reign of King Charles I, and to the*

20th year of King Charles II. It covered cases decided in 1625 through 1668 and appeared in 1693. (1668 was the twentieth year of King Charles II because, notwithstanding the political realities, he was regarded as having become king immediately after the death of Charles I in 1649.) A second volume in the same series, covering decisions from 1668 to 1693, appeared in 1694. The third and final volume, covering 1660 through 1688, appeared in 1716. *Cases Argued and Decreed in the High Court of Chancery* was published in 1697. It contained cases from 1662 to 1679. A second part appeared around 1700, containing cases up to the third year of the reign of James II (1687–1688). Nottingham kept detailed records of his own reasons for judgement, but they were not published until long after his death. See note 35 of this chapter.

64 For example Bodin 1576, Book I Ch. 8, Book II Ch. 5; Holdsworth 1937: 11–13, 273–276.

65 Holdsworth 1937: 20.

66 Chief justice of Common Pleas, 1606 to 1613; chief justice of Kings Bench, 1613 to 1616. James removed him as a judge in November 1616, but until 1628 Coke used Parliament as his forum to oppose James's views concerning the prerogative.

67 Holdsworth 1945: 428–440.

68 Holdsworth 1937: 112–119.

69 Charles II appointed Lord Keepers of the Great Seal during the period of his exile, but the Great Seal had been captured and destroyed by Parliament on 11 August 1646, and it was only upon the Restoration in 1660 that Charles's appointees actually exercised power.

70 An illustration of the departure from the ordinary course of the law, which attracted the ire of the parliamentarians, was that Lord Keeper John Finch in 1641 remarked that an order of the Council was sufficient ground for issuing a decree in the Chancery: Yale 1965: 9.

71 Holdsworth 1956: 431–434 gives some detail of the attack on the Chancery during the Commonwealth. A further attempt to limit the jurisdiction of the Chancery, in 1690, also failed because it would produce great hardships and injustices: Holdsworth 1956: 464.

72 Holdsworth 1937: 162.

73 Holdsworth 1937: 162.

74 His appointment may well have been motivated by Henry VIII needing to mollify the dissatisfaction of the common lawyers (who were an influential element in the House of Commons) at the high-handed way in which Cardinal Wolsey had overridden the common law in the name of conscience and to gain the support of the common lawyers for his matrimonial and ecclesiastical policy: Holdsworth 1945: 219–220.

75 Hake 1953, cited in Klinck 2010: 102.

76 A law of 1538 required every parish to purchase a copy of the Bible in English and to keep it available for anyone to read. King James commissioned a new English translation of the Bible, which was published in 1611 and was widely used.

77 Klinck 2010: 184ff.

78 Tillotson 1688: 96, cited in Klinck 2010: 185.

79 In *Tatham v Huxtable* (1950) 81 CLR 639 at 649 Fullagar J referred to the law as "a system which has never regarded logic as its sole inspiration."

80 Spence 1846: 414. In *Lenthall v Waring* (1676) Case 466 at Yale 1957: 333 Nottingham said "all courts which hold pleas in equity by any kind of pretence administer the conscience of the King, for the subject is not to be governed by the conscience of a corporation, or of any court but what depends upon the King . . ."

81 The letter is advice about what Williams should say in his speech upon becoming Lord Keeper. Given that context, the references to "his Lordship" and the "he" in "rules he means ever to hold" are references to Williams in his role as Lord Keeper.
82 Klinck 2010: 124–125.
83 Condren 2006: 172 describes seventeenth-century casuistry as "too complex phenomenon to be seen as any single school, or doctrine of moral reasoning. It was, rather, a constellation of propensities sharing the recognition that principles under-determine conduct."
84 And still continues to be.
85 In the sense used by Bacon. See Holdsworth 1945: 240.
86 Klinck 2010: 216; Leites 1988b: 119–120.
87 Aspects of this course of procedure can be found much earlier (see Spence 1846: 338, 345, 349, 362), but what matters for the development of principles is that the procedural frame in which litigation occurred was already in place by the reign of Elizabeth (see Spence 1846: 379), which was before the detailed development of many of the principles.
88 Story 1853: 99.
89 Nottingham, *Prolegomena*, Ch. VII [2], in Yale 1965: 213; Nottingham, *Prolegomena*, Ch. VIII [1], in Yale 1965: 217.
90 The defendant's equity could in turn be qualified or obliterated by still further facts that the plaintiff established. For example, facts that by themselves would suffice to establish a defence of *laches*, could fail to do so if the defendant had encouraged the plaintiff to delay in bringing his claim. In principle, this countering of one party's equity by an equity of the other party could go on indefinitely, though in practice it hardly ever went beyond two or three steps. Other alternatives open to a defendant were to file an *answer*—a document that disputed one or more of the facts that the plaintiff alleged, and stated what the defendant asserted was the truth concerning that matter—or a *disclaimer*—a statement that he had no interest in the matter (Spence 1846: 372). However, answers and disclaimers had no consequences for the development of equitable principles.
91 *Dering v Earl of Winchelsea* (1787) 1 Cox 318 at 321; 29 ER 1184 at 1185.
92 *Sedgwick v Evan* (1582–1583) Choyce Cases 167, 21 ER 97 (mortgagor not entitled to recover possession of mortgaged land when mortgagee had been in possession for 40 years), *Sibson v Fletcher* (1632–1633) 1 Ch. Rep 59, 21 ER 507 (mortgagee, not in possession of the mortgaged land, who had made no claim for payment for 17 years, refused relief), *Hales v Hales* (1636–1637) 1 Chan Rep 103, 21 ER 520 (similar, but neither interest paid nor claim made for payment of principal for forty years); *Garford v Humble* (1618) Tothill 27, 21 ER 113 ("ancient bonds" ordered to be cancelled when holder attempted to sue on them), *Simons v Lee* (1613) Tothill 129, 21 ER 144 (action for payment of legacy fails when not pursued for sixty years).
93 The Commissioners' rules 4, 7; Lord Clarendon's rules 12, 17; and Lord Nottingham's Rules title Pleadings, rule 36 (in Yale 1965: 101). Even earlier than these rules, in the advice to Bishop Williams about what to say in his induction speech in 1621, Sir Francis Moore suggested that he say that "if any man demur upon a bill presented in Chancery and rest upon the demurrer, his Lordship mindeth to have the bill and demurrer brought before him and to take conusance thereof himself," Yale 1965: 79. Under earlier Chancery procedure if a defendant demurred to a bill the demurrer would be decided by a Master. It was only if there was an appeal from the Master's decision that the matter would be considered by the chancellor.
94 Sometimes the adequacy of the plea was left to be decided at a hearing. On occasion Nottingham said he would order a defendant to answer the bill "saving the benefit of his plea at hearing, but would not stifle the equity of this case upon a plea": *Shermar v Cox* (1674) in Yale 1957: 76 (Case 124).

95 Explained in paragraph preceding note 18 in this chapter.
96 The *Statute of Uses*, 1536.
97 Yale 1957: lxi–lxii. The chapter heading to Ch. XXV in the *Prologemena* was "He that seeks relief in equity must proceed equitably." For example, the Chancery would set aside a transaction procured by fraud only if the plaintiff refunded whatever he had received under the transaction, with interest: *Vere Essex v Muschamp* (1684) 1 Vern, 23 ER 438.
98 *Rich v Sydenham* (1671) 1 Chan Cas 202, 22 ER 762 ("He who has committed Iniquity, shall not have Equity").
99 *Amand v Brasdbourne* (1682) 2 Chan Cas 138; 22 ER 884.
100 *Ratcliff v Graves* (1683) 1 Vern 196, 23 ER 409, also reported (1683) 2 Chan Cas 152, 22 ER 890.
101 *Morley v Morley* (1678) 2 Chan Cas 2, 22 ER 817.
102 *Townley v Shurborne* (1633) Tothill 88, 21 ER 132.
103 *Capell v Gostow* (1614–1615) Tothill 88, 21 ER 132.
104 *Palmer v Jones* (1682) 1 Vern 144; 23 ER 376.
105 Nottingham, *Prolegomena* Ch. VI [14], in Yale 1965: 211.
106 A statute, in this context, means a bond given with particular formalities in a trading town, pursuant to the statute *De Mercatoribus* 1285 or the *Statute of the Staple* 1353. There were special procedures for enforcing such bonds, designed to provide quick remedies to traders for payment of debts.
107 Nottingham, *Prolegomena* Ch. VI I [1], in Yale 1965: 213. In other words, if a person pays a statute or a debt, and does not take a written acknowledgement that he has paid, and is sued for it, the Chancery will prevent that action from proceeding.
108 Nottingham, *Prolegomena*, Ch. VI I [3], in Yale 1965: 213.

Bibliography

Baker, J. H. (2002) *An Introduction to English Legal History*, 4th edn, Oxford: Oxford University Press.
Bodin, J. (1992 [1576]) *On Sovereignty*, ed. and trans. J. H. Franklin, Cambridge: Cambridge University Press.
Bracton, H. (1922 [1250s]) *De legibus et consuetudinibus Angliae*, vol. 2, ed. G. E. Woodbine, New Haven: Yale University Press.
Browne, D., ed. (1933) *Ashburner's Principles of Equity*, 2nd edn, London: Butterworths.
Bryson, W. H. (1975) *The Equity Side of the Exchequer*, Cambridge: Cambridge University Press.
Cicero. (2003) *Topica*, ed. and trans. T. Reinhardt, Oxford: Oxford University Press.
Condren, C. (2006) *Argument and Authority in Early Modern England*, Cambridge: Cambridge University Press.
Crompton, R. (1594) *L'Avthoritie et Iurisdiction des Covrts de la Majestie de la Roygne*, London.
Durand, G. (1576) *Speculum iuris*, Venice.
Fifoot, C. H. S. (1949) *History and Sources of the Common Law: Tort and Contract*, London: Stevens and Sons.
Hake, E. (1953) *Epieikeia: A Dialogue on Equity in Three Parts*, ed. D. E. C. Yale, New Haven: Yale University Press.
Holdsworth, Sir W. (1937) *A History of English Law Vol. VI*, 2nd edn, London: Methuen.
———. (1945) *A History of English Law Vol. V*, 3rd edn, London: Methuen.

————. (1956) *A History of English Law Vol. I*, 7th edn, London: Methuen.

Klinck, D. R. (2010) *Conscience, Equity and the Court of Chancery in Early Modern England*, Farnham: Ashgate Publishing.

Leites, E., ed. (1988a) *Conscience and Casuistry in Early Modern Europe*, Cambridge: Cambridge University Press.

————. (1988b) "Casuistry and character" in ed. E. Leites 1988a, pp. 119–133.

Plucknett, T. (1956) *A Concise History of the Common Law*, 5th edn, London: Butterworths.

Potter, H. (1948) *An Historical Introduction to English Law and its Institutions*, 3rd edn, London: Sweet and Maxwell.

Richardson, H. G. (1941) "The English coronation oath," *Transactions of the Royal Historical Society*, 23: 129–158.

Rolle, H. (1668) *Abridgement des Plusiers Cases et Resolutions del Common Ley: Alphabeticalment Digest Desouth Severall Titles*, London.

Saint-German, C. (1974) *Doctor and Student*, eds. T. F. T. Plucknett and J. L. Barton, Selden Society, vol. 91, London: Bernard Quaritch.

Selden, Sir J. (1892) *The Table Talk of John Selden*, ed. S. H. Reynolds, Oxford: Clarendon Press.

Snagg, R. (1654) *The Antiquity & Original of the Court of Chancery, and Authority of the Lord Chancellor of England*, London.

Spence, G. (1846) *The Equitable Jurisdiction of the Court of Chancery*, London: V. and R. Stevens and G. S. Norton.

Stephenson, C. and Marcham, F. G., eds. (1937) *Sources of English Constitutional History*, vol. 1, New York: Harper Brothers.

Story, J. (1853) *Commentaries on Equity Jurisprudence, as Administered in England and America*, 6th edn, Boston: Little, Brown and Company.

Tillotson, J. (1688) *The Rule of Faith*, London.

West, W. (1594) "Of the Chauncerie" in *Three Treatises, of the Second Part of Symbolaeographie*, London.

Whitman, J. Q. (2008) *The Origins of Reasonable Doubt: Theological Roots of the Criminal Trial*, New Haven: Yale University Press.

Yale, D. E. C., ed. (1957) *Lord Nottingham's Chancery Cases, vol. 1*, Selden Society, vol. 73, London: Bernard Quaritch.

————. ed. (1961) *Lord Nottingham's Chancery Cases, vol. 2*, Selden Society, vol. 79, London: Bernard Quaritch.

————. ed. (1965) *Lord Nottingham's "Manual of Chancery Practice" and "Prolegomena of Chancery and Equity,"* Cambridge: Cambridge University Press.

3 Alchemical and Chymical Principles

Four Different Traditions

William R. Newman

It is a well-established fact that the *tria prima* (three first things) of early modern chymistry, namely, mercury, sulfur, and salt, descend from the medieval alchemical theory that metals and minerals are composed of mercury and sulfur. There have been few attempts, however, to look at the development of these theories in detail, and to examine the relationship of the main medieval versions to their early modern descendants.[1] If we take the trouble to do so, several interesting features emerge. First, the integration between theory and experimental practice, even in the earliest phase of the sulfur–mercury theory in medieval Europe, becomes obvious. Second, close analysis allows one to understand the grounds for the appeal of these theories. Unlike the four scholastic elements, which were abstract entities sharing only the tactile qualities of heat, cold, humidity, and dryness, the theories of two principles and *tria prima* were based on materials that shared the full range of perceptible qualities (the "secondary qualities" of the scholastics) accessible to the senses. It is, therefore, ironic that one sometimes encounters the claim among historians that the alchemical-chymical principles were "metaphysical," "spiritual," or "ideal" entities, in contrast to the material constituents of things envisioned by eighteenth-century chemistry. According to this historiographical school, the alchemists—conceived of as a monolithic group—acknowledged that their material principles were too pure or "high" to be capable of isolation in the laboratory: they were mere notional entities that could not be handled or touched.[2] Other scholars have recently pointed to the fact that pre-Lavoisian chymistry was home to more than one prominent tradition, however, and that while some early modern chymists may have propounded a view that their principles were of a "spiritual" nature, many historical figures took a more materialist position.[3]

The present chapter extends the notion of multiple traditions in chymistry in two fashions. First, it expands the discussion to include high medieval alchemy alongside the chymistry of the early modern period, in order to illustrate both continuities and diachronic shifts in the concept of material principles distinct from the traditional four elements. And second, this chapter examines three different yet more or less synchronic schools that were all influenced by the medieval claim that metals and minerals consist of sulfur and mercury but which all accepted the third principle of salt that had been introduced by Theophrastus von Hohenheim or Paracelsus in the sixteenth century. The three schools can be viewed respectively as upholders of Hypostatical Principles, Posterior Principles, and Hierarchically Complex Principles. As we will see, there is some overlap among the Hypostatical, Posterior, and Hierarchical positions, and yet they served as rallying points for framing diverse arguments.

The Mercury–Sulfur Theory: Arabic Origination and Medieval European Appropriation

The theory that metals are composed of mercury and sulfur probably originated in the early Middle Ages with the Arabic *Book of the Secret of Creation* (*Kitāb sirr al-Khalīqa*) attributed to a pseudonymous Balīnūs (probably meant to signify Apollonius of Tyana).[4] This text did not receive wide dispersion in the Latin West, however. Instead, the sulfur–mercury theory met its earliest European circulation in Arabo-Latin alchemical texts of a largely practical nature and in the work of a vehement opponent of chrysopoeia (transmutation of base metals into gold), the Persian philosopher Avicenna. Despite his opposition to the *summum bonum* of the aurific art, Avicenna accepted the sulfur–mercury theory as a way of explaining the subterranean generation of metals: in fact, he gave a particularly cogent exposition of it that would enter into the mainstream of scholastic philosophy. Avicenna makes considerable use of the sulfur–mercury theory in his *De congelatione et conglutinatione lapidum*, also known as *De mineralibus*.[5] This short but influential section of Avicenna's *Kitāb al-Shifā'* was translated into Latin at least twice in the Middle Ages.[6] The most important of the translations, done by Alfred of Sareshel at some point in the second half of the twelfth century, often circulated with the translation of Aristotle's *Meteorology* Book IV made by Henricus Aristippus, and sometimes with the newer translation of the *Meteorology* made by William of Moerbeke in the thirteenth century.[7] Because of its manuscript transmission, Avicenna's text looked to contemporaries as though it was the conclusion of Aristotle's *Meteorology IV*; this fact lent the text considerable prestige and provided it with an authority that it might otherwise have lacked.[8]

Although Avicenna famously attempts to rebut *chrysopoeia* in another passage of his *De congelatione*, the text nonetheless accepts that metals are formed within the earth from mercury and sulfur that differ mainly in color and purity. There are dirty and clean sulfurs and mercuries, and these can vary in color as well, according to Avicenna. Only a very clean and pure mercury and sulfur will yield gold; the converse will produce lead or iron. Hence, the metals are obviously compounds, but to Avicenna this is true of the principles as well: sulfur and mercury are themselves compounded from more primitive ingredients. Sulfur is made up of an earth and water that are cooked together underground until they form a "strong commixture" and become unctuous. Eventually this unctuosity finds a cool place within the earth and congeals to form sulfur. Mercury is produced when a "subtle, sulfurous earth" and a subterranean water combine in a "strong mixture": the presence of the earth accounts for the fact that quicksilver can roll around on a flat surface without sticking to it.[9] Avicenna makes no attempt to distinguish his two principles from ordinary brimstone and quicksilver. In fact, he deduces the presence of water in sulfur and earth in mercury from empirical considerations such as the easy fusibility of the former and the latter's tendency to be repelled by other surfaces. Nonetheless, the fact that his sulfur and mercury exist in varying degrees of purity within the metals suggests the possibility that the two principles could themselves be subjected to processes of refining.

This is precisely the approach that numerous Arabo-Latin *practicae* of the High Middle Ages adopted. The *Liber secretorum de voce Bubacaris*, for example, a reworking of a genuine Arabic text by Abū Bakr Muḥammad ibn Zakariyyā' al-Rāzī (d. 925 or 935 CE), advises the alchemical practitioner to wash and sublime quicksilver in order to free it from its excess humidity.[10] Sulfur, on the other hand, must be liberated from its superfluous combustibility and oiliness, for which other alchemical processes are required.[11] The object of these operations is to arrive at a clean, dry form of quicksilver, probably to be identified with the "corrosive sublimate" ($HgCl_2$) of later times, and an unburning sulfur: the *Liber secretorum* then employs these "purified" principles in a variety of operations meant to result in chrysopoeia. From the alchemist's perspective, these refined principles would not be compounds, but rather the mercury and sulfur themselves, which had now been liberated from the undesirable materials previously clinging to them. There is no reason to assume that the *Liber secretorum* has anything else in mind other than a simple refining of the principles to free them from their impurities. Indeed, as anyone who consults the German translation that Julius Ruska made of Rāzī's *Kitāb al-Asrār* can see, the original text is filled with similar recipes for purifying and refining a host of mineral products ranging from sal ammoniac and orpiment to boraxes and alkalis.[12] To reiterate, there can be little doubt that the native quicksilver and brimstone of the Rāzīan tradition consisted of the ordinary materials that we know by those terms today, although the products arrived at by means of alchemical purification and refining could differ radically from our modern elemental mercury and sulfur. At any rate it is clear that mercury and sulfur were treated as ordinary materials to be operated on in the alchemical workspace, not as hypothetical bearers of properties.

While remaining true to its empirical roots, the sulfur–mercury theory underwent an important and little-studied development in the course of the thirteenth century at the hands of its European recipients. Writing under the influence of the rather obscure Nicolaus Peripateticus, Albertus Magnus exercised important modifications to the theory in his *Liber mineralium* stemming from the early to mid-1250s.[13] In his chapter on the matter out of which metals are made, Albert argues that the two principles, like many other things, contain a twofold unctuosity or oiliness. Sulfur, for example, has an extrinsic unctuosity containing a burning, feculent earthiness, yet the same principle also holds an intrinsic, noncombustible humidity at the roots of its essence, which cannot be removed by fire.[14] It is important that Albert uses the distillation of ethanol from wine to justify these comments, pointing out that wine has "a supernatant, flammable unctuosity" that is accidental and can be extracted from it, leaving behind the second, noncombustible "unctuosity" in the less volatile part of the wine.[15] Just as wine contains a highly volatile, combustible material that can be distilled off (ethanol) and a less volatile component that is not combustible, so too does the metallic principle sulfur. Albert therefore continues to say that normal sulfur contains a burning unctuosity that blackens and burns metals when it is fused and dropped on them. For this reason, Albert adds, alchemists advise that sulfur be washed in acid solutions and sublimed until its unctuosity is removed and the sulfur can withstand fire without burning. As for quicksilver, Albert first paraphrases Avicenna's claim that it contains a liquid component along with a subtle earth. In conformity with his theory of a twofold unctuosity, however, Albert describes the moist component of mercury as a "subtle, unctuous, humidity" (*humidum subtile unctuosum*) rather than a "water."[16]

In short, Albert's theory postulates that mercury and sulfur each contain two unctuosities or humidities, one volatile and one fixed. Since the metals are themselves composed of the two principles, it follows that each metal should itself contain a fixed and a volatile component, though the degree of fixity and volatility will vary with the particular metal. To summarize Albert's position, then, each of the metals is composed of the two principles, sulfur and mercury. But because sulfur and mercury contain their own extrinsic, volatile components, and also intrinsic, permanent ones, the metals themselves are also volatile or permanent in varying degree. Moreover, Albert insists that a clever metallurgist or alchemist can determine the presence or absence of the principles by examining the characteristics of the given metal when subjected to an intense fire, as in the case of silver, which releases a sulfurous stench when it is calcined at the fire. And as we have seen, Albert thinks that alchemists can remove the extrinsic components of sulfur, at least, by operations involving sharp waters and sublimation. Albert's claim that quicksilver and sulfur have their own extrinsic and intrinsic components, and also his observation that these can be detected by heating the metal, would exercise considerable influence in subsequent alchemy, above all in the *Summa perfectionis* of Geber and the tradition stemming from it.[17]

The *Summa perfectionis*, composed around the last third of the thirteenth century, employs fire analysis to detect both fixed and unfixed principles, which are merely new terms for the radical and extrinsic principles of Albertus Magnus. The *Summa* devotes extensive chapters to determining the fixed and volatile constituents of the materials then known as metals, namely gold, silver, tin, lead, copper, and iron, as well as those of mercury and sulfur themselves. In addition to "manifest" tests such as weight and color of the metals, the *Summa* also extensively employs analysis by fire, that is, calcination, which causes the metals to release sulfurous vapors and leave behind calces of varying fixity and color. Other tests are also used, such as the respective metals' ease of amalgamation with quicksilver (easier amalgamation indicates that metal contains more mercury), their ability to melt before or after glowing with red or white heat in the furnace, their resistance to the vapors of "sharp things" (mostly dilute corrosives), and so forth. Even the standard assaying tests of cupellation and cementation are employed in order to determine the nature and quantity of the metals' mercury and sulfur.[18]

The *Summa perfectionis* exercised a powerful influence on later medieval alchemy, forming a point of departure for the most important texts of both the pseudonymous corpus ascribed to Arnald of Villanova and to that of Ramon Lull. In point of fact, the calcination tests found in the *Summa perfectionis* and prefigured by Albertus Magnus provided the origin of the fire analysis that is usually associated with the Paracelsian school of the sixteenth and seventeenth centuries, though these medieval sources far exceeded Paracelsus in terms of clarity and comprehensiveness. Although it would be a very worthwhile project to follow the fate of this materialist tradition into the fourteenth and fifteenth centuries, we must now turn to the early modern period, where we will see both appropriation and radical reorientation of the sulfur–mercury theory.

Paracelsus and the Origination of the *Tria Prima*

Let us now pass from these sober theories of metallic composition to the exuberant imagination of Paracelsus, who is famous for adding a third principle, salt, to the preexisting dyad of sulfur and mercury, and for asserting that *all* things—not just metals and minerals—are composed of three principles. Although the choice of salt was conditioned by empirical considerations, Paracelsus's addition of a third principle was clearly the product of theological concerns. Indeed, his entire system has been aptly described by a recent scholar as "Bible-based science," for he tried to derive the major features of his cosmology directly from scripture.[19] Indeed, in his *Book of Meteors*, Paracelsus argues that all things are made "from three" because the *fiat lux* by which God created the world was necessarily threefold, being the expression of the Father, Son, and Holy Spirit.[20] The three principles exist in the outer reaches of the cosmos just as they do in the sublunary realm. Hence, the Sun is composed of a white, diaphanous salt; the Moon, of a red sulfur; and the "darkness" of the night sky is actually a type of mercury. But precisely in this same locus we also learn that there are four elements, which are "heaven," air, water, and earth. Paracelsus argues that these elements act as "mothers" to produce the material world. They are stamped with the Trinitarian activity of the three principles.

In a closely related work, the *Book of the Generations and Fruits of the Four Elements*, Paracelsus relates that these four immaterial mothers produce four distinct physical layers in the cosmos, the totality of which he calls *Domor*.[21] These elemental layers are paired so that the outer one consists of air or "chaos" on its external side and fire on the internal, while the inner pair is made up of earth on the outer and water within. This is somewhat artificial, however, for Paracelsus admits that products of fire exist in the air, such as the Sun, which "swims" in the "chaos." Similarly, brooks and streams obviously are found on the Earth's surface. The outer borders of each element are referred to as its Yliadum. At the internal surface of each sphere, the material element is produced by its "mother," which is evidently intangible, whereas the external Yliadum is the point at which the element "dies" and returns to the immaterial. In typically Paracelsian fashion, however, the Yliadum is not only a physical border; it is also the maturation point of a given elemental "fruit." When a fruit of earth, such as a chestnut tree, grows to the end of Earth's Yliadum, it begins to penetrate the sphere of fire, and so dies. The putrefaction of vegetable matter by the heat of the Sun, which "consumes" it, is thus viewed as the result of a given "fruit's" trespass on the neighboring elemental sphere: "each thing, when it arrives at its Yliadum, is subjected to putrefaction, and rots." Thus, Paracelsus describes the Yliadum of the Earth as not proceeding beyond the surface of the Earth more "than the height of growing things."[22] This area is the "lower chaos," which Paracelsus distinguishes from the "upper chaos" where the element of fire is located.

Paracelsus's frequent denials of elemental mixture may stem from the fact that he equated the elemental layering of the sublunary world with the elements themselves. The observation that there are four distinct layers of air, fire, water, and earth that act as mothers or wombs of material things in the Paracelsian cosmos may be what led his Danish follower Petrus Severinus to argue in his *Idea medicinae* of 1571 that the Paracelsian elements were *loci*—places—rather than material things. Yet the material "fruits" or products of each individual elemental mother ranged from four material elements conceived of much as Empedocles or Aristotle envisioned them to such defined species as metals and minerals, which Paracelsus claimed to be fruits of water. The grafting of three principles onto these four elements surely derives in large part from Paracelsus's overwhelming insistence on the Trinity as a pattern for the cosmos. At the same time, however, Paracelsus built on medieval practices of fire analysis like those of Albertus Magnus and Geber. He frequently appeals to the chymical anatomizing or separation of the three principles in material substances, as in this passage from the *Opus Paramirum*, recently translated by Andrew Weeks:

Begin with wood: It is a body. Let it burn: that which burns is sulphur. What smokes is mercurius. What turns to ash is sal. This combustion shatters the understanding of the peasant; but the physician begins to receive from it the eyes of a physician. Thus are found in it three things, no more, no less; and each separated from the others. Furthermore, it can be said of these three that all things have the three; and even though they do not disclose themselves in one manner to the eyes, art discloses them by bringing things to a certain point and making them visible [*so eröffnets die kunst/ die solchs dahin bringet vnd sichtig macht*]. For that which burns is *sulphur*. Nothing burns but *sulphur*. What appears as smoke is *mercurius*; nothing is sublimated but *mercurius*. That which turns into ash is *sal*: nothing turns into ash unless it is *sal*.

(*Opus Paramirum* in Paracelsus 2008: 319)

From this and many similar passages in Paracelsus's work, it seems clear enough that he viewed the three principles as they existed in the sublunary realm to be material entities that were combined with one another in sensible bodies. In some instances the empirical bases of this theory emerge strikingly: "salt" in the form of potash and lye are indeed leached from the ash of plants, as Paracelsus well knew. But unlike the medieval proponents of the sulfur–mercury theory which we examined earlier, Paracelsus no longer thinks of his *tria prima* as literal brimstone and quicksilver (along with salt): instead, Paracelsus's mercury, sulfur, and salt are more general analogues that receive their names from these apothecary-substances.[23] The art of *spagyria* or chymical analysis merely "opens" (*eröffnet*) the body and reveals its preexisting anatomy of mercury, sulfur, and salt by separating them. Nonetheless, this seemingly straightforward view of chymical analysis contains a host of difficulties, as can be seen in the interpretations of Paracelsus proffered by his early modern interpreters. The multiple attempts to adopt or exploit Paracelsus's system and to salvage the less objectionable parts of it led directly the emergence of the three schools of interpretation that I have labeled the Hypostatical, Posterior, and Hierarchical.

Hypostatical Principles

The term *hypostatical principles* was immortalized by Robert Boyle, who used it throughout his works as a blanket expression for the *tria prima* of Paracelsus and those who followed him. Yet Boyle's usage was in reality quite inexact even if some of his contemporaries also employed "hypostatical principles" broadly.[24] As a term of chymistry, *principia hypostatica* originated from a specifically Trinitarian interpretation of the *tria prima*, which many of those who adopted the material theory of three principles did not share. From a scholarly perspective, we should therefore limit our use of "hypostatical principles" to the particular school that retained the main concepts implied by the term "hypostasis," in particular the ubiquity of the three principles *stricto sensu*. The most famous popularizer of the hypostatical principles was without doubt Joseph Du Chesne, the Huguenot writer who adopted the Latinized *nom de plume* Quercetanus. In his *De priscorum philosophorum materia* of 1603, a work devoted primarily to the medical uses of chymistry, Quercetanus repeatedly employs the term *hypostatical principle* when speaking of mercury, sulfur, and salt. He was not the first to do so, as the expression is already found in Helisaeus Röslin's 1597 *De opere dei creationis seu de mundo*. Indeed, Quercetanus knew Röslin's work well, since his own *Ad veritatem hermeticae medicinae* (1604) incorporates large chunks of the *De opere dei*, including the very passages where Röslin uses the term *principia hypostatica*.[25] Nonetheless, Quercetanus was dependent on numerous sources beyond Röslin, in particular, Severinus.[26] The issue of plagiarism need not detain us here in any event, however, for Quercetanus was certainly the main point of dispersion for the Hypostatical interpretation of the three principles even if he was not the originator.

Much of Quercetanus's *De priscorum philosophia* is devoted to the marvels of saltpeter, a material that contributed alike to the deflagration of gunpowder, the manufacture of nitric acid, the preservation of meats by corning, and the fertilization of crops. His comments on the three principles of niter are revealing:

> The three principles are extracted from nitrous earth, which can be mutually separated from one another and yet all of them consist of one and the same essence. They are distinct in their properties and powers alone, in which [their essence] is manifested. The incomprehensible mystery of three persons in one and the same hypostasis, which make up the divine Trinity, can be referred to this in a certain fashion. The highest Creator has wished to be revealed *unitrinus* or *triunus* not only in that which is found in the nature of the earth, but in all works of the whole creation.
>
> (Quercetanus 1609: 17)

Here we can see that Quercetanus did not employ the term *hypostasis* merely in order to give the sense of the English "underlying nature" as one might reasonably suppose from the Greek origin of the word. His point here is not that the three principles underlie the sensible phenomena of matter, though he would of course have agreed with that statement. Rather, his use of the term comes from the indissoluble union of the persons in the Trinity, whereby Father, Son, and Holy Spirit are distinct, and yet all share the same substance or hypostasis. Moreover, since the Holy Trinity is found in "all works of the whole creation," it would seem to follow that even mercury, sulfur, and salt themselves must each reveal this tripartite structure. In other words, just as the persons of the Trinity are never found alone, so the *tria prima* should each contain one another. In fact, Quercetanus meant this claim to be taken quite literally, as his further remarks on nitrous earth reveal.

In *De priscorum philosophia* Quercetanus describes the refining of saltpeter in some detail since he explicitly wants to base his doctrine of hypostatical principles on the three materials extracted from nitrous earth.[27] He points out, quite correctly, that nitrous earth (or, as we would say, impure niter) often contains common salt (our NaCl), which can be separated out of the raw material by crystallization. For Quercetanus, this constitutes one of the hypostatical principles of nitrous earth. The saltpeter that is isolated by this procedure itself consists of two additional salts, one sulfurous and one mercurial; since these were already contained in the original nitrous earth, they provide its two other hypostatical principles. The sulfurous component reveals itself in the ignition of saltpeter when charcoal is added to it or when it is an ingredient in gunpowder. Hence, in accordance with the Paracelsian concept of the *tria prima*, we have now extracted two components of the nitrous earth, one solid (salt) and the other combustible (sulfur); what remains to be found is a volatile, vaporous constituent (mercury). This third component Quercetanus identifies with "an acid mercurial [part] participating in the nature of sal ammoniac."[28] As he points out, this is usually bound up with the sulfurous component since both are capable of being sublimed, and yet the acid portion can be separated. Quercetanus clearly has in mind the manufacture of "spirit of niter" (nitric acid) from saltpeter by means of destructive distillation (usually with the addition of vitriol). He is making much the same move as the phlogiston theorists of the eighteenth century did when they argued that ordinary sulfur contains an acid spirit (our sulfuric acid).[29] Just as sulfur releases its acid when burned *per campanum*, so saltpeter relinquishes its acid, mercurial spirit when analyzed by distillation. Thus, Quercetanus argues that he has now extracted all three of the *tria prima* from the original impure niter. But what of his claim that the hypostatical principles exist "in all works of the whole creation" and hence should be inseparable from one another like the persons of the Holy Trinity?

The *De priscorum* insists on a strict reading of the term *hypostatical* in several places, as in the following passage:

> Therefore none of these three principles are found alone and simple, not participating in another. For salt by virtue of the two said salts [i.e. the sulfurous and mercurial components of niter] contains an oleaginous substance within itself, and a mercurial one; sulfur [contains] a salty and mercurial substance, and mercury a sulfurous and saline [one]. But each retains the name of that which predominates in it.
>
> (*De priscorum*, Quercetanus 1609: 97)

Unfortunately, Quercetanus provides no further empirical basis for this claim, though one could take it as a simple admission of the fact that all processes of purification are necessarily imperfect (even the highly refined, "chemically pure" products sold by modern laboratory supply companies are usually only brought to 95%–99% purity). More important, the requirement that any material must contain all three principles follows necessarily from Quercetanus's adoption of the Trinity as a model for material being, and this is a consequence that he readily accepts.

This ubiquity of the three principles should be seen as the real hallmark of the Hypostatical school. The same claim is found, for example, in the founding work of the French "chemical text-book tradition," Jean Beguin's 1612 *Tyrocinium chemicum*. After describing in some detail the separation of multiple natural substances into the three principles, Beguin introduces an important concept with a marginal tag: "No principle is entirely simple" (Nullum principium plané simplex est). Beguin's text then goes on to reproduce almost verbatim the passage that we just quoted from Quercetanus's *De priscorum philosophia* (without acknowledgment), indicating that "this must be noted first" before proceeding.[30] Given the multitude of editions, commentaries, and derivative texts that the *Tyrocinium chemicum* generated, we have every right to think of the claim that the three principles are literally ubiquitous as representing a distinct Hypostatical school.[31]

Posterior Principles

It will be a matter of surprise to some that I include Joan Baptista van Helmont among the supporters of the three principles since the Flemish chymist devoted a long and biting chapter of his 1648 *Ortus medicinae* to debunking the "*tria prima chymicorum principia.*" It would seem quite natural for van Helmont, who insisted that water was the primitive material of the world, to deny a primordial status to the three principles. Moreover, van Helmont's attack was the immediate inspiration for Robert Boyle's *Sceptical Chymist*, whose primary target was the three chymical principles and the four Aristotelian elements.[32] A careful reading of van Helmont reveals, however, that the bulk of his attack was devoted to the superficial use made of the three principles by "tyrocinists," that is, "vulgar" beginners and textbook writers like Beguin, who did not, at least in his view, have a profound knowledge of chymistry.[33] Before proceeding to the actual use that van Helmont himself made of the three principles, it will be necessary to look at his debunking of the "vulgar" doctrine.

In his chapter on the three principles, van Helmont provides seven numbered theses attacking the *tria prima*. They can be summarized as follows:

1.   Although three principles are indeed drawn out of many things by means of fire, these are not preexistent entities, but new creations of the fire itself, made by transmutation.
2.   The amount of oil extracted from the sprout of a tree varies widely depending on the time of year the sprout is harvested, ranging from hardly a drachm to as much as two ounces.
3.   Things which were not in a material as initial constituents cannot be "first" (as in the Paracelsian *tria prima*) but are mutually transmuted and arise in a posterior fashion under the influence of *semina*.
4.   Elemental water becomes an oil in animals, vegetables, and minerals, and any oil can be turned back into water with the aid of certain added

ingredients. But the first principles of other things cannot be transmuted into one another or cease to be what they were before.

5. Some bodies do not contain three, but only one or two constituents.
6. There are some bodies that never have been dissolved into three principles equaling the original body in weight, at least not by means of the common chymical techniques.
7. There are some bodies that are completely immutable and indivisible, as they lack any principle of "duality" (meaning "heterogeneity").

Let us first consider the theses that depend on empirical observation. As one can see, van Helmont begins by ruling out fire analysis as a reliable way of separating three preexistent principles, since the fire can just as easily create new substances as decompose pre-existent ones. He then provides various arguments against the tyrocinists' claim that the three principles are universal: sometimes oil (sulfur) is present in great quantity in plants, sometimes not at all; some materials reveal only one or two constituents by analysis; some materials are completely resistant to decomposition. Even when the vulgar chymists break materials down into their supposed three principles, the latter do not always equal the original material in weight: this discrepancy in mass means that something is amiss. As one can see, these arguments provide evidence against commonplace forms of analysis, but they do not in themselves rule out the possibility that matter, or at least some matter, consists of three principles. As for the more theoretical Theses 3 and 4, these explicitly rule out the three principles as initial ingredients of bodies, but not necessarily as posterior constituents. As we will see, this is a point of primary importance for van Helmont, for it opens up the possibility of the three principles existing in fully formed materials even if those materials did not originally derive from sulfur, salt, and mercury. As we shall see, van Helmont capitalizes on this eventuality, and with his usual love of paradox, he extends it to bodies that he calls homogeneous.

At various points in the *Ortus medicinae* van Helmont insists that both water and "the mercury extracted from metals" are absolutely homogeneous.[34] His point is that neither of these substances contains the *tria prima*, despite the claims to ubiquity made by the believers in hypostatical principles. The very fact that van Helmont believed in a mercurial component found in metals, however, illustrates his own adherence to some form of principle theory. He even offers that this internal mercury of the metals, once deprived of the "stain" of its metallic sulfur, becomes entirely indissoluble and incapable of division.[35] Obviously van Helmont believed that metals are normally composed of sulfur and mercury and that the mercury can be isolated free of sulfur. Such mercury liberated from its "external sulfur" can no longer be precipitated into an earth by long heating (we would say oxidized).[36] And yet van Helmont speaks of another "internal sulfur" that cannot be removed from the mercury by any means whatsoever.[37] All of this is, in fact, an attempt to interpret Geber's theory of fixed (essential) and volatile (accidental or superfluous) metallic principles in the light of van Helmont's own understanding of matter. As if to drive his debt home, van Helmont explicitly refers to Geber at this point, saying that the author of the *Summa perfectionis* understood the absolute homogeneity of the pure metallic mercury. More than that, van Helmont adds that he himself only learned of the true nature of water under the stern switch made from the caduceus of Mercury (*sub ferula, ex caduceo Mercurii parata*).[38]

Despite the fact that van Helmont often adduces water as an example of a purely homogeneous material, he sometimes speaks as though it contained the three principles. This is particularly evident in his "Gas aquae" and "Progymnasma meteori," dense chapters of the *Ortus medicinae* where van Helmont discusses the process whereby water vaporizes. He envisions the water as consisting of tiny particles made up of shells that correspond to mercury, sulfur, and salt. These three cannot be separated in water, but they can exchange places under the proper conditions. Hence if the water is heated, its salt, which cannot tolerate the heat, is driven upward, and since the mercury and sulfur cannot be divided from it, they are dragged along. If the water vapor continues to rise into colder regions, the warmer sulfur forms a skin over the particle and the resulting inversion of layers results in the conversion of the vapor into Helmontian "Gas." This complicated theory may suggest that the water is far from being homogeneous, but van Helmont explicitly rejects this implication. After affirming that elemental water contains the three principles, he says,

> I suppose these things just as the astronomers do their eccentrics, so that one may proceed in an expedient fashion, thanks to our weakness in understanding.
>
> (van Helmont 1648: 74, #8)

Just as the astronomers postulate eccentrics and other geometrical tools that allow them to perform calculations without assuming the physical reality of those models, so van Helmont speaks of the three principles of water. Elsewhere he describes this employment of the principles as "analogical" and adds the illuminating comment that the *tria prima* in water are not "anterior principles of composition" but, rather, principles "of heterogeneity." Moreover, this heterogeneity is only "conjectural" and does not imply that the three principles in water can actually be separated.[39]

It is clear that the three principles serve an important function for van Helmont despite his reduction of their scope. Precisely because he recognized that the evaporation of his professedly homogeneous water, as well as freezing and sublimation of ice did not involve what we would today call chemical change, van Helmont was forced to propose another mechanism by which such transformations could occur. This is the origin of his theory that water (and other homogeneous material like the "internal" mercury of metals) contained inseparable "heterogeneities" that merely corresponded to the traditional three principles. Given sufficient space, one could show that this operational use of mercury, sulfur, and salt as "posterior principles" was characteristic not only of van Helmont's own work but also that of his followers, such as the New England alchemist George Starkey. Both in his pseudonymous writings composed under the *nom de guerre* of "Eirenaeus Philalethes" and in his openly acknowledged publications, Starkey explains mineralogical and chymical phenomena by means of the three operational principles, unimpeded by his adherence to the Helmontian position that all matter draws its origin from water.[40] Like van Helmont, Starkey belongs to the school of Posterior Principles, though the degree to which he maintained the homogeneity of water and mercury remains to be determined.

Hierarchical Principles

The fourth school of chymical principles shared important features with the third in that both groups rejected the claim of hypostatical ubiquity promoted by Quercetanus and his followers. Additionally, the Hierarchical school accepted that the *tria prima* were derivative rather than primordial constituents of matter, a position similar to that of the proponents of posterior principles. Nonetheless, there were important differences as well. The most important hallmark of the fourth chymical school lay in its attempt to build up a corpuscular hierarchy of composition in which the primordial constituents of matter—be they the four elements or, in Robert Boyle's case, *prima naturalia* composed of a uniform catholic matter—combined to form larger, semipermanent corpuscles, which could in turn recombine to form even larger and more complex agglomerations. The roots of this theory lay once again in medieval alchemy, for Geber's *Summa perfectionis* already argued that particles of the four elements combine in a "very strong juxtaposition" (*fortissima compositio*) to make sulfur and mercury and that these are further combined in a strong union to compose the various metals.[41] This position could hardly satisfy the natural philosophers of the early modern universities, however, who were, for the most part, committed to the position that metals were purely homogeneous substances in which the four elements had ceded their actual existence and undergone a total mixture (*mixis*) under the influence of a unifying substantial form. Hence one finds learned chymists from the early seventeenth century onward trying to make the hierarchical view of chymical combination a philosophically acceptable alternative to the strong view of *mixis* as involving homogeneity *simpliciter*.

One of the most influential early proponents of a hierarchical corpuscular theory involving the three principles was the Wittenberg professor of medicine, Daniel Sennert, who openly professed himself an atomist in his *De chymicorum cum Aristotelicis et Galenicis consensu ac dissensu* of 1619. In the first edition of this work, Sennert rejects the hypostatical view of the three principles and then proceeds to argue that they are *prima mixta* made by God in the initial Creation. As he puts it,

[w]hen God the Optimus Maximus was creating by means of his word, he mixed the elements and imposed their own peculiar forms, seminal reasons, and essences analogous to the heavens on the mixts.

(Sennert 1619: 173)

Sennert then proceeds to offer empirical evidence for the robust character of these mixts, such as the fact that burning sulfur does not yield the four elements but, rather, a highly acidic spirit or oil of sulfur. He admits that the *tria prima* do contain the four elements but that God has granted the principles their own substantial forms, which cannot be reduced to a mere intermixture of the elements: after all, the latter can only produce the four elementary qualities hot, cold, wet, and dry. The net result of this, of course, is a reduction in the practical importance of the four elements since the *tria prima* cannot be returned to them by laboratory means and, as Sennert argues elsewhere, the three principles are responsible for most of the sensory input experienced by humans. In his later *Hypomnemata physica* (1636) Sennert goes so far as to speak of "atoms of their own genus." In this text he is more explicit in his claim that the four elements retain their full being within atoms of "the second genus," namely, the *prima mista*.[42] In still other works Sennert speaks at length about the affinity and tight bonding that can occur among these second-genus atoms in the formation of what we would now call chemical compounds. Here, too, he uses the term *atom* or *atom after a fashion* (*atomus quasi modo*), indicating that particles of the material in question have some degree of permanence. The term *atom* for him is clearly relative rather than absolute: for Sennert, something is fully atomic if it resists decomposition by means of the analytical tools of the seventeenth-century laboratory.[43]

The rough hierarchy of atoms erected by Sennert was appropriated and modified by Boyle, who combined Sennert's empirical approach with other ideas and terminology derived from a variety of sources. Beginning with *The Sceptical Chymist* of 1661, Boyle distinguishes between the *prima naturalia* or *minima naturalia*—the first or smallest particles that cannot be divided further and the composite corpuscles, which begin with the *prima mixta* and continue upward to form more and more complex particles, called "compounded" and "decompounded" bodies.[44] Now it is obvious that a hierarchical theory of this sort admits of the possibility that the *tria prima* could exist at a higher level of composition than the primordial one occupied by Boyle's *prima naturalia*. In fact, Boyle affirms this possibility at some length in his 1680 *Experiments and Notes about the Producibleness of Chymicall Principles*, a work that is marked by its rather sympathetic approach to chrysopoeia. Here he allows that the chymists' sulfurs and mercuries of metals are real, but that they may be "magisteries," namely, products of the metal produced without analyzing it into its parts.[45] Yet Boyle never published a systematic account of his hierarchical matter theory, and it is fair to say that the three principles occupy an even smaller position there than they do in the work of the man whose work inspired *The Sceptical Chymist*, van Helmont. We must turn back to the European continent to see the full flourishing of the Hierarchical interpretation of the *tria prima*.

Johann Joachim Becher, who moved to England in 1679, was an avid reader of Boyle's works; as I have argued elsewhere, his exposure to *The Sceptical Chymist* and other Boylean writings of the 1660s facilitated Becher's transformation from a rather traditional alchemist and projector to the more critical and systematic writer behind his *Physica subterranea* of 1669.[46] Becher tells us in the *Physica subterranea* that material things are composed ultimately of earth and water, which can be mixed together in different ratios.[47] Sulfur, salt, and mercury, on the other hand, are mixed bodies, not ultimate principles. Indeed, Becher stresses that the *tria prima* of the Paracelsians are actually *decomposita* (in the Boylean sense) for no mercury, sulfur, or salt can be extracted from metals unless some other component is added, which results in a further compounding of substances that are already mixts. The same is true of vitriols, *amausa* or metallic glasses, crocuses, butters, and flowers, which thereby have the same right to be called principles as do the Paracelsian *tria prima*. All of these materials are in reality *decomposita*, having been recombined from simpler components.

A superficial reading of Becher might therefore produce the impression that he had altogether abandoned the traditional principles; in fact, nothing could be farther from the truth. What Becher did, in reality, was to return to the medieval concept that the commonplace materials going by the names of mercury and sulfur that could be purchased from apothecaries, were impure forms of the primordial principles of the metals and that the latter were hidden in these unrefined materials.[48] Accepting the third principle salt as well, Becher renamed the three principles "fusible earth" (*terra fusilis*), "fatty earth" (*terra pinguis*), and "fluid earth" (*terra fluida*).[49] Although he ridicules Paracelsus at length for confusing the vulgar *decomposita* mercury, sulfur, and salt with fundamental principles, Becher makes it quite clear that his three earths are merely the *tria prima* by another name. Becher identifies fusible earth with the clear stones found in mines that can be melted at high temperatures: like Paracelsus's salt, fusible earth provides the basis or body to metals; similarly, fatty earth provides combustibility and color, properties typically associated with the principle sulfur, and fluid earth *metalleitas*, meaning primarily the properties of malleability and ductility, like the old alchemical mercury.[50] The three earths are the simple and homogeneous bases that combine in multiple ways to produce the diversity of the mineral realm. Becher argues that their immediate intermixture produces eight different classes of minerals, and that these classes can then recombine to yield 40,320 (8 factorial) types of mixture. Comparing mixtures and *decomposita* to syllables and words, he says that his hierarchical system of composition provides the true "alphabet of nature" (*alphabetum naturae*).[51] Becher's three earths therefore underlie an immense superstructure of mineral composition: although we cannot here proceed into the realm of eighteenth-century chemistry, it is precisely this system that was appropriated by the founder of the phlogiston school, Georg Ernst Stahl.

Conclusion

We have now concluded this brief survey of the chymical principles. It should be clear that the theoretical and practical underpinnings of the principle approach were already well established by the High Middle Ages. Building primarily on the theory of metallic generation found in Avicenna, Albertus Magnus and the Latin Geber had already developed extensive and well-thought-out means of testing for the presence or absence of the principles of mercury and sulfur in metals based on an extension of contemporary assaying practices. Paracelsus was the heir of this theory and its probatory apparatus, but his main emphasis lay in expanding the range of the theory rather than in deepening its analytical techniques. Hence, he consciously employed the Holy Trinity as a model for his principles, which required the addition of a third member, namely salt, in order to reflect the triplicity of Father, Son, and Holy Spirit. There was also a strong empirical element to Paracelsus's thought, but the limited scope of this chapter has not allowed a deep examination of his motives for picking salt as the third principle. As we have seen, the Trinitarian component of Paracelsus's theory was emphasized by Quercetanus and the Hypostatical school, which achieved considerable dispersion, thanks to Jean Beguin and others in the French chemical textbook tradition. The two remaining schools of thought, however, the upholders of the Posterior and Hierarchical views of the chymical principles, were not partisans of the literal ubiquity of the *tria prima* as envisioned by those of the Hypostatical persuasion. Additionally, both the Helmontian position of posterior principles and the elaborate hierarchy established by Becher can be seen as attempts to escape the claim that the *tria prima* were the ultimate, primitive constituents of matter. Seen from the perspective of the *longue durée*, both the medieval and early modern principles were much more than a mere supplanting of four elementary constituents by three. From the High Middle Ages onward, they provided intermediary components that were widely thought to be semipermanent and hence to retain their robust existence within compounds. This was a far cry from Thomistic and Scotist theories of mixture, which denied the continued existence of intermediate components such as the elements within mixts. Additionally, mercury and sulfur were revealed to the senses during analysis in a way that the four elements could not be *ex hypothesi*, since the dominant theories of medieval and early modern scholasticism denied their perpetuation after the substantial form of the mixt had been imposed. In sum, we should view the sulfur–mercury theory and its heir, the theory of *tria prima*, as concessions to the senses. More than that, we can see them as attempts to frame a material theory on the manipulation of matter by means of experiment, above all by means of analysis employing the techniques of the laboratory. The historical importance of this approach requires no further reiteration.

Notes

1 Though one should consult Norris 2006 and 2007.
2 For a rather early example of this viewpoint, see Boas [Hall] 1958: 83–86. See pages 83 and 85–86 for the following passages: "Mercury and sulphur seemed eminently suitable candidates for chemical elements, especially since ordinary quicksilver and brimstone were never confused with the elements of the same names, which were purer and higher than anything ordinarily obtainable in the laboratory. . . . It cannot be too often emphasized that the chemists did mean something different from what we mean today when they spoke in terms of elements. For their elements and principles were ideal substances, real in the sense that they existed in matter, but not real in the sense that they could be handled and observed." A somewhat similar perspective underlies the otherwise laudable book of Holmes (1989). Holmes intended to extend various features associated with the Chemical Revolution, such as "operational interpretations of composition," back to the French Academicians of the seventeenth century. See Holmes 1989: 8–37. In the case of earlier chymists such as Paracelsus, Holmes asserted that "the principles were more spiritual than material"; see Holmes 1971, particularly p. 132; and Holmes 2003, especially pp. 45–46.
3 In particular, see Klein 2014. See also Newman 2014a.
4 For a brief introduction to this obscure figure, see Plessner, *sub voce* "Balīnūs". See also Weisser 1980.
5 In a useful recent article, Jean-Marc Mandosio and Carla di Martino (2006) argue that Avicenna's text should actually be referred to as *De mineralibus* since the name *De congelatione et conglutinatione lapidum* properly refers only to the first chapter of the translated text. For the present chapter, I have kept the name *De congelatione et conglutinatione lapidum*, if only because this title has come to be widely recognized in the Anglo-Saxon world as a result of Holmyard and Mandeville's 1927 edition.
6 Mandosio and di Martino 2006: 406. As the authors point out, the fifth part of Avicenna's *Kitāb al-Shifā'*, which contained his treatment of meteorology, was translated once in full; additionally, two partial translations were made—*De congelatione et conglutinatione lapidum* and *De diluviis*.
7 For the date of this translation, which is somewhat earlier than the traditionally accepted one, see Mandosio 2010: 244–245. I thank Sébastien Moureau for bringing this article to my attention.
8 See Mandosio and di Martino 2006, especially pp. 411–419.
9 Unfortunately, there is no critical edition of the *De congelatione et conglutinatione* available today. Avicenna 1927 merely copies a single base manuscript and provides alternative readings in the apparatus with no attempt to emend the obviously quite defective text. In the passage that I have translated here, I have followed the text provided in Manget 1702: 636–638. Needless to say, this uncontrolled text should also be taken with a large grain of salt. For the Latin passage, see Manget 1702: 637B.
10 Ruska 1935.
11 *Liber secretorum de voce bubacaris,* in MS. Paris, BN lat. 6514, fol. 103ra: "Operatio argenti vivi est lavatio deinde levare eius humiditatem. . . . Operatio auripigmenti et sulphuris est ut dealbentur et liberentur ab ardore eorum et pinguedine quam habent." Also *Liber secretorum de voce bubacaris,* in MS. Paris, BN lat. 6514, fol. 103ra—103rb: "Sublimationis argenti vivi due sunt

manieres quarum una est pro albedine et alia pro rubedine tamen in unaquaque sublimatione sunt duo secreta ut unum est ut recipiatur humiditas aliud est ut facias illud durum et siccum et sic deperditur eius humiditas."

12 Ruska 1937, especially 100–126, and 220–225.

13 Halleux 1982. For Albert's theory of a double humidity and its debt to Nicolaus Peripateticus, see Newman 1991: 218–219. For more recent work on Nicolaus Peripateticus, see Moureau 2012, especially 34–35. A critical edition is found in Wielgus 1973.

14 The standard text of the *Liber mineralium* found in volume five of Borgnet's edition of Albertus's *Opera omnia* is defective at this crucial point, so I correct it here with the addition of a few words found in two medieval manuscripts, Vaticanus palatinus latinus 978, fol. 8rb, lines 20–23, and Marburg Universitäts-bibliothek B20(MS. 26), fol. 140v (9v), lines 10–11; otherwise the passage comes from *Liber mineralium* in Albertus 1890: 61. The only difference between the two manuscripts as far as the addition goes is that Marburg B20 (MS. 26) lacks the "autem" before "est intrinseca": "duplex est unctuositas in multis rebus: quarum una est quasi extrinseca <in mixto terrestri adusto feculento et hec est multum combustibilis et inflammabilis altera autem est intrinseca>, subtilis valde, nihil faetulentum vel cremabile habens admixtum: et haec non est inflammabilis, et intrinseca rei retenta in radicibus rei, ne per ignem possit evelli et epotari: et nos dedimus de hoc exemplum in liquore, qui eliquatur ex vino, in quo una est unctuositas supernatans inflammabilis, et facile astringbilis et quasi accidentalis. Altera commixta toti substantiae liquoris ipsius, non separabilis ex ipsa substantia liquoris, nisi per defectionem substantiae: et haec non est cremabilis."

15 Albertus 1967: 157, note 8. Unfortunately, Wyckoff's translation of this passage reproduced here is defective, as she has used Borgnet's text without consulting the manuscript tradition of the *Liber mineralium*.

16 Later in the *Liber mineralium* Albert takes the interesting step of dividing the metals' principles into three rather than two unctuosities. He is apparently subdividing the "extrinsic" humidity into two types—a combustible and a noncombustible version. For details and quotations, see Newman 2014a.

17 See Newman 1991: 214–223.

18 Newman 1991: 719–738 and 769–783.

19 Daniel 2003.

20 Paracelsus 1933: 134–136.

21 Paracelsus 1972 [1590], vol. 8: 55.

22 Paracelsus 1972 [1590], vol. 8: 126.

23 See *Opus Paramirum* in Paracelsus 2008: 322–324: "Nuhn aber Etwas haben wir durch das Fewr Vulcani/ dardurch wir die drey Ersten erkleren: Nemlich durch den Schwefel/ den Sulphur/ dieweil sie sich vergleichen: Durch das Queck-silber den Mercurium/ auß vrsachen auch eines solchen vergleichens: Durch das Saltz/ Salem/ dann es gibt gleiche wirckung."

24 Joachim Jungius, like Boyle, used the term *principia hypostatica* in an expanded sense. But it is quite characteristic of Jungius to employ creative labels, such as the "hypothesis actupotentialis" for the Aristotelian emphasis on potency and act. See Jungius 1982: 51 and 102 for both terms.

25 Didier Kahn demonstrates this at length in his illuminating article. See Kahn 2004: 676–682 for the issue of Du Chesne's debt to Röslin.

26 For Quercetanus's debt to Severinus, see Hirai 2001.

27 *De priscorum*, Quercetanus 1609: 18.

28 *De priscorum*, Quercetanus 1609: 18.

29 See Eklund 1971: 1–39.

30 Beguin 1612: 31.

31 For editions of the *Tyrocinium chemicum*, see Patterson 1937.

32 Debus 1967.

33 Van Helmont's scorn for texts like Beguin's *Tyrocinium chemicum* emerges quite clearly in the *Ortus medicinae*, van Helmont 1648: 524, #1: "Inprimis norunt Adepti mecum, quantum hinc distent dispensatoria Seplasiae, imo & quam remoti absint Scriptores, qui Basilicas, & Tyrocinia chymica ingenti gloriolae pruritu, adhuc ipsimet Tyrones ediderunt."
34 Van Helmont 1648: 407.
35 Van Helmont 1648: 68, #8.
36 Van Helmont 1648: 70, #14.
37 Van Helmont 1648: 70–71, #17.
38 Van Helmont 1648: 68–69, #8.
39 Van Helmont 1648: 407, #54.
40 Newman 1994: 161–164.
41 Newman 1991: 663–665 (Latin 321–326) and 774–776 (Latin 601–608).
42 Sennert 1636: 107.
43 See Newman 2012.
44 Boyle, *The Sceptical Chymist* in Boyle 1999–2000, 2: 296–297, 347; *The Origine of Formes and Qualities* in Boyle 1999–2000, 5: 326; *The History of Particular Qualities* in Boyle 1999–2000, 6: 270–271, 274–275; *Experiments and Notes about the Mechanical Origine and Production of Volatility* in Boyle 1999–2000, 8: 425; *Producibleness* in Boyle 1999–2000, 9: 114. The term *decompounded* had already been absorbed into discussions of matter theory by the time of Heinrich Cornelius Agrippa von Nettesheim's *De occulta philosophia*, if not earlier. See Agrippa 1992: 91.
45 Principe 1998: 54, note 90.
46 Newman 2014b: 70.
47 For the claim that earth and water are the ultimate principles, see Becher 1738: 58: "In quolibet autem regno specifica sunt semina tanquam principia principiata & singularissima; & ita omnium rerum *principium remotissimum est terra & aqua;* prout in creationis actu extiterunt: *principium propinquum est eadem terra & aqua;* sed post productionem, prout varie mixta; *principium vero specialissimum est id,* unde quodvis corpus tanquam ex *principio principiato,* actualem generationem incipit. Et hoc est cuiusvis corporis specificum semen, nempe *terra & aqua, quatenus disposita sunt, & ex principiis propinquis in specialissima, nempe rerum semina, mutata sunt,* ita omnia ex terra & aqua prodierunt, omnia rursus in terram & aquam ultimatim reduci possunt. Prout autem Sect. I. cap. 4. distinctionem inter mixta fecimus, atque alia ex terra & aqua, alia ex variis aquae speciebus, alia vero ex variis terris constare probavimus; ita hoc loco illorum tantum corporum principia considerabimus, quae pure terrea sunt, seu quae ex terris tantum constant."
48 Becher 1738: 85: "Hactenus de *tribus terris* egimus, quas pro mineralium principiis statuimus; & quas veteres sub *salis, sulphuris & Mercurii* nomine complexi sunt. Non quod *vulgare* sal, sulphur & Mercurius, ipsa *pura* principia existant; sed quod pars quaedam istorum decompositorum, pro principiis subterraneorum habenda sit: atque ita totum pro parte sumserint, denominationem, nempe *a potiori.*"
49 Becher 1738: 60–84.
50 In addition to the chapters devoted to each of Becher's three earths, see Becher 1738: 88,
51 Becher 1738: 95.

Bibliography

Agrippa, H. C. (1992) *De occulta philosophia, libri tres,* ed. V. P. Compagni, Leiden.
Albertus Magnus (1890) *B. Alberti Magni ratisbonensis episcopi, ordinis praedicatorum, opera omnia,* vol. 5, ed. Auguste Borgnet, Paris: Ludovicus Vivès.

————. (1967) *Albertus Magnus: Book of Minerals*, trans. D. Wyckoff, Oxford: Clarendon Press.

Avicenna (1927) *Avicennae de congelatione et conglutinatione lapidum*, eds. E. J. Holmyard and D. C. Mandeville, Paris: Geuthner.

Becher, J. J. (1738) *Physica subterranea*, Leipzig: Weidmann.

Beguin, J. (1612) *Tyrocinium chemicum*, Paris.

Boas (Hall), M. (1958) *Robert Boyle and Seventeenth-Century Chemistry*, Cambridge: Cambridge University Press.

Boyle, R. (1999–2000) *The Works of Robert Boyle*, 14 vols, eds. M. Hunter and E. B. Davis, London: Pickering and Chatto.

Casanova-Robin, H. and Galand, P., eds. (2010) *Écritures latines de la mémoire de l'Antiquité au XVIe siècle*, Paris: Classiques Garnier.

Daniel, D. T. (2003) *Paracelsus' Astronomia Magna (1537/38): Bible-Based Science and the Religious Roots of the Scientific Revolution*, PhD dissertation, Indiana University.

Debus, A. (1967) "Fire analysis and the elements in the sixteenth and the seventeenth centuries," *Annals of Science* 23: 127–147.

Eklund, J. (1971) *Chemical Analysis and the Phlogiston Theory, 1738–1772: Prelude to Revolution*, PhD dissertation, Yale University.

Halleux, R. (1982) "Albert le grand et l'alchimie," *Revue des sciences philosophiques et théologiques*, 66: 57–80.

Helmont, J. B van (1648) *Ortus medicinae*, Amsterdam.

Hirai, H. (2001) "Paracelsisme, néoplatonisme et médecine hermétique dans la théorie de la matière de Joseph Du Chesne à travers son *Ad veritatem hermeticæ medicinæ* (1604)," *Archives internationales d'histoire des sciences*, 51: 9–37.

Holmes, F. L. (1971) "Analysis by fire and solvent extractions: The metamorphosis of a tradition," *Isis*, 62: 129–148.

————. (1989) *Eighteenth-Century Chemistry as an Investigative Enterprise*, Berkeley: University of California Press.

————. (2003) "Chemistry in the Académie Royale des Sciences," *Historical Studies in the Physical and Biological Sciences*, 34: 41–68.

Jungius, J. (1982) *Praelectiones physicae,* ed. C. Meinel, Göttingen: Vandenhoeck and Ruprecht.

Kahn, D. (2004) "L'interprétation alchimique de la Genèse chez Joseph Du Chesne dans le contexte de ses doctrines alchimiques et cosmologiques" in ed. B. Mahlmann-Bauer 2004, pp. 641–692.

Klein, J. (2014) "Corporeal elements and principles in the learned German chymical tradition," *Ambix*, 61: 345–365.

Mahlmann-Bauer, B., ed. (2004) *Scientiae et Artes: Die Vermittlung alten und neuen Wissen in Literatur, Kunst und Musik*, Wiesbaden: Harrassowitz Verlag.

Mandosio, J. M. (2010) "Humanisme ou barbarie? Formes de la latinité et mémoire de l'Antiquité dans quelques traductions médiévales de textes philosophiques arabes" in eds. H. Casanova-Robin and P. Galand 2010, pp. 227–263.

Mandosio, J. M. and di Martino, C. (2006) "La 'Météorologie' d'Avicenne (Kitāb al-Shifā' V) et sa diffusion dans le monde latin" in eds. A. Speer and L. Wegener 2006, pp. 406–424.

Manget, J. J. (1702) *Bibliotheca chemica curiosa*, vol. 1, Geneva.

Manning, G., ed. (2012) *Matter and Form in Early Modern Science and Philosophy*, Leiden: Brill.

Moureau, S. (2012) "Les sources alchimiques de Vincent de Beauvais," *Spicae, Cahiers de l'atelier Vincent de Beauvais*, 2: 5–118.

Newman, W. R., ed. (1991) *The "Summa perfectionis" of Pseudo-Geber*, Leiden: Brill.

———. (1994) *Gehennical Fire: The Lives of George Starkey*, Cambridge, MA: Harvard University Press.

———. (2012) "Elective affinity before Geoffroy: Daniel Sennert's atomistic explanation of vinous and acetous fermentation" in ed. G. Manning 2012, pp. 99–124.

———. (2014a) "Mercury and sulphur among the high Medieval alchemists: From Rāzī and Avicenna to Albertus Magnus and pseudo-Roger Bacon," *Ambix*, 61: 327–344.

———. (2014b) "Robert Boyle, transmutation, and the history of chemistry before Lavoisier: A response to Kuhn," *Osiris*, 29: 63–77.

Norris, J. (2006) "The mineral exhalation theory of metallogenesis in pre-modern mineral science," *Ambix*, 53: 43–65.

———. (2007) "Early theories of aqueous mineral genesis in the sixteenth century," *Ambix*, 54: 69–86.

Paracelsus (1933) *Sämtliche Werke*, series 1, vol. 13, ed. K. Sudhoff, Munich: Oldenbourg.

———. (1972 [1590]) *Bücher und Schrifften*, ed. J. Huser, vol. 4 and vol. 8, Basel, 1590. Reprint Hildesheim: Olms, 1972.

———. (2008) *Paracelsus: Essential Theoretical Writings*, trans. A. Weeks, Leiden: Brill.

Patterson, T. S. (1937) "Jean Beguin and his *Tyrocinium Chemicum*," *Annals of Science*, 2: 243–298.

Plessner, M. (1986) "*sub voce* 'Balīnūs'" in *Encyclopedia of Islam*, 2nd edn, online version, eds. P. Bearman, T. Bianquis, C. E. Bosworth, E. van Donzel, and W. P. Heinrichs, http://dx.doi.org/10.1163/1573–3912_islam_SIM_1146. (Accessed: 4 November 2016).

Principe, L. M. (1998) *The Aspiring Adept: Robert Boyle and His Alchemical Quest*, Princeton: Princeton University Press.

Quercetanus (J. Du Chesne) (1604) *Ad veritatem hermeticae medicinae*, Paris.

———. (1609) *De priscorum philosophorum materia*, Geneva.

Ruska, J. (1935) "Übersetzung und Bearbeitungen von Al-Rāzīs Buch Geheimnis der Geheimnisse," *Quellen und Studien zur Geschichte der Naturwissenschaften und der Medizin*, 4: 153–239.

———. (1937) *Al-Rāzī's Buch Geheimnis der Geheimnisse*, in *Quellen und Studien zur Geschichte der Naturwissenschaften und der Medizin*, vol. 6, Berlin: Julius Springer.

Sennert, D. (1619) *De chymicorum cum Aristotelicis et Galenicis consensus ac dissensu*, Wittenberg.

———. (1636) *Hypomenemta physica*, Frankfurt.

Speer, A. and Wegener, L. (2006) *Wissen über Grenzen: Arabisches Wissen und lateinisches Mittelalter*, New York: De Gruyter.

Weisser, U. (1980) *Das 'Buch über das Geheimnis der Schöpfung' von pseudo-Apollonius von Tyana*, Berlin: De Gruyter.

Wielgus, S. (1973) "Quaestiones Nicolai Peripatetici," *Mediaevalia Philosophica Polonorum*, 17: 57–155.

4 The Two Comets of 1664–1665

A Dispersive Prism for French Natural Philosophical Principles[*]

Sophie Roux

Introduction

In November 1664, a comet—the brightest since the three comets of 1618, say the reports of the time—appeared in the European skies. This comet first moved very quickly from the east to the west, that is, in a retrograde way with respect to the planets; then it reversed its direction and became much slower, to the point that it seemed to be stationary; in February it was not to be seen anymore by the naked eye, and by early March, it disappeared completely. At this very moment, another comet appeared, which was even brighter than the first one, had a longer tail, and stayed among the stars until mid-April. Observations of these two comets were made all over Europe and even beyond. In Rome, they were observed not only by the Jesuits Honoré Fabri, Gilles-François de Göttignies, and Athanasius Kircher but also by Giovanni Domenico Cassini, who happened to be staying there at the time and was solicited by Queen Christina, who wanted him to participate in the observations; in Bologna by Geminiano Montanari and Giovanni Battista Riccioli; in Pisa by Giovanni Alfonso Borelli, who published a book under the pseudonym of Pier Maria Mutoli; in Florence by Lorenzo Magalotti; in Venice by Gaudentius Brunacci; in Lisbon by Antonio Pimento; in Valencia, by Enrique Miranda; in Majorca by Vincent Mut and Miguel Fuster; in Jena by Erhard Weigel; in Leipzig by Christoph Richter; in Ulm by Jakob Honold, Johannes Prateor, and Christoph Schorer; in Dresden by Johannes Philip Hahn and Matthias Dannerwald; in Wittenberg by Michael Strauch; in Nuremberg by Christian Theophile and Johannes Christoph Kohlansen; in Hamburg by Stanislas de Lubienietski; in Würzburg by Caspar Schott; in Franeker by Gravius; in Liège by François-René de Sluse; in Leyden by Christiaan Huygens; in the Hague by Samuel Carel Kachel of Hollenstein; in Rostock by Johannes Quistor; in Dantzig by Johannes Hevelius and his nephew Johannes Hecker; in Magdeburg by Otto von Guericke; in Ratisbonne by Johannes von Rautenstein; in Copenhagen by Erasmus and Thomas Bartholin; in Stockholm by Nicolaas Heinsius; in London by Samuel Pepys, Robert Hooke, and Christopher Wren; in Oxford by John Wallis; and in Cambridge by the young Isaac Newton. Across the Atlantic,

Samuel Danforth published a book of observations that seems to have been one of the very first works of astronomy printed in America; in French Québec, the comets were also seen and reported on by Jesuit missionaries and by Marie Guyard, also known as Sainte Marie de l'Incarnation, a mystic born in Tours who introduced the Ursulines in New France.[1]

Although most secondary literature dedicated to these two comets has been focused on England and Italy, France was not to be outdone in terms of observations, small talk, and publications.[2] The observations of Adrien Auzout, Ismaël Boulliau, Jacques Buot, and Pierre Petit, who lived in Paris, had the best instruments, and were in contact with the international community of the time, are still remembered. They should not make us forget observers living in the provinces who night after night did their best in the midst of a cold and rainy winter to identify and locate these comets among persistent clouds. Among them were Nicolas de Croixmare de Lasson, Pierre-Daniel Huet, and André Graindorge in Caen, Mignot de la Voye in Rouen, Claude Comiers in Embrun, Robert Luyt in Tonnerre, Gabriel de Malapeyre in Toulouse, Montalegre in Lyon. Predictably, in each city of France there was at least one Jesuit ready to send a report on his observations to other Jesuits in other cities—Jacques Grandami in Paris, Jacques de Billy in Langres, Michel Seneschal in Douai, Michel Beaussier at La Flèche, Vincent Léautaud in Embrun, Jean Bertet and Claude-François de Chales in Lyon, Henri-Ignace Régis in Aix, Mercure Verdier in Poitiers, Ignace-Gaston Pardies in Bordeaux, and Louis Nyel in Pont-à-Mousson.[3]

Correspondences a nd private diaries show that these comets also made a great impression in France on what can be called the public of the time. Although her main preoccupations at this time were the last episodes of Nicolas Fouquet's long trial, Madame de Sévigné wrote a few lines on the first comet to Pomponne, the Prince de Condé described it to the Queen of Poland, and the Marquis de Castries warned Jean-Baptiste Colbert about predictions and prophecies that were made by an old astrologer in Montpellier.[4] The physician and *libertin érudit* Guy Patin joked that, whatever the astrologists may say, the first comet would certainly not been followed by a cut in taxes but that, for once, their predictions concerning the diseases that the comet may bring about could turn out to be right, since, in order to contemplate it, everybody got cold by standing on the Pont-Neuf at three in the morning after the midnight Mass.[5] Charles Cotin, who is said to have inspired Molière's Trissotin, wrote a courteous piece on the comet, while Nicolas Boileau, who used to mock Jean Chapelain's old, greasy and filthy wig, ad-libbed during a party with friends a short nonsensical comedy in which this wig was transformed into a comet by Apollo.[6] A *Ballet de la comète* was danced at the Oratorians in Soissons on 9 February, and a *Balet des comètes pour la tragédie d'Irlande* at the Jesuits in Paris on 6 August.[7] Anti-cometary remedies were sold.[8] Even the gender of comets was a matter of inflamed discussion—that is, the question of whether one should say "une" or "un" "comète" in French, considering that the original Latin word is *cometa*, one of those rare masculine words to end by the letter *a*.[9] Confronted with this question, one of the authors who will be studied in this chapter suggested an analogy between the undetermined gender of the word *comet* and the undetermined nature of comets, which are neither planets nor stars nor both at the same time.[10]

In this chapter, I would like to use the books that were published on these comets in France as a dispersive prism to enable us to see the ongoing controversies about which principles should be accepted in natural philosophy in the mid-sixties. Although these comets are less famous than those of 1577 and 1680–1681, their significance for the history of science has been the subject of scholarly attention. The main question concerns how a consensus was built among astronomers, who, indeed, formed for the first time a truly international network but made observations neither from the same places nor at the same times, of nor even in a continuous manner, because of not only uncertain weather but also of the frequent ailments that affected everybody in those days.[11] The purported identification of one comet rather than two different ones, or, on the contrary, of two different comets rather than one, was in these circumstances problematic: How could one decide if discontinuous observations bear on one and the same comet or on different ones? Because of its sudden variation of luminosity and change of location, the first comet was sometimes seen as two different, successive comets. This was the case not only for the anonymous author of *Histoire des comètes qui ont paru depuis peu sur notre horizon* but also (albeit briefly) for expert astronomers such as Boulliau and Wallis.[12] Marie Guyard and Vincent Léautaud went as far as claiming three comets in a row appeared between mid-December and the beginning of February.[13] Conversely, some other observers believed that the two comets were one and the same, as testified by Grandami and Billy, who argued against Guyard and Léautaud.[14] But, because a certain number of questions on comets were still unanswered at this time, they were also food for deep thought and for heated debates at the intersection between astronomy and natural philosophy. It was still disputed as to whether comets were permanent objects, as Seneca claimed, or transitory objects, as Aristotle prescribed. Discussions on their location, trajectory and origin had been refueled by the recent discoveries made by sixteenth-century astronomers. Tycho Brahe established that at least some of them were travelling above the Moon and ran across the paths of the planets, thus demonstrating that the heavens are not incorruptible and that they contain no solid spheres. But the space above the Moon was still so vast that the question of the location of comets and the question of their trajectory remained open. A distinction could be introduced between different kinds of comets, the ones below the Moon, the others above the Moon (Maestlin), and, for those who thought that comets were above the Moon, it was still to be decided whether they were between the Moon and Venus (Tycho) or even beyond Saturn (Descartes). Some argued that the trajectories of comets were irregular, others that they were regular, and, among these, some that these trajectories were circular, such as Maestlin and Tycho; others that they were rectilinear, such as Kepler; and it was still to be decided whether or not their motion was uniform. Intimately linked to these points was the question in the defense of which system of the world comets could be enrolled. Kepler ended up remarking that there were as many arguments in favor of the motion of the Earth as there were comets, since their apparent paths result from the combination of their real rectilinear paths with the motion of the Earth. That the tails of comets were almost always opposed to the Sun was, since Peter Apian, generally taken as the proof that, contrary to what Aristotle believed, they are not ignited bodies but, rather, that they resulted from the refraction of the rays of the Sun through the body of the comet, but the nature of this body was itself debated again and again.[15]

That men of letters and scientists were enthralled with these questions on comets in general, and on these comets in particular, was not specific to France, but there were nevertheless several intellectual and institutional circumstances that gave a specific twist to the French debate. It should be first recalled that the sixties were the first years of the war between Jesuits and Cartesians that was to enflame learned France until the late nineties at least. On the Cartesian side, Claude Clerselier, Descartes's friend, correspondent, translator and executor, had launched a campaign for the propagation of Descartes's works. The *Discours de la méthode* was reissued in Paris, while Descartes's unedited works were posthumously published under the supervision of his followers, who in addition published their own works and organized conferences presenting Descartes's doctrines. On the Jesuit side, Honoré Fabri received from Jean Bertet the letters where Descartes suggested a physical explanation of transubstantiation and censured it *privatim*, that is, unofficially, on 15 April 1660. Following some troubles in Leuven and then the official censures of the *carme déchaux* Agostino Tartaglia and the *somasque* Stefano Spinula, Descartes's works were put on the Index of Prohibited Books *donec corrigantur* in Rome in 1663. This was not to be the end of the controversy between Jesuits and Cartesians: on the contrary, it was only the beginning of a war among French natural philosophers.[16]

The early sixties were also the period of gestation for the future Académie des sciences.[17] The letters of Henry Oldenburg, first secretary of the Royal Society, and of Christiaan Huygens, who after his stay in London during the spring of 1661 remained a correspondent of Robert Moray, made French scientists aware of the experimental commitment of the Royal Society, which rapidly became a model for them.[18] This model became all the more desired because the *Académie Montmor*, founded by the end of 1657, was known to be doomed because of its endless discussions and bitter quarrels—and it, indeed, came to its end in June 1664. As early as July 1663, among various men of letters, Jean Chapelain, Pierre Petit, Samuel Sorbière, Pierre de Carcavy, Marin Cureau de la Chambre, Johannes Hevelius, and Christiaan Huygens were awarded pensions by the King.[19] In 1663, Samuel Sorbière, then secretary of the Académie Montmor, wrote a letter to Colbert in which he showed this already dying society in the most favorable light. Several other competing projects for the future Academy were written. The most interesting one is the project of the Compagnie des sciences et des arts, probably written in 1664 by Thévenot, Auzout, and perhaps Petit, but two other proposals were composed in 1666, probably at the request of Colbert, one by Jean Chapelain and the other by Charles Perrault, both members of the Petite Académie, a council in charge of proposing initiatives to glorify the King. The Académie des sciences was finally founded in 1666.

Finally, it is also during these years that French elites began to discredit systematically the practice of astrology as a popular superstition. Advancing spiritual or temporal arguments against astrology was indeed an old affair. Judiciary astrology, that pretends to predict with certitude what will happen to a given human being (contrary to natural astrology that concerns natural events that will probably happen), had been condemned in 1586 in the *Constitutio contra exercentes astrologiae iudiciariae artem* and in the papal bull *Coeli et terrae Creator Deus*, then again in 1631 in the bull *Contra astrologos iudiciarios*. The official concern of these bulls was to preserve the freedom of will, but they were also linked to the fear of the social troubles that predictions of the death of a pope would imply. The condemnation of the church was relayed during the whole century by French Jesuits, who published treatises against astrological predictions in 1619 (Jean Leurochon), 1649 (Nicolas Caussin), 1657 (Jacques de Billy), 1660 (Jean François), and 1681 (Claude-François Menestrier).[20] As for the several royal decrees that were issued in France against illicit predictions in 1560, 1579, and 1628, they were increasingly inspired by socio-political reasons, rather than by religious and metaphysical reasons. Contrary to the 1560 and 1579 decrees, the 1628 decree does not ground its condemnation of illicit predictions on the "express commandments of God" that Christian princes should respect but, rather, on the fact that these predictions, being "useless and without certain foundations, can only embarrass weak minds who believe in them."[21] Despite all this, it was only during the reign of Louis XIV that the elites began to turn away from astrological practices, which were now perceived as low class and despicable. The sixties might have been a turning point in this respect: in 1668 Jean de la Fontaine published *L'Astrologue qui se laisse tomber dans un puits*, a fable that asserts that the stars cannot reveal us the will of God and that astrology is nothing but a scam; two years later, Molière issued *Les amants magnifiques*, a comedy that depicts an astrologer as a venal charlatan.

The two comets arrived in the midst of this—the debate on comets that had been refueled by Copernicus, Kepler, Tycho, and their followers, the budding confrontation between the Jesuits and the Cartesians, the excitement preceding the establishment of the Académie des sciences and the growing disregard for the elites for astrology. My working hypothesis in this chapter is that, in such circumstances, the two comets of 1664–1665 were an occasion for each camp to advocate publicly its positions about the proper principles for natural philosophy. By natural philosophical principles, I mean here two things. First, ontological principles, that is, what kinds of beings one should have recourse in order to explain comets—to wit, whether one should admit substantial forms, which system of the world one should favor, whether a body would go on forever once it has been moved, etc. Secondly, I mean epistemological principles, namely: what kind of knowledge can be reached about the nature of comets and about the effects that they may bring about—for example, if such knowledge is certain or only probable, if it relates to signs or to causes, if it is to be obtained mostly through the senses or through reason, and so on.

It is also to be noted that, if some of the authors I study are explicitly speaking in terms of principles and indeed conceive principles as the source of all knowledge—those who have good principles will reach true knowledge, while those who have bad principles will reach only an appearance of knowledge—others do not describe their epistemic chances to attain some kind of knowledge in these terms. In the case of the latter, it is us who introduce the question of principles in texts written by authors who certainly had some principles that organized their beliefs but did not consider that it was important to make them explicit. Two of the questions explored in this chapter are to determine who speaks in terms of principles and who does not and what are the implications of these differences.

With my questions clear, I would like to add a few words on the corpus dealt with in this chapter. It consists of all the books on comets, whether they were written in Latin or in French, published in France in 1664 and in 1665 that I could recover by a systematic search in contemporary catalogues. Though systematic, such a search is not without bias: as scientific correspondences like those of Ismaël Boulliau, Giovanni Domenico Cassini, Johannes Hevelius, Christiaan Huygens, or Giovanni Battista Riccioli make clear, astronomical observations were not circulated through books but through letters.[22] But it would take another paper to reconstruct how observations were gathered, circulated, and compiled. As for the organization of the maze of books that were published, it appeared to me that the most efficient way to present them in this chapter is the following: I first report on those that were mainly concerned with astrological predictions; then I analyze the philosophical opposition that developed between Jesuits and Cartesians; in the third place I turn to those whom I call the Montmorians, because at some point or another, they participated in the Académie Montmor; finally, I say a few words on a few books that are unique for one reason or other.

Aristotelian Principles for Uncertain Astrological Prognostications

The first group I want to consider are those books that, according to their titles, aim to predict the effects that the comets would have on natural and human affairs, or at least to determine if such predictions are possible. In chronological order of publication, they are sieur de Montalegre's *Discours sur le comete*, D. de Vaissey's *De novo cometa*, Julio Giustiniani's *L'explication de la comète*, Henry de Leschener's *Traité des comettes*, the anonymous *Figure de la derniere estoille*, Robert Luyt's *Questions curieuses sur la comète*, Sombreval's *Advertisssement du ciel*, and Claude Comiers's *La nature et présage des comètes*. Even before analyzing the opinions defended in these books, it is to be noted that, apart from Comiers's book—which was not published until the fall of 1665, had probably a greater circulation than the other books, and was an exception in many respects—they were the very first books to be published, as their licenses to print indicates.[23] This chronological primacy is counterweighed by two kinds of marginality, one rather social and the other rather intellectual.

While the authors belonging to the other groups resided in Paris, the authors of the first group, at least those who are known, were either foreigners or provincials who did not publish much, or at least did not publish much on such matters. Montalegre's name indicates that he was probably from the south of France, and some passages of his book imply that he lived in Lyon.[24] Vaissey's letter introducing his verses was written in Chalon-sur-Saône, a town in Burgundy. Although "Giustiniani" is the name of an aristocratic family from Venice—Marc Antonio Giustiniani was the Venetian ambassador in Paris from 1665 to 1668—Julio Giustiniani was probably a Spaniard attached to the queen mother Anne of Austria, since his book is dedicated to her. The title of Leschener's book mentions that its author is of German origins, and its dedication suggests that he had been close to Prince de Condé since the successful military campaigns that the latter led against the Spaniards before being their ally during The Fronde.[25] Luyt was a doctor in theology, preacher-in-ordinary to the King, who had lifelong ecclesiastic charges at the church of Tonnerre, another small town in Burgundy. His book on comets was his only incursion in "scientific" matters, his other publications being those befitting a devout scholar who had a particular interest in retracing the history of local saints and local noble women.[26] Sombreval is presented as "aumonier de la Reine," but I have not found any indication which queen; he probably lived in the vicinity of Lyon, though, since his book was published by François Larchier, the printer from Lyon, who three months earlier printed Montalegre's book to which he answers and the only extant copy of his book is to be found in the Municipal Library of Bourg-en-Bresse. Comiers was to be associated with Paris's learned circles from the early seventies on and to collaborate extensively to different learned journals, from the *Journal des savants* to the *Mercure galant*, passing by the *Nouveautés journalières concernant les sciences et les arts* and the *Journal des nouvelles découvertes*. However, at the time of his book on comets, which was his very first, he was merely provost of the chapter of Ternand, a village north of Lyon, a charge he had obtained ten years earlier from his first patron, the Marquis de Saint-André Montbrun.[27]

This social marginality goes along with what we would perceive today as a kind of intellectual marginality. Leschener is the only one to have heard about the conference at the collège de Clermont, which may of course be explained by the fact that the books of this group were the first to be published.[28] But the scarce references to sixteenth-century astronomers and seventeenth-century natural philosophers clearly indicate that these authors were intellectually isolated. The author of *Figure de la derniere estoille extraordinaire,* Vaissey, Giustiniani, and Luyt refer to nobody at all. The most frequent references of the other writers, Montalegre, Leschener, Sombreval, and Comiers, are the ancients, whether fathers from the church, philosophers, writers, or poets. Leschener mentions in addition Copernicus, Tycho, Kepler, and Sombreval refers to Riccioli, Fromondus, Alsted, and Gassendi, but this is only in passing.[29] Comiers is something of an exception: he refers not only to Copernicus, Tycho, and Kepler but also to Longomontanus, Regiomontanus, Fromondus, Galileo, and Descartes.[30] Moreover, it is to be noted that Montalegre, Vaissey, Giustiniani, Luyt, Leschener, Sombreval, and Comiers are not mentioned in the books of the other groups. Some of them address criticisms to astrologists but in quite general terms and, with the exception of Pierre Petit's *Dissertation sur la nature des comètes,* not by referring to specific books.

Although I group these books together because they were concerned with predictions, Giustiniani, Vaissey, and the anonymous author of *Figure de la derniere estoille extraordinaire* are the only true astrologists in the group. Their books are quite short leaflets, Giustiniani's ten pages, on which I shall focus, being the longest and the most detailed. The immediate observable qualities of the earlier comet—its color, its position among the astrological signs and its celestial orientation—are briefly described. From this description, it is immediately and categorically inferred that the comet will bring along dreadful events: its Saturnian color predicts many deaths; its position near the Hydra means that these deaths will happen on waters, and in naval battles rather than because of tempests; its head pointing towards the Orient is a sign that "these bad accidents are firm and true" (Giustiniani 1665: 6).[31] The only thing about which doubts are expressed is the location where these accidents will happen, but the conjecture is that they will not happen in France because of the great piety of its King. In a word, Giustiniani is not interested in discussing principles, whether ontological or epistemological. He does not say a word about the ontological principles necessary to explain the nature of the comet, and he does not reflect on the epistemological modalities of his predictions—as I just said, he is happy with the description of observable qualities, taken as immediate signs of the misfortunes to come. This is all the more true for the seven pages of Vaissey's *De novo cometa. . . carmen prognosticum* and of the really quite brief *Figure de la derniere estoille extraordinaire et professie mistique sur l'apparition du dernier comete,* which does not even describe the particulars of these two comets but only asserts that a comet is a "spark of God's wrath, which is threatening you with the eternal fire, against which the sinners can not guard, except if they convert early" (*Figure de la derniere estoille*: 5).

The other books of this first group, however, are different. First, opinions on the nature of comets are expressed. Except for Comiers, their authors all endorse Aristotle's opinion that comets are exhalations and vapors that rise from the Earth into the air. Montalegre falls in with this opinion by convention, noting simply that, although there are several opinions on the nature of comets, this one is the most common. He adds, however, that there are not only terrestrial comets, resulting from exhalations of the Earth, but celestial comets as well, resulting from exhalations from the other planets.[32] Without giving any explanation at all, Luyt admits that, as Aristotle and all the other philosophers after him would have claimed, comets are meteors that belong to the highest region of our terrestrial air.[33] Sombreval does not take sides with any opinion on the nature of comets, but it seems reasonable to conclude that he shares Aristotle's opinion since he explains their effects by referring to Aristotle and does not mention the anti-Aristotelians of the day.[34] Leschener's view is more elaborated than those of Montalegre, Luyt, and Sombreval: he is not satisfied with the idea that Aristotle's position is to be accepted because it is the most common but gives arguments to support it. According to him, natural philosophy came back to childhood in its elderly age: maturity was reached during Aristotle's time, when it was thought that heavens are incorruptible and that comets are terrestrial exhalations enflamed by the Sun when reaching in the highest region of the air; with the modern opinion that comets are celestial phenomena, we would have regressed to childish pre-Aristotelian opinions.[35] Leschener knows that Tycho Brahe called upon the absence of parallax to place the comets above the Moon, but he opposes this conclusion with the argument that it never happened that two astronomers situated at different places were able to observe the same comet at the same time.[36] He then discusses if comets are clouds illuminated by the Sun or perhaps real fires inflamed by the Sun, to finally declare himself favorable to the former.[37] Curiously enough, he still seems to prefer the Copernican system of the world.[38] As for Comiers, he presents himself as theologian by profession, who for nine years has only been able to exercise his talent as a mathematician during his leisure (he did go on to publish extensively and successfully books dealing with fashionable mathematical and physical subjects, like cryptography, optical illusions, and, by the end of his life, the divining rod).[39] After presenting the different existing opinions on comets, he presents his own theory, that comets are planets that were invisible but now became visible by reflecting the light of the Sun. What characterizes his way of proceeding is a rare combination of deference to ancient authorities and use of experiments: as in the First Day of Galileo's *Dialogue*, which is briefly quoted for its view on comets, common optical and chemical experiments made on Earth are used to understand what happens in the heavens.[40]

If Montalegre, Luyt, Leschener, Sombreval, and Comiers have a differ-ent story to tell than those I called the true astrologers, it is also because of their ambivalent epistemological stance toward astrological predictions. On one hand, they indisputably make predictions; on the other, they distance themselves from these predictions in all sorts of ways, rehearsing some of the already common arguments against astrology. First, reporting on the distinction between natural and judiciary astrology, they stress that their predictions are only natural predictions about some natural causes which will probably bring about some effects. For example, the dryness of the exhalations may bring bad crops that could cause famines, which could pos-sibly, in turn, be causes of social turmoil. These are not judiciary predictions that state what will certainly happen to a given human being.[41] Although such a distinction was often linked to the distinction between comets as causes and comets as signs, Luyt and Sombreval are the only ones who men-tion it. Adopting an argument put forward by Claude Pythoys, Luyt notes that predictions are uncertain, not only because it is difficult to know where, when, to which, and to whom they will apply, but also because the relation between a comet and its effects, whether the death of a prince or a war, is not a natural and necessary relation as the causal relation between heat and burning wood, but only a conventional relation as the relation between white and peace, so that it is only the experience of many centuries that shows us that there is indeed a relation between the appearance of comets and unfortunate events.[42]

Second, they make the point again and again that astrological beliefs being typical of children, women, and the lower classes who are unable to listen to the learned, their own predictions are just specious fantasies, railler-ies, and mockeries only proposed in order to entertain their readers.[43] Third, they doubt if the effects of comets are positive or negative. Although the tra-ditional view was that comets foreshadow misfortunes, natural calamities, and human tragedies, Montalegre advocated as early as December 1664 that comets in general were either only natural signs or that, if they announce something, it is only joyful events. He does not give many arguments but explains that such an opinion "pleases his mood, which amounts to avoid increasing purposely my sorrows and voluntarily frightening myself." There are a few examples, though: the defeat of the Barbarians, the defeat of the Turks, and the defeat of the "Diabolic Sect" and "Infernal Cabal" of Protes-tants were announced by the comets that appeared in, 405, 725, and 1618, respectively.[44] As for the comets of 1664–1665, in particular, they would be the first signs of a new golden age when peace will be guaranteed by Louis XIV and abundance furnished by the many treasures discovered in the Isle of Madagascar.[45] Sombreval wrote his *Advertissement du ciel* to defend against Montalegre the traditional view "that comets are ominous presages in the physical and in the moral world, namely dangerous alterations in nature and upheavals in the republics and the states; that they do not bring any good except by accident, because the unhappiness of one is always followed by the happiness of the other" (Sombreval 1665: 4). Sombreval relates Montalegre's mood to his name, Montalegre's literally meaning "joyful mount," and opposes to it the authorities "of the Ancients, of The Fathers and of the Church itself" (Sombreval 1665: 3).[46] Luyt stands somewhere between Montalegre and Sombreval: he notes that, in general, comets forebode calamities, but adds that it can happen that they are used by God to give a premonition of future prosperity. As for the 1664–1665 comet, he thinks that it announces rather a joyful event.[47] Comiers went farther than Luyt and proposed what he himself called a "problematic treatment" of the presages of the comets: in order to leave to his readers the freedom to judge if comets are bad omens, good omens, or no omen at all, he presented in almost three hundred pages the facts in favor of each of these three different opinions successively.[48] But since he devotes more than two hundred pages to the circumstances in which the first opinion has been confirmed, only twenty pages to the second opinion, and ten pages to the third opinion, it is tempting to conclude that he believes that comets are indeed bad omens. This is confirmed by his "General conclusion on the presages of comets," where he asserts that, "after having presented reasons from both sides and left the question as if it was undecided," he abandons this reserve: his own belief, shared by Augustine, Kepler, Carolus Magnus, and Sennert, is that comets are signs sent by God to make his wrath known to us (Comiers 1665: 432–436).[49] Comiers's treatise could have stopped here, but he still deals with another question; whether the misfortunes associated with these two comets are the harbingers of the end of the world announced by the Antichrist. In order to answer this question, he enters the delicate interpretation of religious prophecies and finally concludes that the only relevant prophecy here is that the present King of France will establish a universal empire.[50]

To sum up, if there were still Aristotelians in late seventeenth-century France as far as the question of nature of comets is concerned, they were to be found among those who, even if not astrologers in the strict sense of the word, were nevertheless concerned with predictions and, in the case of Comiers, with prophecies. However, I have shown that not all of them were Aristotelian in this respect and that among the three of them who were, Montalegre and Luyt were such by adherence to an established orthodoxy rather than because of careful arguments. As far as heliocentrism is concerned, both Leschener and Comiers imply that, were it not for the church's condemnation, they would subscribe to Copernicus's principles.[51] As for their epistemological commitments, except for *L'explication de la comète* and for *Figure de la derniere estoille extraordinaire*, these books were not only distancing themselves socially from the vulgar astrological beliefs of women and members of the lower classes, they were also expressing epistemological doubts about the possibility of using comets to make any prediction at all. Although some of them present traditional arguments against astrology and discuss them, they do not engage in any polemics against the authors of these arguments; rather, the only existing controversy among them seems to be on whether comets are good or bad omens—a question that, in its own way, indicates that astrological beliefs were disappearing. Whether one considers ontology or epistemology, it is to be noted that these authors do not express themselves in terms of principles, even when there is some disagreement between them, as is the case between Montalegre and Sombreval. As we will see in the next section, the situation was different in the case of the Cartesians and the Jesuits.

Cartesians and Jesuits: A Predictable Confrontation on Ontological Principles

Cartesians and Jesuits are treated together in this chapter not only because they are the most well-defined groups, both in social and in epistemological terms, but also because of the controversy developing between them, or at least because of the confrontation between some of them. It all began during the learned ceremony—or courtly workshop, or scientific drama—that took place at the collège de Clermont as early as 10 January 1665, that is, only one month after the appearance of the first comet in Paris and at a moment when only prognostications by Giustiniani, Vaissey, and Montalegre were in circulation. The most detailed report of this event is to be found in the newly born *Journal des savants*, which presents "an admirable summary of all that happened at this conference" (*Journal des scavans*, 28 January 1665, in Sallo 1665: 66–71).[52] In the presence of the Prince de Condé; his first male born, the Duc d'Enghien; his brother, the Prince de Conti; and of many prelates and noble persons from the Royal Court, four opinions concerning comets in general were successively defended.[53] A local Jesuit, Nicolas d'Harouys, first argued that comets were occasional clusters of small wandering stars, which were unseen beforehand. Then Gilles-Personne de Roberval, professor of mathematics at the neighboring Collège royal, said that he subscribed to Copernicus's system and that the comets were exhalations of what he called the elementary sphere, that is, the sphere that extends from the Earth to the Moon. The third one to speak was the physician Vincent Phelippeaux, at the time residing in Paris after he had caused some trouble in Leuven because of his Cartesian positions on the human body. He explained Descartes's opinion, according to which a comet is an obfuscated sun which does not belong to any vortex in particular but passes from one vortex to another.[54] Finally, another local Jesuit, Jacques Grandami, repeated what he had already asserted at the occasion of the comets of 1618, that they are parts of the celestial matter condensed by the action of the sun.[55] Still according *Journal des savants*, night fell and the conference came to an end before a third and last local Jesuit, Jean Garnier, could present his opinion, that comets are fires gathered in the air.[56] The order of the speeches was important: the Jesuits obviously wanted to open and conclude a ceremony that was set up to organize the debate by bringing about the true doctrine while a variety of opinions were displayed. They were unsuccessful in the sense that the different protagonists left collège de Clermont without having reached a consensus. But they were successful in the sense that, thanks to the report of their meeting that was immediately published in *Journal des savants*, they set the agenda for most of the books published thereafter.[57] Moreover, the fact that these, together with Denis's book, are nowadays much easier to find in French libraries indicates that they most probably had a greater circulation than the other books in my corpus. In what follows, I first analyze some of the opinions that were presented in this conference; then I examine the Jesuit reaction to Jean-Baptiste Denis's anonymously published book defending Descartes's opinion on comets; finally, I report on Cartesians and Jesuits who stayed out of this confrontation: Jacques Rohault, on one side and, on the other side, Jacques de Billy's in Langres and Ignace-Gaston Pardies in Bordeaux, whose books had a much smaller circulation and are still more difficult to find today than those of the Parisian Jesuits.

Grandami's opinions are detailed in his March 1665, *Le cours de la comète*. In *Le parallèle de la comète* that he published two months later, he only argues that the two comets, the one of December 1664–March 1665 and the one of March–April 1665, are different.[58] *Le cours de la comète* begins by presenting the numerical tables describing the trajectory of the first comet; it then includes the more philosophical *Traité de la comète*, which, according to the dedication to Prince de Condé, reiterates what Grandami said in January on the nature of comets in general. The first point to be noted is that Grandami is no Aristotelian as far as the nature of comets is concerned; as Denis should ironically conclude in his criticisms of the Aristotelian position on comets, "at this famous conference held at the *collège de Clermont* . . . not even one philosopher was to be found, who but dared to suggest Aristotle's opinion" (Denis 1665: 37–38). Like most Jesuits from the beginning of the seventeenth century, Grandami adopts Tycho's main claims: he believes that the Sun revolves around the Earth while the other planets revolve around the Sun, that the celestial matter is fluid, that comets are situated above the Moon and even above the Sun, that they have a circular path, and that comets are made of incorruptible celestial matter, like stars and planets, although they are themselves corruptible.[59] When he discusses the location of the first comet, Grandami makes explicit that, contrary to Copernicans and following Riccioli's criticism of Kepler, he does not believe that comets move along straight lines; even if they had such a motion, this would be no more a proof of heliocentrism than the motion of Mars around the Sun is a proof of the motion of the Earth around the Sun.[60] He finally refers to an experiment he made a long time ago at the Bourdelot Academy and a book he published more than twenty years earlier to establish that, magnetic bodies being immobile and the Earth being a magnetic body, the Earth is immobile.[61]

The main question that this Tychonian Jesuit had to deal with was how a corruptible comet could appear from such incorruptible matter and if this implied the apparition of a new substantial form. Grandami's solution to this problem amounts to saying that a comet appears when the fluid celestial matter is condensed in such a way that it becomes able to reflect the rays of the Sun that it receives and to transmit them. Such a condensation does not imply a change from one substantial form to another—the corruption of a substance and the generation of another substance—but "only an accidental alteration, as the one occurring between flowing and iced water; or between milk and blood and clotted milk and blood, and finally between soft juices and the same things when they have hardened" (Grandami 1665a: 2).[62] So far we have the material cause and the formal cause. As for the efficient cause that condenses celestial matter into comets, Grandami thinks that it is the natural virtue which the heavens, the Sun, the Moon, and the planets received from God: "Nobody doubts that they condense clouds in the sky, that they harden gold and gems in the earth and pearls in the sea: why would they fail to have the virtue to condense and thicken comets in their own womb?" (Grandami 1665a: 7). To this, Grandami adds the swift motion of the celestial matter, which helps the compression of comets exactly as the wind helps with the condensation of clouds.[63] Finally, dealing with the final cause of comets leads Grandami to condemn astrology in quite classical terms.[64] If Grandami is no longer Aristotelian as far as the nature of comets is concerned, he is completely Aristotelian in the categories that structure his thought: the four causes, for example, and the kind of questions that he asks—for example, whether a comet and the celestial matter that surrounds it have the same substantial form or not.

To present in more details the other positions at stake at the January conference is difficult, if not impossible: there is no trace of Vincent Phelippeaux's discourse; the only known books by Nicolas d'Harouys are rhetorical pieces; as for Roberval, his only published writing dealing with comets is a chapter of the *Aristarchi Samii De mundi systemate, partibu. . . libellus* that he published twenty years earlier.[65] However, if one takes into account the theses that were defended at the collège de Clermont that year and Jean-Baptiste Denis's *Discours sur les comètes*, a coherent story emerges, in which a war was waged against Descartes's natural philosophy. Three theses in mathematics were defended at the collège de Clermont on comets that year. (Without going into the details of what defending a thesis in a Jesuit college implied, it should be mentioned that Jesuit teachers wrote the theses of their students, that the oral defense consisted in resolving some problems mentioned in the written thesis, and that, for several institutional reasons, theses in mathematics allowed more space for discussion than theses in philosophy).[66] On 29 January, less than three weeks after the January conference, and at a period of the year most unusual for engaging oneself in such an exercise, Jérôme Tarteron defended a thesis called *De cometa ann. 1664 et 1665. Observationes mathematicae*.[67] Six months later, Louis Ragayne de La Picottière defended two theses in a row, *De duplici cometa vero et ficto positiones mathematicae* on 12 June and *De hypothesi cartesiana positiones physicomathematica* on 13 June.[68] Even if Nicolas d'Harouys officially stopped teaching mathematics at the collège de Clermont in 1664, he was the one who wrote all these theses with the intention, first to expand on the view that he presented at the January conference that comets are clusters of small stars that happen to gather occasionally for a while, then to answer Denis's book, and, finally, to counterattack Denis with a general criticism of Descartes's principles on natural philosophy.[69] This confrontation could have turned into a protracted controversy if Denis had answered; he did not however, nor did any of the contemporary followers of Descartes. This is why I describe it as a "confrontation" rather than as a "controversy."

The first thesis, *De cometa ann. 1664 et 1665. Observationes mathematicae*, begins with the remark that, having discussed comets in general, it remains to deal with a particular comet.[70] Considering the chronology, the fact that, according to *Journal des scavans*, comets were discussed only in general during the January conference, and the resemblance between the opinion on comets defended in this thesis and d'Harouys view, it may very well be that at the January conference, d'Harouys presented his opinion on comets in general.[71] As for the body of the thesis, it consists of twenty-four sections titled "observations." Each of these sections starts by presenting an observation in the strict sense of the word—that a comet appeared on our horizon, that it was seen out of the Zodiac, that it was seen for almost two months, that its tail changed its form, that its tail grew progressively before diminishing, and so on. It is then asserted that this observation cannot be explained by any of the competing opinions, except by the opinion that the comets are made by a cluster of innumerable small stars; a problem related to the observation at stake is finally formulated, probably to be solved by Tarteron during the defense.

In general, the authors of the competing opinions on comets are not identified by name, with the exception of Descartes, whose name appears in three of the problems that the young Tarteron was supposed to solve. Observations 12, 14, and 15 ask the defendant to "demonstrate that according to the first principle of the Cartesian doctrine (*e primario Cartesianae doctrinae principio*), which is quite necessary to explain the phenomena of comets, there would not be a cow, an elephant, or another beast that would not be carried away by all its effort through the skies," to "prove that, according to the principles of the Cartesian philosophy (*ex principiis Cartesianae Philosophiae*), a comet created by God cannot maintain itself by itself," and to demonstrate that, "according to the famous law of the Cartesian philosophy (*ex celebrimo Cartesianae Philosophiae decreto*), it can happen that a comet, once set in motion, even slightly, will not stop moving before a chimera is produced" (Tarteron 1665, Observatio 12, 14, 15: 7–8). At stake are Descartes's "principles" that, although comets are more solid than planets, they move more quickly; that no creature can exist without God's concurrence; and, finally, that, once something has been set in motion, it will continue to move forever.[72] It is not clear why d'Harouys highlighted these specific propositions, but it is clear that, already in the

January thesis, Descartes's philosophy is characterized by its principles. D'Harouys's aim was not only to fight against the explanation of comets proposed by Descartes but, taking literally his claim that the whole system of the world depends on a few principles concerning matter and motion, to also oppose these very principles because of the paradoxes that they imply. However, this is a relatively light opposition compared to what we shall see in the two June theses.

At some point in the spring, the young Jean-Baptiste Denis published anonymously a *Discours des comètes selon les principes de Mr. Descartes*.[73] Like d'Harouys in the *Observationes mathematicae*, Denis proceeds dialectically. Descartes's view is that comets are old stars that have been so obscured by their spots that they stop having their own vortices and begin to wander from one vortex to the other. Denis intends to establish it by refuting the other opinions, first Aristotle's and then the four that were presented during the January conference. D'Harouys's position, which is introduced as first formulated by Democritus and Anaxagoras, is the one on which Denis is the more negative.[74] Interestingly enough, he uses comparisons that do not appear in *Observationes mathematicae*, but that could have been used orally by d'Harouys during the January conference: comets are like a swarm of gnats flying together or like a company of travelers riding together; their tails are like footmen that stand sometimes before the carriage and sometimes behind it.[75] His judgement on d'Harouys is particularly harsh: he would make as many suppositions as there are questions; his position would be contradictory.[76] According to Denis, while the others instituted "some principles that have no other use except for explaining the phenomena of comets [*ont pose des principes qui n'avoient aucun autre usage, que pour expliquer les Phœnomenes des Cometes*]," Descartes has recourse only to principles "that are used to explain all the physical motions, mostly the motions of planets, fixed stars, and other celestial bodies. So that, if somebody finds a difficulty in that, it won't come from the thing itself, but from him who, because he does not know well enough the principles of this philosophy, will have difficulties applying it [*celuy qui, n'estant pas verse dans les principes de cette Philosophie, aura peine à en faire l'application*]" (Denis 1665: 77–78). Denis adds that such principles might appear paradoxical to "those who had never a glance at another philosophy than the one of Aristotle" but that, if one does not accept them, one is obliged to stay with "occult qualities, secret influences, attractions, sympathies, antiperistases, and the other words that, because they fill in only the mouth and leave the mind void, are now reputed to be an asylum and refuge for ignorance" (Denis 1665: 105–106). In a word, Denis insists that the only existing alternative is between being an Aristotelian and being a Cartesian and that this choice is a choice that does not only concern comets but the general ontological principles that are to be defended in natural philosophy.

In June, d'Harouys replied to Denis's attack in two theses written as short treatises: in particular, unlike most theses in mixed mathematics they do not include problems to be solved by the defendant. While in the first of these theses, *De duplici cometa*, d'Harouys concentrated on comets; in the second one, *De hypothesi cartesiana*, he launched a much more general attack on Cartesian natural philosophy, explicitly formulated in terms of, and focusing on, principles. In the foreword of *De duplici cometa vero et ficto positiones mathematicae*, d'Harouys explains that he had to write it because "the hypothesis on the matter of comets that he proposed and publically defended was recently attacked by a Cartesian (*Cartesianus quidem*) who forgot the Cartesian method, which seems to have this particularity, that it establishes its own claims as far as possible, but does not attack others' claims" (Ragayne de la Picottière 1665a, sig. A1r).[77] This unnamed Cartesian is Denis, not only because his book was published anonymously and because it actually proceeds by attacking the other opinions but also because *De duplici cometa* is a point-by-point answer to *Discours de la comète*, based on the opposition between two comets, the true, or at least verisimilar, comet, which is the comet inasmuch as it is described according to d'Harouys's principles, and the false and fantastic comet, on the other hand, which is the comet inasmuch as it is explained by "the Cartesian." The first part (§§1–38) answers Denis's book paragraph by paragraph, and even line by line; the second part (§§39–61) counterattacks by criticizing it, here again following it quite closely.[78] Last, but not least, is a list of paradoxes that would follow from Descartes's philosophy (*consectaria e principiis Cartesianae Philosophiae*). This list is interesting because, except for one consequence concerning Cartesian doubt (§31, "What is doubted by many is better known than what is doubted by nobody"), it concerns only natural philosophy, but also because it reveals which principles in Descartes's natural philosophy were difficult to digest for d'Harouys. First, there are paradoxes coming from the identification of matter and extension (§§1, 2, 3, 10, 11, 13, 27; e.g., §2, "The infinite in act is not only possible[79] in material things, it is also necessary," or §13, "God can not add or subtract anything to this universe, not even an insect or an atom"). To these first paradoxes, some paradoxes on magnitudes (e.g., §4, "The part is not smaller than the whole") may be related, the idea probably being that there is an infinite number of actual parts in a smaller chunk of matter as well as in a bigger chunk of matter. Second, there are paradoxes arising from the definition of motion that Descartes gave in *Principia philosophiae*, like §5, "There is not a particle of this universe that can change place," §6, "A boat following the course of the Rhône is as steady as a rock on the earth," §29, "In order to say that a body is at rest, it is not enough to say that it is immobile," or §32, "In the Copernican hypothesis, the Earth is much steadier than in the Tychonian hypothesis." Third, there are paradoxes arising from the reduction of qualities to matter and motion, for example, from the reduction of solidity (§§14, 15, 20, 33), of hardness (§8: "If the parts of this air that we inhale would rest, they would be harder than any metal"), of attraction (§34), or of life (§§16, 22: "a clock can fall ill and die just like a horse"). For d'Harouys, a paradox is not something that contradicts a particular philosophical doctrine, let alone a theological one, but, true to the etymology, what goes against received opinions and revolts against common sense.[80] In particular, d'Harouys neither alludes to substantial forms and real accidents, nor speaks of the Eucharist, which were at the heart of the condemnation of Descartes in 1663.[81]

The process of "principlization" triggered in this list—starting with a discussion on the comets, one ends up with an examination of the principles of Descartes's natural philosophy in general—is still amplified in *De hypothesi cartesiana positiones physicomathematicae* that was defended the next day. Moreover, while in *De duplici cometa* "mathematical positions" were at stake, in *De hypothesi cartesiana* "physico-mathematical positions" are presented. More precisely, *De hypothesi cartesiana* considers first the principles of "the Cartesian hypothesis" in general (part I), then one by one (part II), and finally with regard to their consequences (part III). D'Harouys obviously read *Principia philosophiae* very carefully: he takes numerous expressions *expressis verbis* from it. The first kind of problem that he encountered in Descartes's natural philosophy comes from its use of fictional hypotheses. D'Harouys points out repeatedly that Descartes allowed himself a certain *libertas fingendi* and that he presented his natural philosophy as a hypothesis, a fable, and a fiction.[82] The title of the thesis itself, *De hypothesi cartesiana*, makes sense when one reads paragraphs 14 and 15 of the first part, according to which Descartes is only a mathematician who makes hypotheses and forges fictions, while Aristotle is a philosopher who advances true theses and explains the reality of things.[83] In another passage, d'Harouys refers to the astronomical machines that he built for teaching the different "astronomical hypotheses (the Ptolemaic system, the Copernican system, the semi-Copernican system, the Tychonian system, the Harouysian system, and all the others)" and makes the point that, although one can build machines full of gears to explain the motions of the stars, no sane person will assume that these machines can show how things actually happen in the sky.[84] Hence, the criticism that d'Harouys addresses to Descartes from a methodological point of view amounts to reestablishing the traditional division of labor between the mathematician and the natural philosopher and condemning Descartes for being only a mathematician. The second kind of problem that Cartesian natural philosophy arouses according to d'Harouys comes from Descartes's ontological principles. As in the thesis of the previous day, it is common sense that d'Harouys defends against the paradoxes that, according to him, would be implied by the identification of matter and extension,[85] the existence of infinitely divisible parts in matter,[86] the confusion of animals and machines or the assimilation of nature and art,[87] the conservation of motion,[88] the transmission of motion from one body to another body,[89] the persistence of rectilinear motion,[90] and so on. Contrary to the thesis defended the day before, however, the final nail is put in the coffin: although brief, it is said that, the one who follows Descartes's philosophical principles does not need substantial forms and that he cannot explain the conversion of the bread and wine into the body and blood of Christ.[91]

The confrontation between d'Harouys and Denis thus presents three characteristics. First, there is what could be called a dialectization of the debate on comets. While Grandami presented his own opinions without discussing his contemporaries, except to mention their observations, both d'Harouys and Denis proceed dialectically, establishing their respective positions by refuting the opposing ones. Second, a polarization of the debate is also to be noted. While there were four positions at the January conference, even five if Garnier is taken into account, and much more if we consider all the books on comets published in 1665, Denis and d'Harouys ended up polarizing the debate between two and only two positions: the only choice is between being Cartesian or being a Jesuit. Third, and perhaps most important, there is a radicalization, or perhaps "principlization," of the debate, in the sense that it is no longer opinions on comets that are important but, much more generally, the kind of ontological principles that are assumed in natural philosophy. In this respect, two things are relevant. First, this principlization manifests itself in words: d'Harouys and Denis use the term *principle* in every single line. Second and foremost, principles enable these natural philosophers to draw boundaries and make the distinction between friends and foes. Friends are not those with whom one agrees on everything but those with whom one agrees on principles. Some amount of disagreement was possible between the Jesuits, as long as it did not concern ontological principles: Grandami and d'Harouys did not agree with Aristotle, and not even between themselves on the origin of comets—d'Harouys said that they come from the gathering of small stars, while Grandami claimed that they result from the condensation of parts of celestial matter—but they nevertheless agreed on basic ontological assumptions, for example, that the nature of a material entity like a comet depends on its location in the universe, that there exist real qualities in nature, and that every natural being is associated with a substantial form. These basic assumptions were the principles that they would not abandon and that defined the boundaries between their camp and the Cartesians. In a confrontation like this, Cartesians and Jesuits sometimes pretend that their principles were what everybody agrees on, although they were only what defined their intellectual identities against others'.

These three characteristics—dialectization, polarization, principlization— are often to be seen in protracted controversies, but as I noted at the beginning, the confrontation between Denis and d'Harouys did not develop after the June theses. Denis left the ground and there was no Cartesian to take over the fight against the Jesuits. Indeed, the notes taken by a lawyer on the lecture on comets that Jacques Rohault gave in his *Mercredis* that very year, contain neither allusion to the January conference nor to any kind of controversy about the nature of comets between Cartesians and Jesuits or, more generally, "Ancient philosophers," as they were called. Rohault was apparently happy to establish that comets are neither stars nor planets, to put forward overused arguments against Aristotle, and, finally, to briefly introduce the theory of comets that Descartes presented in *Principia philosophiae*.[92] On the Jesuit side as well, there was more variety in the field than one might have suspected. Without doubt, contemporaries took d'Harouys's theses as the official Jesuit answer to the Cartesian camp. Petit announced to Huygens, "Denis was mistreated at the public disputes by the Jesuits because this same author has refuted Father Darouys [*sic*], and as you know these Gentlemen (*ces Messieurs*) do not forgive anything. They wanted to take their revenge by defending even their bad opinion that comets are the gathering of several stars" (Petit to Huygens, 7 August 1665, Huygens 1888–1950, 5: 433). Similarly, Oldenburg, who received the theses from Thévenot, immediately reported on them to Boyle, going as far as recopying most of the paradoxes with which the first thesis ends as well as the concluding paragraph of the second thesis.[93] It is striking, however, that, among the Jesuits, there were different ways of conceiving the interplay between natural philosophy and astronomy than those illustrated by the Jesuits from collège de Clermont in Paris. This is particularly true of Jacques de Billy and Ignace-Gaston Pardies, both of them mathematicians and both of them settled in the provinces, the first in Dijon and the other in Bordeaux.

Jacques de Billy published two books on the 1664–1665 comets.[94] Like Grandami's *Parallèle des deux comètes*, the short *Traité de la comète* only attempts to establish that the first and the second are different comets; it ends with some brief considerations against judiciary astrology, against which Billy had already written a book.[95] The longer *Crisis astronomica de motu cometarum*, which was published in 1666, that is a year later than the other books I discuss in this chapter, is a technical astronomical book, devoted to refuting the opinion that the apparent motions of comets can be explained if one attributes to them rectilinear motions, in which case, as Kepler argued in his *De cometis libelli tres* in 1619, their variations of speed and their curvilinear paths would only be appearances resulting from the annual motion of the Earth, exactly as the stations and retrogradations of planets. Billy claims, rather, that comets follow the arc of great circles on the celestial spheres, as Tycho believed. He does not refer to Kepler and Tycho, however, but to two French astronomers: Adrien Auzout, whom I shall come back to in the next section, and Pardies, on whom I concentrate now.[96]

Like Billy, Pardies published two books on the 1664–1665 comets, one shorter in French and a longer one in Latin.[97] The shorter *Remarques sur les comètes et autres phaenomènes extraordinaires de ce temps* includes first some reports on recent, well-attested extraordinary phenomena, like the ghosts that were seen during an earthquake in Canada or the rain of blood that fell on Apulia and then astronomical observations of our two comets.[98] The longer *Dissertatio de motu et natura cometarum* is a most interesting book. Pardies's intention is to put forward an hypothesis that would reduce the irregular motions of comets to regular motion as, he claimed, the hypothesis of ancient Chaldeans did.[99] Unlike other astronomers, especially Auzout, he is quite explicit on the different instruments for situating a comet among the stars and on the method to be used to calculate its direction and predicting its future positions, a method which relies on precisely this hypothesis.[100] The hypothesis he formulated is that comets have uniform rectilinear motions, an hypothesis already defended by Kepler.[101] Contrary to Kepler, however, Pardies does not associate such a rectilinear path with the transitory nature of comets; instead, he places comets among the eternal works of nature, and this, precisely because he sees rectilinear motion as the perpetual motion *par excellence*.[102] Gassendi had already claimed for comets not only rectilinear paths but an everlasting nature, but, in his case, it was not a problem since he supposed that the universe "begins nowhere and ends nowhere" (Gassendi 1658: 710). It is not clear whether Pardies realized that maintaining both rectilinear paths *and* a perpetual nature was somewhat contradictory in a finite universe, but, as an excursus on Seneca and the Ancients indicates, he may have thought that uniform rectilinear motions were only an approximation of the truth, which was that the comets moved along extremely large circumferences.[103] Later on in the book he defends opinions that cannot but sound Cartesian: the tails of comets would come into being because of a specific kind of refraction, there would be a celestial matter moving in vortices and carrying around the different planets, and the direction of the tails of comets should be explained by the centrifugal effort that pushes away the lighter and more subtle parts, which would raise the question why planets do not have tails.[104] By the end of his *Dissertatio* anyhow, probably because Kepler, as we saw, linked the opinion that comets have rectilinear paths to the defense of Copernicus, Pardies takes care to distance himself from heliocentrism.[105] Neither the senses nor reason can make us decide which system of the world is true, but, as theses defended by Chrétien François de Lamoignon at the collège de Clermont in 1663 prudently indicate, "even if it were true that the Earth moves and that the firmament stays immobile, the hypothesis that places every motion in the heavens and that considers the Earth as immobile would be stronger," and this because "the one who puts the world in motion and who alone understands which of the two is moved, said so clearly, so distinctly and so expressly: *the Earth stands for ever: the Sun rises and sets, and changes it course at noon; sometimes it stands still, obeying God through the voice of a man*" (Pardies 1665b: 61–74, 69).[106]

Thus, although the confrontation between d'Harouys and Denis was perceived by their contemporaries as part of the controversy on principles that was then developing between Cartesians and Jesuits, it was not representative of all the attitudes available for Jesuits and Cartesians during this period. Rohault avoided entering this confrontation. Contrary to Grandami and d'Harouys, mathematician Jesuits like Billy and Pardies were practicing an observational and mathematical astronomy, which, at first sight, was not as concerned with establishing ontological principles as with finding the true locations and paths of the comets. However, the question which system of the world was to be adopted could not but lurk in the background of their enterprise, and it seems that Pardies was well aware of Descartes's doctrine concerning comets, and even somehow favorable to it.

Between Gassendism and Experimentalism: The Montmorians

The next group is constituted of members of the Montmor Academy, which came to a definite end in June 1664: its former secretary Samuel Sorbière, Caen's erudite Pierre-Daniel Huet, the astronomer Adrien Auzout, and the Intendant des fortifications Pierre Petit. Contrary to the Jesuits and to the Cartesians, among whom one finds at least theoretically a certain uniformity of doctrine, inasmuch as they refer to an authoritative text, be it Descartes's or Aristotle's, the Montmorians do not defend the same opinions, adopt the same attitudes, or even develop friendships among themselves. In order to carry out astronomical and anatomical observations, Petit, Auzout, and Thévenot began to meet independently from the other Montmorians, whom they reproached for being idle talkers, while Sorbière criticized them for their radical commitment to observations and experiments.[107] In this respect, one finds among Montmorians a distinction similar to the distinction to be found among Jesuits: some of them were astronomers actually committed to observations; the others were natural philosophers who, rather than making observations, referred to observations in a kind of rhetorical way. But in the case of Montmor Academy, there were personal enmities as well: while Sorbière condemned Petit for his crude manners, explaining, for example, to Hobbes that he is "apt to destroy orderly arrangement and philosophic moderation" and that "we know no one who is more of a troublemaker among the ex-Peripatetics or exponents of a more refined philosophy" (Sorbière to Hobbes, 5 January 1663 [?], Hobbes 1994, 2: 551–553), Petit wrote to Oldenburg to accuse Sorbière of having unduly pretended to represent the Montmor Academy at the Royal Society.[108] Following the chronological order of their publications, I begin with the natural philosophers who were taking inspiration from Gassendi, Sorbière, Chapelain, and Huet, and continue with Auzout and Petit who conceived themselves as astronomers, founding their work on observations and mathematics.

Samuel Sorbière was a multifaceted man of letters who in 1654 converted from Protestantism to Catholicism out of careerism; he consequently succeeded in becoming one of the royal historiographers, in getting the patronage of Montmor, and in being appointed secretary of Montmor Academy when it was officially founded in 1657.[109] In 1664, however, he wrote such a deprecating presentation of some of the most illustrious members and patrons of the Royal Society, which he had visited in 1663, while the Montmor Academy was already nearing its end, that, to calm down the ensuing diplomatic tensions between France and England, he was stripped of his charge of historiographer and sent to lower Brittany for a few months. On returning to Paris from his short exile, he published in 26 January 1665 a *Discours sur la comete* in the form of a letter to Claude Auvry, bishop of Constance.[110]

Sorbière mentions Boyle's experiments on the void that he may have seen at the Royal Society, but there is no other allusion to English scientists. His booklet is rather placed under the intellectual patronage of Gassendi: not only is it written in the form of a letter to Claude Auvry, who asked Gassendi his opinion on the 1654 lunar eclipse, but its explicit purpose is also to say what Gassendi would have said on the 1664–1665 comets had he been alive.[111] However, Sorbière does not mention any specific observation concerning these particular comets, and he refers only in passing to the opinion developed in the *Syntagma philosophicum*, according to which comets would travel with uniform rectilinear motion. Instead, he uses Gassendi to mark off the limits of Aristotle's, Galileo's, and Descartes's opinions on comets.[112] The reference to Gassendi is used as an occasion to get to grips with Descartes's philosophy in general, including his metaphysics. Sorbière recalls the reasons why one should resist Descartes's claim that void does not exist: void is a necessary principle in natural philosophy, without which it would be impossible to account for motion and for the distinction between bodies.[113] However, Sorbière's most important reproach to Descartes does not concern ontology but, rather, his lack of certain intellectual virtues: contrary to Gassendi, who used to suspend his judgment, Descartes was imprudent and presumptuous.[114] Thus, Sorbière contrasts Gassendi's modesty to Descartes's dogmatism: while the first expressed clearly that, because of the imperfection of the instruments, the negligence and lack of exactitude of those who use these instruments, and the absence of communication between observers, we can reach only conjectural knowledge when dealing with comets, the second pretended to "become the head of a party, or the founder of a sect, and to impose on the half-learned by the bold efforts of a fertile and strong imagination" (Sorbière 1665: 18; see also 4, 17–18). Although Sorbière uses the little parallax to refute Aristotle's stance that comets are sublunary meteors, he also insists that, while it is difficult to trust celestial observations in general, it is almost impossible to say anything certain about parallaxes, considering that we do not know the exact distances between two places and that we are not sure that the observations were made at the same time exactly.[115] Last but not least, Sorbière condemns Descartes for starting from unproven hypotheses that have no other guarantee than "some metaphysical thoughts that are more embarrassing than persuading." Starting from such hypotheses, Descartes cannot prove that it is necessary that things happen in such a way, but he only tells a story about how things could have happened (Sorbière 1665: 10–16, 11 for the quotation). In a word, what characterizes Sorbière is a certain use of the reference to Gassendi: his point is not to expand his doctrine on comets, or to oppose to Gassendi's principles to Descartes's, but to assert that what matters is the moral attitude that one adopts with regard to the discovery of truth.

The same kind of attitude is to be found in the letter Pierre-Daniel Huet sent to Chapelain by early March 1665, a letter that was published post-humously in his *Dissertations sur diverses matières de religion et de philologie*.[116] Although Huet did not come regularly to the meetings of Montmor Academy, he was introduced into this circle by Chapelain, took part in its meetings when he happened to be in Paris, and even presented a talk on glass-drop in 1661.[117] In 1662 or 1664, together with André Graindorge, he instituted a scientific circle in his house in Caen, which devoted most of its studies to anatomy but which, between dissections, found time enough to comment on the 1664–1665 comets, if not to observe them.[118] In the letter he sent to Chapelain on the subject, he first presents seven different opinions, among which the five opinions presented at collège de Clermont, including the one that Garnier could not put forward because of the falling night, then the opinion that they are planets revolving around other suns that happen to pass by, and finally the opinion that they are inflamed bodies, which was proposed by Isaac Vossius. Then he unveils dialectically the difficulties that each of these opinions raises, in order to finally detail his own view, which he himself presents as an eclectic combination of what is the most valuable in the other opinions.[119] Even if Huet does not agree with Sorbière on what comets are and does not mention him either, he agrees with him in two other respects. First, he casts doubts not only on Descartes's opinion on comets but on the principles that are supposed to ground this opinion: "Those principles [of Descartes] being so uncertain, and subject to so many contestations and objections, from which his supporters will not be able to extricate themselves, the consequences that he draws from them to establish the nature of comets are no more certain" (Huet 1665: 236).[120] Second, he adopts a skeptical stance that was then considered as the main feature of Gassendism, concluding in particular that "he is not in love [with his hypothesis] enough, for not being ready to change it when something more verisimilar will have been found" (Huet 1665: 245).[121] Thus, like Sorbière, Huet identified Descartes through his principles; contrary to the Jesuits, he does not want to juxtapose him against other principles but against another moral attitude to the discovery of truth.

That this was a general characteristic of Gassendism in this period can be confirmed briefly through Chapelain. His own view on comets is that they are a kinds of planets that revolve around a remote sun, so that we see them only when they happen to pass close to our solar system.[122] His judgment on Sorbière's book is harsh, but it is probably out of personal enmity rather than a skeptical attitude towards it.[123] In a letter to Huet, Chapelain indicates "in a question as problematic as the one on the nature and the motions of comets, it would be imprudent to give one's definitive opinion and, according to me, it is only permitted to fall in with the most verisimilar opinion until another opinion comes up, where more probability is to be seen" (6 April 1665, Chapelain 1880–1883, 2: 393).

Auzout and Petit were different from the Gassendist Montmorians; they emphasize not so much the uncertainty of all knowledge than the necessity to develop expert observational and mathematical skills. To put their texts in proper context, it should be remembered that, as early as in the spring of 1662, Petit, Auzout, and "a bit," Thévenot, who felt themselves as lone partisans of experiments among the Montmorians, began to carry out various astronomical observations on their own.[124] One year later, in the spring of 1663, Huygens being in Paris, a "general assembly for telescopes" was organized at the home of Auzout: the power of various telescopes was tested by Auzout, Huygens, Petit, Monconys, and Étienne d'Espagnet, who used a new process to make the lenses for some of these telescopes.[125] The issue of the day in 1663 was also to establish "new laws and ordinances" for the Académie Montmor, or, rather, to refound it in a new form, giving it a much more experimental and practical orientation. The project written on this occasion, known as the project of the Compagnie des sciences et des arts has a very Baconian orientation: it insists on linking together the sciences and the arts, on writing histories of nature and of the arts, on making discoveries useful to the public, on testing secrets, and so on.[126] The attitude of this group is well summarized by the quite negative verdict that Boulliau returned when Lubiniezky happened to ask about Denis's book, which I discussed earlier, and *Le Courrier de traverse ou le tri-comète*, which I will discuss in the following: "since both of these authors do not pay attention to the astronomical foundations that exact observations are, I think that they have neither understood this matter correctly nor defined it properly" (Boulliau to Lubiniezky, 14 August 1665, in Lubiniezky 1667: 533).

Of the astronomers, Auzout was particularly active on an international level, exchanging observations with foreign correspondents, like Huygens, Cassini, Hevelius, and, through Oldenburg, with the English virtuosi.[127] He was respected for that and his name is mentioned with praise by all protagonists of the debate, sometimes associated with those of Pardies and Cassini because the three of them proposed ephemerides.[128] He published successively *L'éphéméride du comète* (at the end of January), *L'éphéméride du nouveau comète*, and the *Lettre à Monsieur L'abbé Charles sur le Ragguaglio di due nove osservationi, etc., da Guiuseppe Campani* (both of them by mid-April). Auzout did not take a stand on the nature of comets in any of these, and his central claim is rather that, with only three observations at his disposal, he was able to find out what he calls the "law" of the comet. However, he formulates explicitly neither what we call a law nor the method that would have allowed him to deduce it from observations. The only thing he provides his readers with are tables predicting the positions of the comet from early December 1664 to early February 1665.[129] In fact, *L'éphéméride du comète* was out only *after* the dates of his alleged predictions. Auzout himself claimed that it was the fault of the printer and called upon Huygens, Thévenot, and Petit to give a testimony that his predictions preceded the publication of the book.[130] But in a letter to Thévenot, Huygens notes "that they should be suppressed rather than to raise the suspicion of a falsification" (Huygens to Thévenot, 29 January 1665, in Huygens 1888–1950, 5: 210). If this suspicion was so easy to have—Huygens undoubtedly appreciated Auzout—one can ask oneself why Auzout was so keen on boasting to have found out the law of the comet.

One of the obvious reasons was that he wished that the King, to whom his first *Éphéméride* is dedicated, would support the construction of large telescopes and an astronomical observatory. As he wrote in the *Lettre à Monsieur L'abbé Charles*, he did not want to publish any book, knowing that being an author carries unwanted consequences, but he changed his mind when the comet appeared, because, as he was the first to predict its day-by-day progression, it gave him the opportunity to make the King understand that Paris was lacking the wherewithal for exact observations and that a proper place with suitably qualified men and up-to-date instruments was called for. Consequently, Auzout's *Éphémérides* rest on two almost contradictory constraints: on one hand, he has to brag about an important discovery based on exact observations; on the other hand, he has to make the point that he made his observations with only ropes, rulers, set squares, and sticks and that he would have done much better with proper equipment. He makes it no secret that, in this way, he wanted to contribute to the establishment of the Compagnie des sciences et des arts, the glory of the King, and the reputation of France demanding that such a company would be supported and subsidized by the king.[131]

Besides this question of patronage, the law of the comet was important in at least two other respects for Auzout. First, it helped him make the point that subjecting comets to laws and showing that they are natural bodies that have regular motions demonstrates that they have no signification at all, astrological or other, and if one comet has no signification, neither do two or three.[132] Second, Auzout initially hoped that comets might help to decide if the Earth moves or not.[133] Although his method for predicting the future positions of comets presupposes that comets have rectilinear paths, and although it was known to his correspondents that such was his conviction,[134] he does not mention it in his first books, perhaps because it was difficult to express such a conviction in works dedicated to the King, because it had been linked by Kepler to heliocentrism. In his third book, however, Auzout openly declares his Copernicanism and writes quite explicitly that comets might give us reasons to believe that the Earth moves, not "by a mathematical and metaphysical conviction" but "by a conviction as reasonable as the conviction which makes us judge that the Sun and the planets do not revolve around Jupiter and Saturn, but rather all the planets including Jupiter and Saturn around the Sun" (Auzout 1665c: 17–18, 54 for the quotation). Quoting and following the Jesuit Honoré Fabri, Auzout explains then that the condemnation of heliocentrism by the church was meant to be only provisional, "in order to prevent the scandal that novelty causes or could cause," with respect to what the common opinions were at a certain time (Auzout 1665c: 49–56).[135]

Although Auzout was known in these years to oppose the Cartesians, especially Rohault, he does not allude here to Descartes's opinion on comets, probably because he was not interested in saying anything on the nature of comets but in finding their laws.[136] It is not the same with Petit, who was a lifelong opponent of Descartes.[137] In certain respects, however, Auzout and Petit have similar positions. Like Auzout, Petit dedicates his book to the King and also has the ambition to discover some regularity in the appearance of comets. In Petit's case, the idea is that each comet describes a very large circle and appears at regular intervals.[138] In the case of the 1664–1665 comet, it would be every forty-six years. This number comes from Petit's conjecture that the 1664 comet is the same as the 1618 comet. He then notes that 1664 minus 1618 equals forty-six and continues that 1618 minus 1572 also equals forty-six. As the reader might know, what appeared in 1572, according to Tycho's observations, was not a comet but a star. Yet Petit argues that the brilliant star discovered by Tycho might have prevented us from seeing a less brilliant comet, and then searches in various historical reports for proofs of the apparition of a comet every forty-six years.[139] The last point that Petit has in common with Auzout, but not the least, is that he was a Copernican, or at least, as Huygens would say, a "semi-Copernicist," since he accepted the daily rotation of the Earth but not its annual rotation.[140] To the usual arguments against the condemnation of Galileo—for example, that the church may not have authority concerning natural matters—Petit adds a Gallican argument, according to which the Church of Rome has no authority to promulgate a decree in France until it has been approved by the Parliament and the faculty of theology of Paris.[141]

Apart from his determination to find some regularity in the paths of comets and his commitment to Copernicanism, Petit is different from Auzout, if only for the length and confusion of his book, which piles appendix upon appendix, each written when Petit happened to read a new book on comets—as he explained at some point, he had to "write hastily, at the same time as his book was printed" (Petit 1665: 218). In one of these appendices, he criticizes not only astrology in general but also Leschener, in order to show that "he is not feigning enemies in order to fight against them" (Petit 1665: 149).[142] Another appendix, the longest of all with almost seventy pages, is entirely devoted to refute the opinion of Denis, whom Petit did not identify: he is explicit that he does not know who hides behind the initials "I.D.P.M" under which Denis published his book.[143] At the beginning of his book, Petit reported Descartes's opinion on comets, but, rather than criticizing it in detail, he only indicated that "according to many, to report it and to refute it is the same thing" (Petit 1665: 14–16). But, when reading Denis's book, he explains, he decided to make public his judgment on Descartes's explanation of comets: the hypothesis that spots would completely cover the Sun is "visionary" and "does not conform to the laws of nature and mechanics about which those who defend Descartes's opinions speak so much"; how heavy comets could be moved by vortices of subtle matter is not comprehensible; Descartes's explanation of the tail of comets is "an enigma"; it is "evidently false" that comets are not going faster when they appear than afterward (Petit 1665: 235–250, and 211–215, 213, 215). Once again, however, the question at stake is not only the question of comets, but, much more generally, Descartes's natural philosophy: subtle matter was only invented as an *ad hoc* explanation of Torricelli's quicksilver experiment; "sense and experiments" refute the "nice words and arguments drawn from the obscure, arrogant, and presumptuous principles" through which Descartes pretended to explain how fire comes from flintstone; Steno proved that most of Cartesian anatomy is a fantasy.[144] And all these false explanations derive from fundamental errors: contrary to what Denis claimed, Descartes introduces particular principles for the phenomena he wants to explain; his very notion of explanation relies on a sophism, which consists of believing that one should consider that an hypothesis is proved if it is sufficient to account for an effect.[145] To conclude, quoting one of the many passages where Descartes says that, if one of his explanations is false, then his whole philosophy is false, Petit takes him at his word and concludes that, indeed, the whole of his philosophy might be false.[146]

Interestingly enough, such a criticism of Descartes is not grounded in a simplistic opposition between reason and observations. In several places, Petit insists that unfortunately amateurs are meddling in "scribbling paper and making printers sweat in winter," the result of which is that we have too many reports that contradict each other, preventing the establishment of the truth and wearing down the trust that good observations should inspire, hence both the necessity and the difficulty of distinguishing between skilled astronomers, who have the best instruments, and those who, with only "bad paper astrolabes" and not versed in astronomy, mistake latitudes and declinations, indicate meridian heights that are impossible and contradictory, and believe that they have seen the two comets at the same time.[147] Discriminating between reliable and unreliable testimonies does not concern only astronomical observations but also the historical reports on which astrological pretentions were grounded. As Sorbière notes, the astrologers used to argue that "there is no better reason and no rule more assured, than the experience, about which the historians give evidence. And if in politics, medicine, and civil life, this kind of proof is received, it should not be rejected from astrology and the predictions made thanks to comets" (Petit 1665: 1). This is why Sorbière's refutation of astrology does not only concern unreasonable beliefs, such as the confusion between antecedents and causes, but also false, approximate, or credulous testimonies.[148]

The Montmorians can be considered as a group not only because they happened to participate to Montmor Academy at some point but also because of their commitment to a natural philosophy based on observations and experiments and because of their anti-Cartesianism. Although they might have some ontological reasons for not appreciating Descartes's way of philosophizing—Sorbière argued, for example, that there is necessarily some void in nature—their main reasons for objecting to it were rather linked to what they saw as the primary intellectual virtues: they thought that Descartes was too much of a dogmatist, imbued with his own ideas. Thus, contrary to d'Harouys, it was not in terms of ontological principles that they argued against Descartes; they rather objected to him because of his moral attitude towards truth and its discovery. However, not all of them had exactly the same attitudes toward truth: while men of letters like Sorbière, Huet, and Chapelain were advocating a mild skepticism, in which observations are called upon rather than actually done, Petit, and Auzout, as practitioners of mixed mathematics, were committed to the discovery of natural truths through observations. Strangely enough, while both of them thought that comets might be enrolled in the defense of Copernicanism, they did not attribute to them the same kind of trajectory, Auzout thinking that it is rectilinear, and Petit that it is circular.

Some Isolated Voices

The groups that I have discussed so far were more or less closely knit, both socially and in their members' natural philosophical beliefs. It remains to analyze briefly three books published on the comets of 1664–1665 that do not fall under my previous categories and appear somewhat isolated. I have no coherent story to tell about these books, but I want to include them in order to present a picture as systematic as possible and because of the originality of some of them.

Le Courrier de traverse ou le tri-comète was written by Nicolas de Croixmare de Lasson, an aristocrat from Rouen, residing near Caen, member of its Académie des sciences and interested in the fine arts as well as the sciences. It was, however, published under the name of M. Vortfischer, presented as the translator of this piece from English to French.[149] Considering the Anglomania that affected Graindorge and Huet, this pseudonym was probably de Croixmare's way of satirizing English virtuosi and their French followers.[150] At first sight, Le Courrier de traverse appears as a fantasy, making a mockery of geometric language, for example, when the "prostasphereses" and "apocastastes" of the tri-comet are said to be identified thanks to a "spherical-spherical."[151] This explains why Journal des scavans assumed that Le Courrier de traverse was published to "laugh at false savants and at those who enjoy celestial observations and the vain curiosities of astrology" (Journal des scavans, 30 March 1665, in Sallo 1665: 88–89). At the same time, however, it claims to rediscover an ancient truth, already known to Seneca, Ptolemaeus, Lokon the Indian, the Babylonians, and the Chaldeans, according to which comets are primordial stars made of the primitive light.[152] Considering that de Croixmare was implicated in alchemical studies that he kept secret, he might have seriously believed in this ancient truth, in which case Huygens's comment would be more exact than the one of *Journal des savants*: "It seems to me that the author of the Tricomete speaks seriously and consequently that he is insane" (Christiaan Huygens to Constantyn Huygens, 2 April 1665, in Huygens 1888–1950, 5: 301–302).[153] In other words, there is clearly a parodic tone in *Le Courrier de traverse*, but this does not mean that de Croixmare did not believe in all that he wrote.

This ambivalence is all the truer of the second book, the anonymous *L'esprit du sage*. As with many others we have followed in this chapter, its author presents the opinions of the "cometists" as they were expressed during the January conference, criticizes them on one ground or another (although with more vivid expressions than in other texts, d'Harouys's students being depicted as "tamed parrots," while Descartes is presented as thinking that "a gang of small corpuscles as black as Africans pounce on a star, conceal it, embrace it so strongly that it cannot anymore accomplish its ordinary function"), and finally proposes still another opinion, which to modern eyes is not that different (*L'esprit du sage*, Ch. 2: 19–40; Ch. 4–5: 68–89). The first and last chapters are, however, distinctive. The first details how the truth on comets was revealed to the author: after a long day of work in his study, he began to walk along a stream and then climbed to the top of a mountain; he had barely reached it when his senses fell dormant and his spirit was transported through the air until he arrived in the middle of a thousand suns, where an astounding voice revealed to him the secrets of the world.[154] It is only in the third chapter that these secrets are displayed. They are important enough to be repeated in the very last chapter, written in smaller characters and with a separated pagination: the Earth moves around the Sun; fixed stars are as many suns; the world is infinite and eternal; "there is only one soul which exists in different degrees in all the Creatures, according to their needs, which makes the variety of so many things that is to be found in each species; there is nothing which is not a soul, even the most insensible and wretched things; this soul is placed at the highest degree of perfection in man, which makes him the most considerable of creatures" (*L'esprit du sage*, Ch. 6: 3–4).[155]

Finally and most intriguing is the third and last book of this improbable series, Gabriel de Vendages de Malapeyre's *De la nature des comètes*.[156] His foreword is typical for an *honnête homme* of this time: de Malapeyre explains that he was not looking for anything except a quiet life but was obliged to write a book on comets to answer many solicitations from his friends; that the radiant lights of philosophy are now dissipating the public terrors that comets inspired and that the darkness of our imaginations favored; that "the attachment that most learned men have for an Author, and the oath of fidelity that they took in favor of a Master, being an insuperable obstacle to the possession of the truth," he will take what he finds pleasant and reject what he finds unpleasant in each author. Malapeyre concludes that his book is neither the book of an astrologist, since he does not make horoscopes, nor of an astronomer, since he does not propose an almanac or an ephemerides but a physicist's book, dealing with the nature of celestial bodies (Malapeyre 1665: 1–8). What is surprising considering this beginning is the amount of information that *De la nature des comètes* displays on natural philosophy: Malapeyre refers accurately to parallax and to the ephemerides published by Auzout and Pardies; using an argument from the relativity of motion, he takes a stand for Copernicanism; he denounces attraction and other "terms full of vanity and ignorance"; his main references in natural philosophy are Galileo, Gassendi, and Descartes.[157] Concerning comets, in particular, he is the only one to understand how paradoxical it was that, in his controversy with the Jesuit Horatio Grassi, Galileo defended a quasi-Aristotelian position, according to which, comets being just luminous reflections of atmospheric exhalations, to calculate their locations, one cannot use parallaxes that apply only to real and permanent objects. He attributes Galileo's move to psychological motivations, whether, as "a generous enemy . . . satisfied to have put Aristotle to the ground a thousand times, he testifies now that the only source of their quarrels and of his aversion was his reason" or as a mighty dialectician "to show that, in the worst cases, he is capable of inventing better reasons than would their most obstinate defenders" (Malapeyre 1665: 101, 104).[158] His account of comets is more generally the most systematic one to be found in all the books that I have read: he first presents Aristotle's opinion (Ch. 7), then Galileo's opinion (Ch. 8), and then the opinions of those who think that comets are ephemeral bodies, Malapeyre distinguishing among seven different positions (Ch. 9) and finally the opinions of those who consider that comets are as ancient as the world, among them Descartes and Gassendi (Ch. 10). It is only then that Malapeyre reports on his own opinion, which relies on what he finds more solid in each of the other opinions, that comets are planets that abandon their sun and pass through our solar system.

Conclusion

The maze of books that the 1664–1665 comets gave rise to disappeared almost as rapidly as the comets themselves: the only book that was published on these comets after 1665 was Billy's *Crisis astronomica de motu cometarum*. Not only were these publications on comets ephemeral, but with no significant truth discovered, they seem to have also been forgotten as soon as they were published—even if it remains to be determined whether any of the arguments we encountered here did not contribute to the debate on the 1680–1681 comets. Last, but not least, they appear to have occasioned one of the fool's games so common to intellectual life: every natural philosopher expresses his views and eventually defends them by indicating the weaknesses of all other opinions but is not, for all that, able to take into consideration what the other natural philosophers said. If the 1664–1665 comets did not induce the discovery of new truths or the interaction between different natural philosophers, they nevertheless reveal the various antagonisms existing in the field of natural philosophy in France in the 1660s, and it is in this sense that they can be considered as a prism.

Seventeenth-century France is often associated with Descartes's legacy, Descartes himself being often associated with a kind of speculative and anti-experimental philosophy. Although I have not discussed Descartes himself in this chapter, such a judgment obviously depends on which work of Descartes one looks at. As far as Descartes's legacy is concerned, I have shown that the situation was more complicated than this common picture might suggest. There were groups in France who defended natural philosophical opinions that were not Cartesian, and among them at least two groups were defending an explicitly *anti*-Cartesian agenda. One such group consisted of Jesuits who attacked Descartes's followers because they would grab hold of wrong ontological principles. The other group included people with different positions but united in their rejection of Cartesians, whom they perceived as unduly dogmatic and insensible to observations and experiments. In a word, if there was ever something like a Cartesian legacy in France, it was at least as much because Descartes was contested as because he was followed.

But the most important issue in a study like this is to know how diverse opinions were expressed in the field of natural philosophy. It is indeed remarkable that while there was a diversity of opinions on comets and more generally on natural philosophy, such diversity did not always instigate open conflicts and that, when such a conflict did occur, it was not always expressed in terms of principles. While astrologists were not criticizing the others, astrological beliefs and practices were condemned by all the others. They were in a sense condemned even by those interested in astrological predictions—as we have seen, they doubted their own predictions. But this does not mean that those who condemned astrology read contemporary astrologers and openly quarrelled with them—except for Petit, they were content with an abstract condemnation of astrology in general, *in absentiam astrologorum*. It is only between Cartesians and anti-Cartesians that an open conflict developed. However, as we have seen, Gassendi's followers did not reproach Descartes so much for his principles as for his lack of certain intellectual virtues. Thus, the only open confrontation that developed was between the Jesuits and the Cartesians, and this because they proposed two systems of the world founded on explicitly formulated principles between which a choice was to be made. In other words, in the situation described in this chapter, principles are not self-evident propositions on which everybody agrees, as philosophers would like them to be, but controversial statements that delineate different social groups according to their different intellectual commitments. By contrast, observations concerning the successive positions of the comets were traveling both from one social group to another and from one country to another. Though these observations travelled mostly through letters, they are also present in some books, like those of the Jesuit mathematicians (de Billy and Pardies) and of some Montmorians (Auzout and Petit), who were able to communicate beyond the boundaries of their social, intellectual, and national groups precisely because they were not reasoning in terms of principles. In other words, the paradoxical nature of principles is that, while they are presented as what everybody should agree on, they are in fact what we are endlessly arguing about.

Notes

* I would like to thank Ofer Gal, Catherine Goldstein, Isabelle Pantin, and Koen Vermeir for encouragements and helpful comments on this chapter; Hugues Chabot for providing me with a copy of Sombreval 1665; and Peter Anstey and Ofer Gal (again) for proofreading my English. Unless otherwise indicated, translations are mine.

1 Marie Guyard to her son, 28 July 1665, in Guyard 1681: 601; Le Mercier 1666: 105–117.

2 In the case of England and northern Europe, see Hetherington 1972; Jensen 2006: 51–110; Shapin 1994: 266–291; Yeomans 1991: 69–93. In the case of Italy, see Aricò 1998 and 1999; Boschiero 2009; Campinoti 2006: 217–228; Cassini 2003; Lugli 2004, Ch. 10. In the case of France, the literature is sparse; see, however, Atkinson 1951; Drévillon 1996.

3 Grandami 1665a: 10; Billy 1665: 6. As is well known, the Jesuit network was one of the first scientific networks to circulate such observations. Two qualifications are however necessary. First, most of the time, we know that a given Jesuit received observational reports from his fellow Jesuits, but we have no information on the content of these reports. Second, the expression "Jesuit network" should not give the impression that one has to deal with a network isolated from the other networks: on matters like comets, Jesuits were integrated in the larger network of astronomers.

4 Mme de Sévigné to Pomponne, 17 and 22 December 1664, in Sévigné 1853: 65 and 70; Prince de Condé to Marie-Louise de Gonzague, Queen of Poland, 19 and 24 December 1664, in Condé 1920: 114–116; René Gaspard de La Croix, Marquis de Castries, to Jean-Baptiste Colbert, 29 December 1664, in Depping 1855: 682–683.

5 Patin to Falconnet, 25 and 30 December 1664, in Patin 2015, L. 804 and L. 805.

6 Cotin 1665; Boileau, "La Métamorphose de la Perruque de Chapelain" in Boileau 1772: 218. The rationale behind this comedy was that, in early modern French, the tail (*chevelure*) of comets was also sometimes designated as their wig (*perruque*), *perruque* meaning not only artificial hair but long and abundant hair (*chevelure*) and *chevelure* being the French word for "tail." In a similar genre, see also the anonymous *Les effets ridicules du comète, envoyez à Lysandre malade*.

7 *Ballet de la comète*; Diez 1665.

8 Patin to Falconnet, 2 January 1665, in Patin 2015, L. 807.

9 *Journal des scavans*, 2 February 1665, in Sallo 1665: 94–95; Patin à André Falconet, 23 January 1665, in Patin 2015, L. 809; Lantin, n.d., fol. 33; Comiers: 7.

10 Malapeyre 1665: 55, 60–62.

11 Shapin 1994: 267–287.

12 Lubiniezky to Boulliau, 21 February 1665 and Boulliau to Wallis, 6 March 1665, in Lubiniezky 1667: 412, 472. Wallis to Oldenburg, 24 December 1664/3 January 1665 and 21/31 January 1665, in Oldenburg 1965–1986, 2: 339.

13 Pardies 1665a: 18; Marie Guyard to her son, 28 July 1665, in Guyard 1681: 601; Léautaud 1665: 13–18. Petit 1665: 206–208, attributes this opinion to Comiers, but it seems to me that this comes from a misunderstanding.

14 Billy 1665b; Grandami 1665b.

15 For a general, standard account on comets in the sixteenth and seventeenth centuries, see Ruffner 1971; Yeomans 1991: 1–68, *passim*. On Tycho, see Lerner 1996–1997: 39–66; on Kepler, see Boner 2013: 105–134; on Galileo and Grassi, see Gal and Chen-Morris 2011. For an overview of recent historiographical trends on comets, see Mosley 2014.

16 Roux 2013 presents in detail the controversy between Cartesians and Jesuits on natural philosophy.

17 On the circumstances preceding the foundation of the Académie des sciences and the story of Montmor Academy, see the pioneering work of Brown 1934; Roux 2014: 58–72.

18 The following letters from 1661 are particularly telling: Chapelain to Huygens, 30 May, Huygens to Chapelain, 14 July, Chapelain to Huygens, 20 July, in Huygens 1888–1950, 3: 273, 295, 299.

19 The complete list of those who received pensions is given in Huygens 1888–1950, 4: 405–406.

20 Drévillon 1996: 40, 57–61, 185–186, 211, 228–230.

21 These decrees are to be consulted in Decrusy *et al.* 1821–1833, 14: 71, 390–391, 15: 215–216. Drévillon 1996: 65–66, 100. For similar considerations in the case of early modern England, see Schechner 1997: 70–87.

22 An important source for such a project would be Lubiniezky 1667.

23 Montalegre 1664: License to print, 25–26 December 1664; Vaissey 1665: Letter dated from 4 January 1665; Giustiniani 1665: License to print, 25 February 1665; Luyt 1665: License to print, 12 March 1665; Sombreval 1665: Approbation from Doctors and Prosecutor of the King, 13–26 March; Comiers 1665: Dedication, 28 April 1665; Approbation from Doctors and Prosecutor of the King, 15–17 September 1665. Although *Figure de la derniere estoille extraordinaire* has no license and Leschener 1665 no license and an undated Dedication, since they mention only the first comet, it may be inferred that they were published in March at the latest.

24 Montalegre 1664, Dedication, sig. A1r, Observations, sig. B1r.

25 "Composé par Henry de Leschener, Allemand." In the context of a controversy on the spelling of Boulliau's name, Granet 1738: 264, mentions that Boulliau gave his approbation to this book and signed it "Boulliau." But this approbation cannot be anything but the "Avis d'un particulier au Libraire sur le Traitté des comettes," which is signed by the letters "F.H.L. de Paris." Moreover, Nellen 1994 does not mention in which circumstances Boulliau would have written this approbation. No other publication from Leschener is known.

26 On Robert Luyt (1609 or 1619–1667), see Matton 1979. His other books include *L'antiquité renouvelée, ou l'Éclaircissement d'un raisonnable doute, par lequel on recherche de scavoir si les nouvelles opinions sont moins recevables que les anciennes*, 1647; *Table généalogique des seigneurs de la maison de Clermont en Dauphiné, comtes de Tonnerre*, 1648; *La plus éminente sagesse du christianisme. Jésus-Christ enfant, traite de dévotion en deux livres*, 1648; *La Régence des reynes en France, ou les Régentes*, 1649; *La princesse charitable et aulmoniere ou l'histoire de la reyne Marguerite de Bourgogne*, 1653; *La découverte d'un saint caché en la ville de Tonnerre, ou l'histoire de saint Micomer*, 1657; *Le plus illustre ornement de la noblesse, les ordres de chevallerie institués par les rois et princes souverains*, 1661.

27 On Claude Comiers (?–1693), see "Comiers, Claude," *Dictionnaire des journalistes*, to be consulted on line at http://dictionnaire-journalistes.gazettes18e.

fr/journaliste/189-claude-comiers. Once in Paris, Comiers resumed his contributions to scientific journals in his books, among which *La duplication du cube, la trisection de l'angle, et l'inscription de l'heptagone régulier*, 1677; *Le pantographe physico-mathématique*, 1677; *Nouvelles instructions pour réunir les Eglises prétendues réformées à l'Eglise romaine*, 1678; *Lettres de Monsieur Comiers [. . .] à Mgr le Marquis de Seignelay, sur l'excellence et usages de la nouvelle pompe*, 1682; *La médecine universelle*, 1687; *L'art d'écrire et de parler occultement et sans soupçon*, 1690; *Traduction polyglotte du verset du psaume 112*, 1691; *Traité de la parole, langue et écritures*, 1691; *La baguette justifiée et ses effets démontrez naturels*, 1693; *Factum pour la baguette divinatoire*, 1693; *calendrier perpétuel et invariable*, 1693; and *Pratique curieuse, ou les oracles des sibylles*, 1694.

28 Leschener 1665: 5.

29 Leschener 1665: 4–7, 10, 12; Sombreval 1665: 12–14.

30 Comiers 1665: 29, 31, 34–35, 40, 58, 63, 100, 102, 131–132, 329, 340, 398, 432.

31 On the different signs to be taken into account in the practice of prognostication, see Schechner 1997: 51–65.

32 Montalegre 1664, Questions curieuses concernant les Comettes, sig. D2r—v.

33 Luyt 1665, Question 3 and Question 6: 7–10 and 14–18.

34 Sombreval 1665: 11–12.

35 Leschener 1665: 1–5.

36 Leschener 1665: 4–5, 7. This argument against the use of parallax had been put forward by Giambattista Riccioli in his *Almagestum novem*, Bologna, 1651. On the consequences Parisian textbooks draw from parallax, or from its absence, see Ariew 1999b: 110–112.

37 Leschener 1665: 9–12.

38 Leschener 1665: 10, 12.

39 Comiers 1665, Preface: 3.

40 Comiers 1665: 35, 132 for reference to the *Dialog*; 75–78, 84–88, 95–99, 104–105, 130–132, 135–136, 143–145 for terrestrial experiments.

41 Comiers 1665: 355–356; Leschener 1665, Avis d'un particulier au libraire, n.p; Luyt 1665, Au Lecteur, sig. a3v. Drévillon 1996: 32–36 notes rightly that the exact boundary between natural and judiciary astrology was in practice difficult to draw.

42 Luyt 1665, Questions 21–22: 49–55; Sombreval: 5.

43 Comiers 1665: 356; Leschener 1665: 16, 32, 46; Montalegre 1664, sig. B2r–v. Mme de Sévigné to Pomponne, 17 December 1664, in Sévigné 1853, I: 65, notes as well: "at the beginning, it was announced only by women, and it was mocked; but now everybody saw it [the first comet]." On the distinction of beliefs on comets in high and low cultures during the early modern period, see more generally Schechner 1997.

44 Montalegre 1664, "Présages et prophéties qu'on doit attendre du comete present," sig. D1r—v. In the same vein, Indovino 1665 wrote that the 1664–1665 comets announced the defeat of Jansenism.

45 Montalegre 1664, "Présages et prophéties qu'on doit attendre du comete present," sig. C2v—D1r. Louis XIV founded the *Compagnie des Indes orientales* in September 1664, a first expedition to Madagascar was prepared during the winter 1664 and the ships sailed off in March 1665.

46 *Sombreval*, meaning "dark valley"; it is tempting to think that it was a pseudonym answering "Montalegre," but I can not find any confirmation of this hypothesis.

47 Luyt 1665, Questions 18–20: 40–49.

48 Comiers 1665: 147–151.

49 *Pace* Drévillon 1996: 186, who, probably because he did not go farther than the dedicace, writes that, according to Comiers, comets are "innocent."

50 Comiers 1665: 466–480. On both the distinction and the complementarity of astrological predictions and religious prophecies, the first ones resulting from a human science using observation and reason, the second ones being given by God through direct revelation, see Boudet 1990. It is not clear whom Comiers was targeting; treatises on prophecies that he enumerates pp. 453–458 date from the sixteenth century. That there were such treatises at the time is, however, exemplified by Courcelles 1665.

51 Leschener: 10, 12; Comiers 1665: 110.

52 This event is reported with fewer scientific details but with more insistence on the greatness of the assembly, in *La Gazette*, 17 January 1665, in Renaudot 1665: 67; it was also described in verses in *Gazette rimée*, in Loret 1857–1891, 4: 299.

53 Louis II de Bourbon-Condé, also known as Le Grand Condé (1621–1686), was at the time Prince de Condé; his son Henri-Jules de Bourbon-Condé (1643–1709) was Duc d'Enghien; his brother Armand de Bourbon-Conti (1629–1666) was Prince de Conti. On their scientific patronage, see Béguin 1999; on their education at Jesuit colleges, see Chérot 1896; on the scientific education that was more especially to be given to Louis de Bourbon (1669–1710) and to Louis-Henri (1692–1740), grandson and great-grandson of Le Grand Condé, see Mormiche 2011.

54 On Descartes's theory of comets, see Camerota 2002.

55 Grandami 1665a: 16 mentions the 1618 observations. *Journal des scavans*, in Sallo 1665: 70, and Denis: 51–52, insist that Grandami was only rehearsing what he wrote fifty years beforehand.

56 On Jean Garnier (1612–1681), see Kane 1940 and "Garnier, Jean," *Scholasticon*, to be consulted on line at http://scholasticon.ish-lyon.cnrs.fr/Database/Scholastiques_fr.php?ID=585. Garnier first taught humanities, philosophy, and theology at collège de Clermont. Traces of his teaching are to be found in his textbook, *Organi philosophiae rudimenta, seu Compendium logicae aristotelicae, traditum a J. G. P. S. J.*, 1651, in his *Physica*, Paris BNF, Ms. Lat. 11257 (on which Brockliss 1995: 203–204 comments briefly) and in the theses in philosophy that he had defended (*Theses peripateticae de logica philosophiae organo, propugnatae a nobilibus adolescentibus in collegio claromontano S.J.*, 1650; *Theses de philosophia morali morum magistra, propugnatae a nobilibus adolescentibus in collegio claromontano Societatis Jesu, a kalendis martiis anni 1650 ad kalendas julias ejusdem*, 1651). After a few years spent in Bourges, he came back in Paris and became the librarian of collège de Clermont, probably in 1674. In 1678, he published a *Systema bibliothecæ collegii parisiensis Societatis Jesu*, which is an outline of how to classify the books of this library, and, more generally, of any library.

57 The January conference is mentioned in *L'esprit du sage*: 21–35; Croixmare 1665: 6; Denis 1665: 17–149; Huet 1665: 233–234; Leschener 1665: 5, who adds two other opinions on the nature of comets.

58 Jacques Grandami (1588–1672), also known as Grandamy, occupied several important positions among the Jesuits, including those of rector at the colleges of Bourges, Rennes, Tours, and Rouen. Although he never taught mathematics, he had a lifelong interest in astronomy and cosmology: he observed the comets of 1618 and his first-known publication was *Nova demonstratio immobilitatis terrae petita ex virtute magnetica*, 1645. From the mid-sixties on, probably in retirement, he published extensively in astronomy (*Deux eclipses en l'espace de quinze jours. La premiere de lune horizontale le 16. de juin. La seconde de*

soleil le 2. juillet. Supputées suivant les tables astronomiques de Kepler, du P. de Billy, & du P. Riccioli, 1666; Tractatus de eclipsibus solis et lunae, ex parte secundâ chronologiae christianae, 1668) but also in sacred chronology (De die supremo et natali Christi quaestio evangelica. In qua asseritur perfecta consensio annorum Christi et Aerae communis, in Ecclesia a mille et amplius annis usu, 1661; Tractatus euangelici. De summa dei gloria in Christo Iesu domino nostro. Ad materiam et formam concionum accomodati, 1664; Chronologia christiana. De Christo nato et rebus gestis ante et post eius nativitatem, 1668).

59 Grandami 1665a. Grandami's positions on comets are also analyzed in Ariew 1999b: 115–119. On the adoption of Tycho's cosmology by the Jesuits, see more generally Lerner 1995.

60 Grandami 1665a: 17.

61 Grandami 1665a: 19–21. On Grandami's magnetic demonstration that the Earth is immobile, see Pumphrey 1990.

62 Translation from Ariew 1999b: 117.

63 Grandami 1665a: 8.

64 Grandami 1665a: 21–23.

65 Roberval 1644: 114–137, according to which comets are exhalations of the Earth, which is provided with a sensible soul that draws air and rejects vapors and exhalations. L'esprit du sage: 36, confirms that Roberval did not publish the presentation he made at collège de Clermont: "It would be ungraceful to print something against somebody that only gave a talk (qui n'a parlé que de vive voix) and I would fear to attribute to him what he does not think."

66 On these details and on the content of the theses defended at collège de Clermont between 1637 and 1682, see Collacciani and Roux forthcoming.

67 Jérôme Tarteron (1644–1720) was only twenty-two in 1666. He was to be a Jesuit teacher in humanities and rhetoric and to translate into French the Latin works of Persius, Juvenal, and Horace. According to Catalogue collectif de France, this thesis is also attributed to Louis Prou; it was not unusual for different students to defend the same thesis. The letter from Guy Patin to Charles Spon, 1 January 1665, in Patin 2015, L. 806 may allude to this thesis: "It [the taxes] will be worse than the comet, which does not show up anymore. The Jesuits made a very dry thesis on that [the comet], where there is almost nothing to learn." In this case, there would be an error on the date of this letter, but it is anyhow a surprising date because the first comet was still around at this time.

68 I cannot find any information on Louis Ragayne de la Picottière, but there were some Ragayne de la Picottière who were bourgeois in Sées, a small city in Normandy. According to Catalogue collectif de France, these theses were defended by Louis de la Bletonnière and Louis Prou as well.

69 Nicolas d'Harouys (1622–1698) was to be rector at the Jesuit colleges of Rennes and Nantes. According to Dainville 1954: 111, he taught mathematics at collège de Clermont from 1661 to 1664 and was then replaced by Michel Beaussier, who came from La Flèche. However, La Gazette, 17 January 1665, in Renaudot 1665: 67, when presenting the conference of 10 January, called him "Professor of Mathematics in this college." His known works are two Latin tragedies played at the collège de Clermont in 1659 and 1661 and, in between, a Panegyric to the Queen Marie-Theresa of Austria, who married Louis XIV in 1660. Although internal evidence is enough to attribute the three theses to him, this attribution can be externally confirmed by the letters from Huygens to his father, 15 January 1665 and from Petit to Huygens, 7 August 1665, in Huygens 1888–1950, 5: 195 and 433.

70 Tarteron 1665, sig. A1r.

71 Journal des scavans, 28 January 1665, in Sallo 1665: 70.

72 Descartes, Principia philosophiae, III 126–127, I 51, II 35, in AT VIIA 174–177, 24 and 62–63.

73 On Jean-Baptiste Denis (ca. 1640–1704), see "Denis, Jean-Baptiste," *Diction-naire des journalistes*, to be consulted on line at http://dictionnaire-journalistes.gazettes18e.fr/journaliste/220-denis-jean-baptiste. Nothing is known for sure on Denis's initial formation, but around 1664 he began to give conferences on scientific subjects on Saturdays at his home in Paris. His anonymously published book on comets was his first book; in 1667–1668 he performed experiments on the transfusion of the blood for which he was to be famous. In 1672–1673, he published a journal competing with *Journal des savants*, called first *Mémoires concernant les arts et les sciences*, then *Conférences présentées à Monseigneur le Dauphin*. It is in this journal that, on 1 April 1672, while another comet was passing by, he admitted that he was the author of *Discours sur les comètes* (*Sixième Mémoire concernant les arts et les sciences*, in Denis 1682: 95). However, it was known that he was the author since, on 1671, the famous satirical *Requeste des Maistres es arts, Professeurs, & Regens de l'Université de Paris* required that "Sir Denis will be obliged to fix immediately and is own expense all the gaps and crevasses that he introduced in the heavenly vault in order to make a way for the last comets that appeared in 1664 and 1665" (Boileau, Despreaux and Bernier 1671: 7). Petit, Auzout, and Cassini are then mentioned as attempting to undermine Aristotle's authority.

74 Denis 1665: 56–57 does not name d'Harouys by his name but mentions "the professor that proposed [this opinion] a few months ago."

75 Denis 1665: 59, 64–65, 70–71. The comparison with insects flying together is also employed in Petit 1665: 17, in reference to d'Harouys's opinion.

76 Denis 1665: 62, 72.

77 A few lines later, the assailant is accused of retaining his name purposely.

78 Ragayne de la Picottière 1665a, §21, 27 and 31: 7–8, complains that Denis did not read the theses on comets and on magnetism that were defended one year earlier.

79 The text read "In rebus materialibus, non tantum impossibile est infinitum actu, sed etiam necessarium," but one should read "possibile" rather than "impossibile."

80 Ragayne de la Picottière 1665a, §38: 10, explicitly says that, while some of Descartes's "first dogmas" contradict the light of nature, others contradict the rules of art, and almost all of them contradict common sense.

81 After many others, I told the story of this condemnation in Roux 2013: 63–65. Although it is improbable that such a condemnation was not known by the Jesuits teaching at the collège de Clermont, it is still to be established how it was diffused in France. For example, *Le Journal des savants* mentions it only in passing on the occasion of a review of Plempius's *Fundamenta medicinae* published in 1665 in Louvain, 1 February 1666: 61: "He says that several articles from this doctrine were censured by the Faculty of Theology in Louvain, and that some books were condemned in Rome by the Inquisition, and reports its decree."

82 Ragayne de la Picottière 1665b, I §2, I §18, III §16: 3, 6, 15. This theme was already present in Ragayne de la Picottière 1665a, §9: 5. It might refer to *Le monde*, in AT XI 31, 48, where Descartes presents his cosmology as a fable, in which case, as Domenico Collacciani pointed out to me, d'Harouys would have been a quite early reader of this book, which was first published in 1664. Note, however, that in the preface that he wrote in 1647 for the French version of *Principia philosophiae*, Descartes recommended to his readers that they read the book "first in whole as if a novel (*tout entier ainsi qu'un roman*)," that is to say, all at once, not interrupting the reading, and suspending the question of truth (AT IXB 11). On the critique of Descartes's physics as being only a novel, that is an unjustified hypothesis, see Roux 2014: 77–84.

83 Ragayne de la Picottière 1665b, I §14–15, §19: 5–6. The distinction between mathematical hypotheses and physical theses appeared already in the former thesis, but more briefly, see Ragayne de la Picottière 1665a, §3, 4, 39: 3–4, 11.

84　Ragayne de la Picottière 1665b, I §7: 4. Garnier 1678: 118, gives a brief description of d'Harouys's machines and explains that, because of their number and magnitude, they could not be stored in the rooms of the library, not to speak of the smaller rooms devoted to the museums, but had to be kept in a large room of their own. They were to be seen by eminent travelers, see, for example, Huygens's *Journal*, in Huygens 1888–1950, 22: 545; André de Graindorge to Huet, 9 May 1665 and 5 August 1665, in Graindorge 1942: 267, 303.

85　Ragayne de la Picottière 1665b, I §3–4: 3.

86　Ragayne de la Picottière 1665b, I §4–5, §9, II §18: 3–4, 11.

87　Ragayne de la Picottière 1665b, I §9, I §23, II §3: 4, 7, 9.

88　Ragayne de la Picottière 1665b, I §3, I §22: 3, 7.

89　Ragayne de la Picottière 1665b, II §14: 10.

90　Ragayne de la Picottière 1665b, II §15: 10.

91　Ragayne de la Picottière 1665b, II §2–4, III §22: 8, 16.

92　Rohault 1660–1661, fol. 83r–91r.

93　Oldenburg to Boyle, 14 July 1665, Oldenburg 1965–1986, 2: 431–432.

94　On Jacques de Billy (1602–1679), see Romano 1999: 564–565. Billy spent part of his career at collège des Godrans, Dijon's Jesuit college, where he filled various spiritual functions between 1638 and 1678; when a chair in mathematics was established in Dijon in 1666, he was the first professor to occupy it. He may have aimed at a wide readership through some of his short books that were written in French (*Le siège de Landrecy dédié au Roy*, 1637; *Abrégé des préceptes de l'algèbre*, 1637; *Le tombeau de l'astrologie judiciaire*, 1657). Most of them are, however, longish books written in Latin for specialists in pure and mixed mathematics. In astronomy, he published *Tabulae Lodoicoecae, seu universa Eclipseon doctrina tabulis, praeceptis ac demonstrationibus explicata*, 1656; *Le tombeau de l'astrologie Judiciaire*, 1657; *Opus astronomicum*, 1661; *Discours de la comète qui a paru l'an 1665 au mois d'avril*, 1665; and *Crisis astronomica de motu cometarum*, 1666. In mathematics, apart from *Nova geometriae clavis algebra*, 1643, and *Tractatus de proportione harmonica*, 1658, Billy is known for his editions of Diophantus (*Diophantus Geometra*, 1660; *Diophanti Alexandrini arithmeticorum libri sex*, 1670; *Diophanti redivivi, pars prior*, 1670).

95　Billy 1657 and 1665;

96　Billy 1666: 1, 8, 42–44.

97　On Ignace-Gaston Pardies (1636–1673), see Ziggelaar 1971. Born in Pau, brilliant mathematician, Pardies was professor at La Rochelle and Bordeaux, then in Paris (1670–1673). Already suspected for his "strange opinions (*opinions étrangères*)" while in La Rochelle, he was accused of Cartesianism after he published the *Discours du mouvement local* (1670); he wanted to wash away this accusation with his *Discours de la connaissance des bêtes* published in 1672, but this presented Descartes's thesis that animals have no soul with such verve that Pardies was definitively identified as a crypto-Cartesian.

98　Pardies 1665a.

99　Pardies 1665b, §1, 20: 5, 47.

100　Pardies 1665b, §7–11: 16–24. On page 47, Pardies notes that Auzout probably relied on this method. Huygens is clear that the first one to propose this method was Kepler; see the minutes of the letter that he sent to Auzout, 12 February 1665, in Huygens 1888–1950, 5: 230.

101　Pardies 1665b, §1–2: 5–7. Kepler conceded that the rectilinear motion of comets was not uniform at the beginning and at the end of their trajectories, see *Astronomia pars optica*, in Kepler 1937–, 2: 287, quoted in Ruffner 1971: 181.

102　Pardies 1665b, §20: 48–49, to be compared with *De cometis libelli tres*: 93–94, in Kepler 1937–, 8: 213, quoted in Ruffner 1971: 181.

103 Pardies 1665b, §20: 50. Toward the end of his work, §30: 65–66, he attributes to them a motion in spirals.

104 Pardies 1665b, §21–24: 51, 52–53, 53–54, 55, to be compared with Descartes, *Principia philosophiae*, III, 134–135, 30, *passim*, 58 *sqq.*, 139, in AT VIIIA 173–174, 92 *sqq.*, 109 *sqq.*, 191.

105 Kepler, *De cometis libri tres*: 98, quoted in Ruffner 1971: 180. See also the letter to Herwart von Hohenburg, 7 October 1602, in Kepler 1937–, 14: 283, quoted in Ruffner 1971: 181.

106 On the political context of Lamoignon's thesis, see Lerner 2001: 534–537.

107 Petit to Huygens, 8 March and 5 May 1662, in Huygens 1888–1950, 4: 73, 127; Petit, Auzout, and Thévenot are mentioned meeting on Tuesdays in the letter from Petit to Huygens, 17 October 1664, in Huygens 1888–1950, 5: 124; they met Christopher Wren when he came to Paris a few years later (Oldenburg to Boyle, 24 August 1665, in Oldenburg 1965–1986, 2: 480). For more details, see Roux 2014: 67–69. As for Sorbière's judgment on them, see the letter to Colbert of 1663, commented in Roux 2014: 63–65.

108 *Proceedings of the Royal Society*, the entry dated 21 October 1663.

109 On Samuel Sorbière (1615–1670), see Sarahson 2004. Born Protestant in Languedoc, he was in Paris by the early forties, where he converted to Catholicism. In his letter to Charles Spon, 25 November 1653, in Patin 2015, L. 332, Patin placed his conversion among "the miracles of our ages which are rather political and economic rather than metaphysical." Believed to have fostered the quarrel between Gassendi and Descartes in the forties, he was a translator and lifelong friend of Hobbes.

110 Sorbière to Hobbes, 3 February 1665, in Hobbes 1994, 2: 668–669.

111 Sorbière 1665: 1. Claude Auvry (1606–1687) became a favorite of Mazarin, who made him Évêque de Coutances (1646–1658) and Trésorier de la Sainte-Chapelle (1653–1687).

112 Gassendi 1658: 700ff.

113 Sorbière 1665: 11.

114 Sorbière 1665: 11. See Roux 2014: 79–84, for other texts illustrating this "moral" criticism of Descartes.

115 Sorbière 1665: 4–5, 7.

116 Chapelain acknowledges receipt of Huet's letter in the letter from 14 March 1665, in Chapelain 1880–1883, 2: 389–390. On Pierre-Daniel Huet (1630–1721), see Shelford 2007.

117 Huet 1853: 106–108. The date of 1661 can be inferred from the letter from Chapelain to Huet, September 1661, Chapelain 1880–1883, 2: 152–153; Tolmer, in Graindorge 1942: 256, refers to two manuscripts corresponding to Huet's talk. Huet 1853: 107, had the same judgment as others on Montmor's secret commitment to Cartesianism, see, for example, Chapelain to Heinsius, 22 September 1667 and to Bernier, 16 February and 26 April 1669, in Chapelain 1880–1883, 2: 530, 622, 640. For more references to testimonies according to which Montmor's agenda when he established his academy was to propagate Descartes's doctrine, see Roux 2014, note 42.

118 Huet 1853: 143 asserts that Caen's academy was instituted in 1662; Huet 1706: 173, that it was at the occasion of the 1664 comet. Huet 1665: 145 confesses that because of bad weather and an eye ailment, he was not able to observe it; observations are, however, mentioned in Huet 1853: 145, 198–199. These and other contradictions are examined and solved in Lux 1989: 22–28.

119 Huet 1665.

120 On Huet's anti-Cartesianism, see Shelford 2007: 133–136, 163–183.

121 On Huet's Gassendism and skepticism, see Shelford 2007: 120–126, 136, *passim*.

122 Chapelain to Graindorge, 3 April 1665, in Collas 1912: 336.
123 Chapelain to abbé de Francheville, 16 March 1665, in Chapelain 1880–1883, 2: 39: "there are several places where he [Sorbière] stepped aside from what he [Gassendi] meant, not to contradict him, but for want of understanding him. He speaks easily, but he does not bite into things, and I do not know why he takes the risk of treating them, while his genius is so improper. The book of this Gentleman (*ce Monsieur là*), which was celebrated, does not sell. If it was ever published, it was at the expense of his purse."
124 Petit to Huygens, 8 March and 5 May 1662, in Huygens 1888–1950, 4: 73 and 127. From the beginning of his correspondence with Huygens, Petit complained of the way in which, in France, people of quality neglected mechanics, see Petit to Huygens, 18 October 1658, in Huygens 1888–1950, 2: 257. Petit, Auzout, and Thévenot are mentioned meeting on Tuesdays in the letter from Petit to Huygens, 17 October 1664, in Huygens 1888–1950, 5: 124. The same three would meet Christopher Wren when he came to Paris a few years later (Oldenburg to Boyle, 24 August 1665, in Oldenburg 1965–1986, 2: 480).
125 Christiaan Huygens to Lodewijk Huygens, 6 April 1663, in Huygens 1888–1950, 4: 324–325. Christiaan Huygens to [Constantyn Huygens], 20 April and 4 May 1663, in Huygens 1888–1950, 4: 333, 338. Contrary to what the editors of Huygens's *Œuvres complètes* claim, the d'Espagnet who appears here cannot be the chemist Jean d'Espagnet (1564–1637?), first president of the parlement of Bordeaux: it is more likely his son, Étienne d'Espagnet, counselor at the same parlement.
126 Huygens 1888–1950, 4: 325–329, commented on in Roux 2014: 69–71.
127 On Adrien Auzout (1622–1691), see Brown 1934: 138–141 and "Auzout, Adrien" in Gillispie et al. 2008: 341–342. Auzout was an aristocrat born in Rouen who in the late forties contributed to Pascal's experiments on the vacuum, in particular with the experiment of the vacuum in the vacuum. He worked as an astronomer with Jean Picard at the Académie des sciences and at the Royal Observatory, of which he was briefly a member (1666–1668) before retiring to Italy and England, for having criticized Charles Perrault's translation of Vitruvius.
128 Billy 1665: 4; Denis 1665: 49–50 and 122–123; Grandami 1665a: 9–10 and 1665b: 6; Petit 1665: 194–195, 303–308, *passim*.
129 Auzout 1665a: 7–8. On p. 1, he says that he will later on "explain" the law if it happens to be true. As I mentioned earlier, his method was used already by Kepler and explicitly formulated by Pardies.
130 Auzout 1665a: 6.
131 Auzout 1665a, Au Roy, n.p.; Auzout 1665c, Au lecteur, n.p.
132 Auzout 1665b: 7. The opposition between being a natural cause and being a divine sign is constant in the literature on comets.
133 Auzout 1665a: 6 and 1665b: 5.
134 Huygens to Thévenot, 29 January 1665 and to Auzout himself, 12 February 1665, in Huygens 1888–1950, 5: 210, 230; Chapelain to Huet, 24 May 1665, in Chapelain 1880–1883, 2: 396.
135 This is resumed in *Journal des savants*, 11 January 1666, in Salo 1666: 22. See Lerner 2001: 536–545, for other restrictions by French theologians and scientists concerning Galileo's condemnation.
136 Huygens 1888–1950, 22: 543, mentions, for example, "a dispute between Rohault and Auzout." Auzout 1665c: 22–23, does, however, attack Descartes's proposition for grinding hyperbolic lenses for being only a theory, that can not be "reduced in practice" and "reduced in use."
137 Pierre Petit (1598–1682), born in Montluçon, resided in Paris from 1633 on as Commissaire provincial de l'artillerie. When *Discourse on Method* and its essays were published, he wrote objections against Descartes's metaphysics and

his explanation of refraction, which Descartes treated contemptuously; a few years later, Petit communicated Torricelli's experiment to Pascal and helped him to make barometric experiments. He became Intendant général des fortifications in 1649; since he was part of various scientific circles in France, it was a bitter blow to him that he was never appointed member of the Académie des sciences (see the letter from Boulliau quoted by Brown 1934: 138). The explanation is perhaps to be found in his character as well as in the confusion of his writings; see the cruel portrait made of him in a letter from Sorbière to Hobbes quoted *supra*, note 102; and Christiaan Huygens to Lodewijk Huygens, 28 September and 9 November 1662, in Huygens 1888–1950, 4: 241, 256, *passim*. In April 1667, however, he was elected to the Royal Society. His book on comets was commissioned by the King, see Petit 1665: 347; Patin to André Falconnet, 30 December 1664, in Patin 2015, L. 805. According to Petit to Huygens, 23 January 1665, in Huygens 1888–1950, 5: 207, it was written "for the Court and the Ladies rather than for Mathematics."

138 Petit 1665: 58, 272, is explicit that a comet can not have a rectilinear motion because, according to him, this kind of motion would be infinite, which would be impossible.

139 Petit 1665: 47–55.

140 Huygens to Petit, 8 October 1665, in Huygens 1888–1950, 5: 499; on 6 November: 530–531, Petit answered that he was in fact a full Copernican but that "he did not want to show it for fear of looking like a ridiculous fool in our Court and among most of the honorable people in France." In the same letter, he added that he and Auzout now doubted how to deduce Copernicanism from the motion of comets; the reason for this doubt might be that, contrary to the motion of the first comet, the motion of the second one could not be rectilinear, see Huygens to Moray, 29 May 1664, in Huygens 1888–1950, 5: 361. According to Ariew 1999b: 114, Garnier defended such a semi-Copernican system in his 1651 course.

141 Petit 1665: 151–155. On this argument, see Lerner 2001: 517–519, 531–533, 536.

142 Petit 1665: 153–182 for the criticism of Leschener. Comiers is criticized only in passing, in Petit 1665: 206–209.

143 Petit 1665: 184.

144 Petit 1665: 222–223, 252–254. Nicolas Steno (1638–1686) spent winter 1664–1665 at Thévenot's house, where the remnants of Montmor Academy met: he performed many dissections for Thévenot's guests.

145 Petit 1665: 230–231, 251–252.

146 Petit 1665: 255.

147 Petit 1665: 76, 259–260, 278.

148 Petit 1665: 96–101 and 87–100.

149 On Nicolas de Croixmare, sieur de Lasson (1629–1680), see Collas 1912: 455–458; Lux 1989: 48–49, 60–62, 110–111. According to Graindorge to Huet, 24 June and 13 August 1665, in Graindorge 1942: 203, 287; Huet 1853: 147–148; Brown 1934: 146, 158–159, he was not only working at the construction of a huge metallic mirror but also trying to grind lenses according to the method prescribed at the end of Descartes's *Dioptrics*. Huet 1706: 429 notes cruelly: "he would have been a greater man if he had had fewer talents."

150 This is suggested in Brown 1934: 229. For manifestations of such an Anglomania, see Graindorge 1942. De Croixmare was also probably the author of the report on worms who ate stones published in *Journal des scavans*, 9 August 1666. Boulliau to Lubiniezky, 4 September 1665, in Lubiniezky 1667: 534 explains that neither he nor Auzout know who Vortfischer is but that he is probably not an English man.

151 De Croixmare 1665: 5.

152 De Croixmare 1665: 3, 6.
153 For indications that de Croixmare was involved in secret alchemical researches, see the letter from Graindorge to Huet, 28 November 1667, quoted in Brown 1938: 150, and Huet 1706: 429.
154 *L'esprit du sage*, Ch. 1: 11–18.
155 See also *L'esprit du sage*, Ch. 3: 41–66.
156 Gabriel de Vendages de Malapeyre (1624–1702), an officer at *Présidial* in Toulouse, was an active figure in the academic circles of his city: he was a member of société des Jeux Floraux, he founded the society of Lanternistes and finally he contributed to the foundation of Académie des sciences, inscriptions et belles-lettres of Toulouse. A devotee of the Virgin Mary, he instituted a prize to the best poem written in her honor and had the Chapelle de Notre-Dame du Mont-Carmel built, of which he gave a written description. His only known book in natural philosophy is *De la nature des comètes*, which is dedicated to Virgin Mary.
157 Malapeyre 1665: 97 and 70, 52–53 and 173, 85, 48–49 and 142–150.
158 He may have known the controversy between Galileo and Grassi through Gassendi 1658: 702b–703a.

Bibliography

Manuscripts

Billy, Jacques de (1665a) *Observations sur la comete par le P. de Billy*, Bibliothèque municipale de Toulouse, Res. C XVII 67 (5).
Lalane, sieur de (1664) *Discours des comètes par le Sieur de Lalane*, Bibliothèque nationale de France, MS 19947.
Lantin, Jean-Baptiste. (n.d.) *Lantiniana, ou Recueil de plusieurs choses dites par M. Jean-Baptiste Lantin, conseiller au parlement de Bourgogne et remarquées par M. Pierre Le Goux Conseiller au même Parlement*, Bibliothèque municipale de Dijon, MS 962.
Rohault, Jacques (1660–1661) *Conférences, recueillies par M. F. avocat qui y a, dit-il, ajouté du sien*, Bibliothèque Sainte-Geneviève, MS 2225.

Primary Sources

[anon.] (1665a) *Ballet de la comète, divisé en deux parties et dansé à Soissons le 9 Fevrier*, Soissons: Nicolas Asseline.
———. (1665b) *Figure de la derniere estoille extraordinaire et professie mistique sur l'apparition du dernier comete*, [n.p.].
———. (1665c) *Histoire des comètes qui ont paru depuis peu sur notre horizon, l'un sur la fin de l'année dernière et l'autre au commencement de la présente avec leur observation, figure et pronostic par un Mathématicien*, Besançon: Nicolas Couché.
———. (1665d) *Les effets ridicules du comète, envoyez à Lysandre malade*, Paris: [n.p.].
———. (1665e) *L'esprit du sage, informé sur le sujet des deux comètes de cette année où sont réfutés tous ceux qui en ont parlé jusqu'à présent, par une nouvelle hypothèse*, Paris: Pierre Josse.
Auzout, Adrien (1665a) *L'éphéméride du comète*, Paris: Jean Cusson.

————. (1665b) *L'éphéméride du nouveau comète*, Paris: Jean Cusson.

————. (1665c) *Lettre à Monsieur L'abbé Charles sur le Ragguaglio di due nove osservationi, etc., da Giuseppe Campani*, Paris: Jean Cusson.

Billy, Jacques de (1657) *Le tombeau de l'astrologie Judiciaire*, Paris: Michel Soly.

————. (1665) *Discours de la comète qui a paru l'an 1665, au mois d'avril*, Paris: Sébastien Cramoisy and Sébastien Mabre-Cramoisy.

————. (1666) *Crisis astronomica de motu cometarum*, Dijon: Pierre Palliot, and Paris: Helias Josset.

Boileau Despréaux, François and Bernier, Nicolas (1671) *Requeste des Maistres es arts, Professeurs, & Regens de l'Université de Paris presentée à la Cour Souveraine de Parnasse: en semble l'Arrest intervenu sur ladite Requeste*, Delphes: ocitée des Imprimeurs de la Cour de Parnasse.

Chapelain, Jean (1880–1883) *Lettres de Jean Chapelain*, ed. Philippe Tamizey de Larroque, 2 vols, Paris: Imprimerie Nationale.

Comiers, Claude (1665) *La nature et présage des comètes. Ouvrage mathématique, physique, chimique et historique enrichi des prophéties des derniers siècles et de la fabrique des grandes lunettes*, Lyon: Charles Mathevet (also published under the title *La nouvelle science des cometes et histoire générale de leurs presages*, [n.p.]).

Condé, Prince de (1920) *Le grand Condé et le duc d'Enghien. Lettres inédites à Marie-Louise de Gonzague, reine de Pologne, sur la cour de Louis XIV (1660–1667)*, ed. Émile Magne Paris: Émile-Paul.

Cotin, Charles (1665) *Œuvres galantes tant en vers qu'en prose. Contenant . . . Galanterie sur l'astrologie judiciaire. Galanterie sur la comette*, Paris: Estienne Loyson.

Courcelles, F. de (1665) *Le desabusement sur le bruit qui court de la prochaine Consommation des Siecles, fin du Monde et du Jour du Jugement Universel. Contre Perrieres Varin qui assingne ce Jour en l'année 1666. Et Napeir Ecossois qui le met en l'année 1688*, Rouen: Laurens Maurry.

Croixmare, Nicolas de, sieur de Lasson [Vortfischer] (1665) *Le Courrier de traverse ou le tri-comète observé à Oxfort en Angleterre depuis le 22 novembre jusqu'au 28 janvier 1665, traduit de l'anglais de M. Vortfischer*, Paris: Jacques Bouillerot.

Decrusy, Isambert, François-André, Jourdan, Athanase-Jean-Léger, and Taillandier, Alphonse-Honoré (1821–1833) *Recueil général des anciennes lois françaises, depuis l'an 420 jusqu'à la Révolution de 1789*, 29 vols, Paris: Belin-Leprieur.

Denis, Jean-Baptiste [I.D.P.M] (1665) *Discours sur les comètes suivant les principes de M. Descartes où l'on fait voir combien peu solides et mal fondées sont les opinions de ceux qui croyent que les comètes sont composées ou d'exhalaisons terrestres ou des sueurs de toute la Sphère Elémentaire ou de quelque manière céleste plus condensée qu'à l'ordinaire ou d'une très grande quantité de petites étoiles qui se sont jointes ensemble*, Paris: Frédéric Léonard (but also: Jean Guignard, Pierre Promé, Charles Savreux; reprinted in 1672 with another title, *Divers sentimens sur les comètes, comparez avec celuy de Descartes, avec l'explication de plusieurs difficultez touchans le système du monde*, Paris: Frédéric Léonard.)

————. (1668) *Discours sur l'astrologie judiciaire et sur les horoscopes, prononcé par J. Denis . . . dans une des conférences publiques qui se font chez luy tous les samedis*, Paris: Jean Cusson.

————. (1682) *Recueil des Mémoires et Conférences sur les arts et les sciences présentées à Monseigneur le Dauphin*, Amsterdam: Pierre Le Grand.

Depping, G. B., ed. (1855) *Correspondance administrative sous le règne de Louis XIV, recueillie et mise en ordre par G. B. Depping. Tome IV et dernier. Travaux publics—Affaires religieuses—Protestants—Sciences, lettres et arts—Pièces diverses*, Paris: Imprimerie nationale.

Descartes, René (1996) *OEuvres de Descartes*, eds. C. Adam and P. Tannery, 11 vols, Paris: Vrin.

Diez, François [also attributed to François Ridelle] (1665) *Balet des comètes pour la tragédie d'Irlande sur le théâtre du collège de Clermont de la Cie de Jésus le 6 aout*, [n.p.].

Garnier, Jean (1678) *Systema bibliothecae collegii parisiensis Societatis Jesu*, Paris: Sébastien Marbre-Cramoisy.

Gassendi, Pierre (1658) *Opera omnia in sex tomos divisa. . . Tomus primus quo continentur Syntagmatis philosophici pars prima, sive logica, itemque partis secundae, seu physicae sectiones duae priores*, Lyon: Anisson.

Giustiniani, Julio (1665) *L'explication de la comète qui apparut sur la fin de l'année 1664 et au commencement de 1665, présentée à la Reine mère en langue espagnole, & d'espagnol traduite en françois*, Paris: Alexandre Lesselin.

Graindorge, André (1942) "Une page d'histoire des sciences. Vingt-deux lettres inédites d'André de Graindorge à P.-D. Huet" in ed. Léon Tolmer, *Mémoires de l'Académie nationale des sciences, arts et belles-lettres de Caen. Nouvelle série, Tome X*, Caen: Charles Le Tendre, pp. 245–337.

Grandami, Jacques (1665a) *Le Cours de la comète qui a paru sur la fin de l'année 1664 et au commencement de l'année 1665. Avec un traité de sa nature, de son mouvement et de ses effets, présenté à Monseigneur le Prince par le P. Iacques Grandamy*, Paris: Sébastien Cramoisy and Sébastien Mabre-Cramoisy.

———. (1665b) *Le Parallele des deux cometes qui ont paru les années 1664 & 1665. Par le P. Iacques Grandamy*, Paris: Sébastien Cramoisy and Sébastien Mabre-Cramoisy.

Granet, François (1738) *Réflexions sur les ouvrages de literature. Tome Troisième*, Paris: Antoine-Claude Briasson.

Guyard, Marie (1681) *Lettres de la venerable Marie de l'Incarnation, Premiere superieure des Ursulines de la Nouvelle-France*, Paris: Louis Billaine.

Hobbes, Thomas (1994) *The Correspondence of Thomas Hobbes*, 2 vols, ed. N. Malcolm, Oxford: Clarendon Press.

Huet, Pierre-Daniel (1665) *XX. Dissertation de la nature des comètes*, in *Dissertations sur diverses matières de religion et de philologie. . . recueillies par Monsieur L'Abbé de Tilladet. Tome second*, Paris: François Fournier and Frédéric Léonard, 1712, pp. 232–246.

———. (1706) *Les origines de la ville de Caen revûës, corrigées et augmentées. Seconde édition*, Rouen: Laurens Maurry.

———. (1853) *Mémoires de Daniel Huet, Évêque d'Avranches*, trans. Charles Nizard, Paris: Hachette.

Huygens, Christiaan (1888–1950) *Œuvres complètes de Christiaan Huygens*, ed. Société Hollandaise des Sciences, 22 vols, La Haye: Martinus Nijhoff.

Indovino, Fortunato (1665) *Le Mystère caché dans le comète de ce temps, ou le Jansénisme agonissant dans ce météore*, Lyon: [n.p.].

Kepler, Johannes (1937–) *Gesammelte Werke herausgeben im Auftrag der Deutschen Forschungsgemeinschaft und der Bayerischen Akademie der Wissenschaften*, München: C. H. Beck.

Le Mercier, R. P. (1666) *La Relation de ce qui s'est passé en la Nouvelle France es années 1664 et 1665*, Paris: Sébastien Cramoisy and Sébastien Marbre-Cramoisy.

Léautaud, Vincent (1665) *Copie d'une lettre escrite par un père jésuite du Collège d'Ambrun à M. de Ponnat, baron de Gresse, conseiller au parlement de Grenoble. Sur le sujet des Cometes apparuës ès mois passés de décembre et janvier*, Grenoble: Robert Philippes.

Leschener, Henry de (1665) *Traité des comettes où on voit leur causes, leur nature, leurs effets, le temps auquel elles se forment, les lieux où elles paraissent, le moyen de les prédire et de connaître non seulement ce qu'elles annoncent en général, mais aussi en particulier. Avis d'un particulier au libraire sur le traité des comèttes par F.H.L. Epistre à S.A.S Mgr le Prince*, Paris: Thierry Denis.

Loret, Jean (1857–1891) *La muze historique ou Recueil des lettres en vers contenant les nouvelles du temps: écrites à Son Altesse Mademoizelle de Longueville, depuis duchesse de Nemours*, ed. C. L. Livet, Paris: Jannet, then Daffis, then Champion.

Lubiniezky, Stanislaw (1667) *Theatri cometici pars prior. Communicationes de cometis 1664 et 1665*, Amsterdam: Frans Kuyper.

Luyt, Robert (1665) *Questions curieuses sur la comète qui a paru en France depuis le XV du mois de décembre l'an 1664: ou le jugement astronomique que l'on en doit former et ce qu'elle pronostique*, Paris: Charles de Sercy.

Malapeyre, Gabriel de Vendages de (1665) *De la nature des comètes*, Toulouse: Arnaud Colomiez chez A. Pélissier.

Molière (1684 [1670]) *Les amants magnifiques*, Amsterdam: John Benjamins Publishing.

Montalegre, sieur de (1664) *Discours sur le comete qui paroit à present; avec sa figure, sa situation dans le ciel, & les bons ou mauvais effets qu'il presage. Suivant les observations de plusieurs scavants astronomes. Recueillies en faveur des curieux par le Sieur de Montalegre, amateur des sciences mathematiques*, Lyon: François Larchier.

Oldenburg, Henry (1965–1986) *The Correspondence of Henry Oldenburg*, 13 vols, eds. and trans. A. R. Hall and M. B. Hall, Madison, Milwaukee and London: University of Wisconsin Press, Mansell, and Taylor and Francis.

Pardies, Ignace-Gaston (1665a) *Remarques sur les comètes et autres phaenomènes extraordinaires de ce temps*, Bordeaux: G. de la Court.

———. (1665b) *Dissertatio de motu et natura cometarum*, Bordeaux: Pierre du Coq.

Patin, Guy (2015) *Correspondance française de Guy Patin*, ed. Loïc Capron, Paris: Bibliothèque interuniversitaire de santé.

Petit, Pierre (1665) *Dissertation sur la nature des comètes. Au Roy. Avec un Discours sur les prognostiques des éclipses et autres matières curieuses*, Paris: Louis Billaine and Thomas Jolly.

Pithoys, Claude (1646) *Traitté curieux de l'Astrologie Judiciaire, ou préservatif contre l'astromantie des généthliaques, auquel quantité de questions curieuses sont resoluës, etc.*, Montbéliard: Jacques Foylet (but also Sedan: Pierre Jannon, 1641 et 1661).

Ragayne de La Picottière, Louis (1665a) *De duplici cometa vero et ficto positiones mathematicae. Propugnabuntur a Ludovico Ragayne de la Picottière in Collegio claromontano Societatis Iesu. Die 12 junii 1665*, [n.p.].

———. (1665b) *De hypothesi cartesiana positiones physicomathematicae. Propugnabuntur a Ludovico Ragayne de La Picottiere Parisino (i.e. Sagiensi) in Collegio Claramontano Societatis Iesu. Die 13. Junii 1665*, [n.p.].

Renaudot, Théophraste (1665) *Recueil des Gazettes nouvelles ordinaires et extraordinaires. Relations et récits des choses avenues tant en ce Royaume qu'ailleurs pendant l'année mil six cents soixante-cinq*, Paris: Bureau d'Adresse.

Riccioli, Giambattista (1651) *Almagestum novum astronomiam veterem novamque complectens observationibus aliorum, et propriis novisque theorematibus, problematibus, ac tabulis promotam, in tres tomos distribuía*, Bologna: Heirs of Victor Benatius.

Roberval, Gilles-Personne de (1644) *Aristarchi Samii De mundi systemate, partibus, et motibus ejusdem, libellus*, Paris: Antoine Bertier.

Sallo, Denis de (1665) *Le Journal des sçavans de l'an MDCLXV*, Cologne: Pierre Michel.

———. (1666) *Le Journal des sçavans de l'an MDCLXVI*, Paris: Jean Cusson.

Sévigné, Madame de (1853) *Lettres de Madame de Sévigné avec les notes de tous les commentateurs*, tome I, Paris: Firmin Didot Frères.

Sombreval, sieur de (1665) *Advertissement du Ciel que Dieu donne aux chrestiens par la Comete*, Lyon: François Larchier.

Sorbière, Samuel (1665) *Discours de M. de Sorbière sur la comète*, [n.p.].

Tarteron, Jérôme (1665) *De cometa annorum 1664 & 1665 Observationes mathematicae propugnatae Parisiis, ab Hyeronymo Tarteron, in aula Collegi Claramontani soc. Jesu. Die Jovis 29. Januarii 1665*, Paris: Edmond Martin.

Vaissey, D. de (1665) *De novo cometa qui apparuit mense decembri anni 1664 carmen prognosticum illustrissimo D. de Vaissey*, [n.p.].

Secondary Literature

Aricò, D. (1998) " 'Res caelestes': Miracoli e osservazioni astronomiche in un carteggio inedito di Giovan Battista Riccioli con altri intellettuali del suo tempo," *Filologia e critica*, 23: 249–294.

———. (1999) "Les 'yeux d'Argos' et les 'étoiles d'Astrée' pour mesurer l'univers: Les Jésuites italiens et la science nouvelle," *Revue de synthèse*, 120: 285–303.

Ariew, R. (1999a) *Descartes and the Last Scholastics*, Ithaca: Cornell University Press.

———. (1999b) "Scholastics and the new astronomy on the substance of the heavens" in ed. R. Ariew 1999a, pp. 97–119.

Atkinson, G. (1951) "Précurseurs de Bayle et de Fontenelle: la comète de 1664–1665 et l'incrédulité savante," *Revue de Littérature Comparée*, 25: 12–42.

Barker, P. and Goldstein, B. R. (1988) "The role of comets in the Copernican revolution," *Studies in History and Philosophy of Science*, 19: 299–319.

Béguin, K. (1999) *Les Princes de Condé. Rebelles, courtisans et mécènes dans la France du Grand Siècle*, Seyssel: Champ Vallon.

Berger, S., Garber, D., and Grafton, A., eds. (forthcoming) *Teaching Philosophy in the Seventeenth Century*.

Boner, P. J. (2013) *Kepler's Cosmological Synthesis: Astrology, Mechanism and the Soul*, Boston: Brill.

Boschiero, L. (2009) "Giovanni Borelli and the comets of 1664–1665," *Journal of the History of Astronomy*, 40: 11–30.

Boudet, J.-P. (1990) "Simon de Phares et les rapports entre astrologie et prophétie à la fin du Moyen Age," *Mélanges de l'École française de Rome. Moyen Age, Temps modernes*, 102: 617–648.

Brockliss, L. (1995) "Pierre Gautruche et l'enseignement de la philosophie" in ed. L. Giard 1995, pp. 187–219.

Brown, H. (1934) *Scientific Organizations in Seventeenth Century France (1620–1680)*, New York: Russel and Russel.

———. (1938) "L'Académie des sciences de Caen d'après les lettres d'André de Graindorge (1666–1675)" in *Mémoires de l'Académie nationale des sciences, arts et belles-lettres de Caen. Nouvelle série, Tome IX*, Caen: Charles Le Tendre, pp. 117–198.

Camerota, M. (2002) "Sidera ex unis vorticibus in alios migrantia. Note sulla teoria cometaria cartesiana" in eds. M. T. Marcialis and F. M. Crasta, pp. 91–105.

Campinoti, V. (2006) "Galileo contro Aristotele nello studio di Pisa: Resoconto di una 'disputatio circularis' di Alessandro Marchetti sulla natura delle comete," *Galilaeana*, 3: 217–228.

Cassini, A. (2003) *Gio. Domenico Cassini, Uno Scienziato del Seicento*, Perinaldo: Comune di Perinaldo.

Chérot, H. (1896) *Trois éducations princières au dix-septième siècle*. Lille: Société de Saint-Augustin, Desclée, De Brouwer et Cie.

Collacciani, D. and Roux, S. (forthcoming) "The mathematical theses defended at *collège de Clermont* (1637–1682): How to guard a fortress in times of war" in eds. S. Berger, D. Garber, and A. Grafton.

Collas, G. (1912) *Jean Chapelain 1595–1674. Étude historique et littéraire d'après des documents inédits*, Paris: Perrin.

Dainville, François de 1954. "L'enseignement des mathématiques dans les Collèges jésuites de France du XVIᵉ au XVIIIᵉ siècle," *Revue d'histoire des sciences*, 7: 109–123.

Dascal, M. and Boantza, V. D., eds. (2011) *Controversies within the Scientific Revolution*, Amsterdam: John Benjamins Publishing.

d'Enfert, R. and Fonteneau, V., eds. (2011) *Espaces de l'enseignement scientifique et technique. Acteurs, savoirs, institutions, XVIIᵉ–XXᵉ siècles*, Paris: Hermann.

Dobre, M. and Nyden, T., eds. (2014) *Cartesian Empiricisms*, Dordrecht: Springer.

Drévillon, H. (1996) *Lire et écrire l'avenir: l'astrologie dans la France du Grand Siècle, 1610–1715*, Seyssel: Champ Vallon.

Gal, O. and Chen-Morris, R. (2011) "Galileo, the Jesuits, and the controversy over the comets. What was The Assayer really about?" in eds. M. Dascal and V. D. Boantza 2011, pp. 33–52.

Garber, D. and Roux, S., eds. (2013) *The Mechanization of Natural Philosophy*, Dordrecht: Springer.

Giard, L., ed. (1995) *Les Jésuites à la Renaissance. Système éducatif et production du savoir*, Paris: Presses universitaires de France.

Gillispie, C. C., Holmes F. L. and Koertge, N., eds. (2008) *Complete Dictionary of Scientific Biography*, vol. 1., Detroit: Charles Scribner's Sons.

Granada, M. A., Mosley, A. and Jardine, N., eds. (2014) *Christoph Rothmann's Discourse on the Comet of 1585: An Edition and Translation with Accompanying Essays*, Leiden: Brill.

Henry, J. and Hutton, S., eds. (1990) *New Perspectives on Renaissance Thought. Essays in the History of Science, Education and Philosophy*, London: Duckworth and Istituto italiano per gli studi filosofici.

Hetherington, N. S. (1972) "The Hevelius—Auzout controversy," *Notes and Records of the Royal Society*, 27: 103–106.

Jensen, D. (2006) *The Science of the Stars in Danzig from Rheticus to Hevelius*, PhD Thesis, University of San Diego.

Kane, W. T. (1940) "Jean Garnier, Librarian," *Mid-America: An Historical Review*, 11: 75–95 and 11: 191–222.

Lerner, M. P. (1995) "L'entrée de Tycho Brahe chez les Jésuites ou le chant du cycle de Clavius" in ed. L. Giard 1995, pp. 145–185.

———. (1996–1997) *Le monde des sphères*, 2 vols, Paris: Les Belles Lettres.

———. (2001) "La réception de la condamnation de Galilée en France au xviie siècle" in eds. J. Montesinos and C. Solís 2001, pp. 513–548.

Lugli, M. U. (2004) *Geminano Montanari*, Modena: Edizioni il Fiorino.

Lux, D. S. (1989) *Patronage and Royal Science in Seventeenth-Century France*, Ithaca: Cornell University Press.

Marcialis, M. T. and Crasta, F. M., eds. (2002) *Descartes e l'Eredità Cartesiana nell' Europa Sei-Settecentesca*, Lecce: Conte.

Matton, A. (1979) "L'œuvre méconnue et l'œuvre inconnue de Robert Luyt," *Bulletin de la Société d'archéologie et d'histoire du Tonnerrois*, 32: 30–42.

Montesinos, J. and Solís, C., eds. (2001) *Largo Campo di Filosofare. Eurosymposium Galileo 2001*, La Orotava: Fundación Canaria Orotava de Historia de la Ciencia.

Mormiche, P. (2011) "Les Condés et les sciences: entre l'académie Bourdelot et l'éducation de Monsieur le Duc (1670–1700)" in eds. R. d'Enfert and V. Fonteneau 2011, pp. 15–29.

Mosley, A. (2014) "The history and historiography of early modern comets" in eds. M. A. Granada, A. Mosley, and N. Jardine 2014, pp. 282–325.

Nellen, H. J. M. (1994) *Ismaël Boulliau (1605–1694), astronome, épistolier, nouvelliste et intermédiaire scientifique*, Amsterdam: APA-Holland University Press.

Pumphrey, S. (1990) "Neo-Aristotelianism and the magnetic philosophy" in eds. J. Henry and S. Hutton 1990, pp. 177–189.

Romano, A. (1999) *La contre-réforme mathématique. Constitution et diffusion d'une culture mathématique jésuite à la Renaissance (1540–1640)*, Rome: École française de Rome.

Roux, S. (2013) "A French partition of the empire of natural philosophy (1670–1690)" in eds. D. Garber and S. Roux 2013, pp. 55–98.

———. (2014) "Was there a Cartesian experimentalism in 1660s France?" in eds. M. Dobre and T. Nyden 2014, pp. 47–88.

Ruffner, J. A. (1971) "The curved and the straight: Cometary theory from Kepler to Hevelius," *Journal for the History of Astronomy*, 2: 178–194.

Sarasohn, L. T. (2004) "Who was then the gentleman? Samuel Sorbière, Thomas Hobbes, and the Royal Society," *History of Science*, 42: 211–232.

Schechner, S. G. (1997) *Comets, Popular Culture and the Birth of Modern Cosmology*, Princeton: Princeton University Press.

Shapin, S. (1994) *A Social History of Truth: Civility and Science in Seventeenth-Century England*, Chicago: University of Chicago Press.

Shelford, A. (2007) *Transforming the Republic of Letters: Pierre-Daniel Huet and European Intellectual Life 1650–1720*, Rochester: University of Rochester Press.

Yeomans, D. K. (1991) *Comets: A Chronological History of Observation, Science, Myth, and Folklore*, New York: Wiley.

Ziggelaar, A. S. J. (1971) *Le physicien Ignace-Gaston Pardies S. J. (1636–1673)*, Odensee: Odensee University Press.

5 Corpuscularism and Experimental Philosophy in Domenico Guglielmini's *Reflections* on Salts[1]

Alberto Vanzo

Natural philosophical *novatores* in late seventeenth-century Italy typically endorsed a corpuscularist view of principles. They claimed that natural philosophy should identify the causes or principles of natural phenomena, and they identified those principles either with corpuscles, or with motion and matter, which, in turn, consists of corpuscles. Yet, several Italian *novatores* were also adherents of early modern experimental philosophy, which recent studies have portrayed as being incompatible with corpuscularism. This raises the question of whether early modern philosophers could consistently endorse both a corpuscularist doctrine of principles and the tenets of experimental philosophy. This chapter addresses this question by examining Domenico Guglielmini's *Philosophical Reflections derived from the Figures of Salts* (1688).[2] In this treatise on crystallography, Guglielmini puts forward a corpuscularist theory, and he defends it in a way that is in line with the methodological prescriptions, epistemological strictures, and preferred argumentative styles of experimental philosophers. The examination of the *Reflections* shows that early modern philosophers could consistently endorse, at the same time, both experimental philosophy and a corpuscularist doctrine of principles.

The chapter starts by explaining what I understand by experimental philosophy and corpuscularism. I then show that a corpuscularist doctrine of principles was widely accepted among late seventeenth-century Italian *novatores*, although most practitioners refrained from highlighting it or defending it explicitly. I then turn to the main corpuscularist claim of the *Reflections* and its methodological preface, that explicitly endorses experimental philosophy. I conclude by discussing how Guglielmini's adherence to experimental philosophy relates to his corpuscularism.

Experimental Philosophy and Corpuscularism

The movement of early modern experimental philosophy emerged in England around 1660 amongst fellows of the early Royal Society such as Robert Boyle and Robert Hooke. It quickly spread to Italy, where it found a favourable reception among the naturalists and physicians who regarded themselves as Galileans.[3] It even influenced those Jesuits, such as Daniello Bartoli and Filippo Buonanni, who were willing to integrate new insights in an eclectic version of Aristotelian-Scholastic *scientia* and to engage with *novatores* on the specific details of their discoveries (e.g., Bartoli 1677: 5–18), rather than rejecting their outlook *a priori* for its metaphysical and theological implications (Torrini 1979b: 20–27). Among other works, Geminiano Montanari's *Physico-Mathematical Thoughts* (1667) and Francesco Redi's *Experiments on the Generation of Insects* (1996 [1668]) endorse central tenets of experimental philosophy.

Experimental philosophers shared a common rhetoric, based on the praise of experiments and the criticism of hypotheses and speculations. They had common heroes, like Bacon, and common foes, especially Aristotelian and, later, Cartesian natural philosophers. But their most important common trait lies in their views on how we can acquire and expand our knowledge of nature. Experimental philosophers held that, before firmly committing oneself to any substantive claims or theories of the natural world, one should gather extensive empirical information by means of experiments and observations. They assigned the same primary role to experiments and observations (Anstey 2014: 105–106): identifying matters of fact which are the basis for developing and confirming theories of the natural world. Experiences (i.e. experiments and observations) reported by others must be critically evaluated, and if possible, their experiments must be replicated. Only once this process of fact gathering and checking is nearing completion will we be entitled to commit firmly to substantive claims or theories (Hooke 1705: 18) and only insofar as they are warranted by experiments and observations (*Defence against Linus,* 1662, Boyle 1999–2000, 3: 12; Sprat 1667: 107).

Seventeenth-century experimental philosophers often claimed that empirical information should be organized in experimental natural histories (Oldroyd 1987: 151–152): large structured collections of experiments and observations on any kinds of items (biological kinds, minerals, diseases, states of matter, counties, and arts). These collections should serve as the preliminary step to the construction of "a Solid and Useful Philosophy."[4] Natural philosophical theories should be derived from empirical information through a process called induction (Glanvill 1668: 87; Hooke 1705: 331; Montanari 1980: 540) or deduction (Newton 1999 [1726]: 943). Yet, seventeenth-century experimental philosophers did not take up Bacon's theory of induction,[5] nor did they develop detailed accounts of how theories can be derived from experiments and observations.[6]

These methodological and epistemological views of seventeenth-century experimental philosophers entail neither the endorsement nor the rejection of corpuscularism. I understand corpuscularism as a view on explanatory natural philosophical principles, namely the view that physical phenomena should be explained in terms of the shape, size, and spatial arrangement of the particles that make up physical bodies, along with the motion of such particles according to the laws of nature.[7] Yet, several recent studies on early modern experimental philosophy distinguish sharply between experimental philosophy and corpuscularism. According to Luciano Boschiero and Marta Cavazza, experimental philosophy was a "purely descriptive" endeavour, "programmatically disinterested" in the "metaphysical causes" of natural phenomena (Cavazza 1998) and independent from "theoretical convictions," "presuppositions and preconceptions" (Boschiero 2007: 1, 9).[8] Corpuscularism did not merely describe phenomena but sought to explain them in terms of causal processes involving unobserved, metaphysically basic entities. For Stephen Gaukroger, on the other hand, experimental philosophy was not purely descriptive. It provided "non-reductive explanations" that avoid any mention of corpuscles (Gaukroger 2006: 254), "as opposed" to corpuscularist attempts to explain phenomena "in terms of some underlying micro-corpuscular structure" (Gaukroger 2014: 28). According to Alan Chalmers, corpuscularist explanations aimed to identify the "rock-bottom or ultimate causes of material phenomena." The explanations of experimental philosophers singled out non-ultimate, intermediate causes, as "opposed to accounts of the ultimate structure of matter" (Chalmers 2012: 551).

The view that experimental philosophy and corpuscularism were sharply distinct and, possibly, in conflict with each other underlies several studies of the authors who committed themselves to both, like Robert Boyle and several Italian natural philosophers. To what degree Boyle's experimental science depends on his corpuscularism is a matter of controversy,[9] as is the view that Boyle suspended his commitment to corpuscularism when articulating his experimental philosophy (Gaukroger 2014: 19). Yet, it is generally agreed that Boyle's "theoretical reflections on the corpuscular hypothesis" belong to his "speculative theory," as opposed to his experimental philosophy.[10] This divide between experimental philosophy and corpuscularism has been portrayed as being so deep that some Italian natural philosophers allegedly endorsed experimental philosophy to conceal their corpuscularism. According to Luciano Boschiero, the members of the Florentine Accademia del Cimento portrayed themselves as experimental philosophers to hide their allegiances to competing Aristotelian and corpuscularist matter theories.[11] And for Marta Cavazza (1990: 145), Bolognese authors followed the model of English experimental philosophers to downplay their matter-theoretical commitments and preserve the freedom of teaching "in the ideological framework of the Catholic Counter-Reformation." According to Cavazza, it is by presenting themselves as experimental philosophers, rather than corpuscularists, that Bolognese practitioners avoided the charges of atheism raised against Neapolitan *novatores* (Torrini 1979a) and the conflicts that led to the imposition of Aristotelianism as the sole natural philosophy to be taught in Florence (Galluzzi 1974, 1995).

This chapter provides a different perspective on the relation between experimental philosophy and corpuscularism by focusing on Domenico Guglielmini's *Philosophical Reflections derived from the Figures of Salts* (1688).

In this work, Guglielmini commits himself to experimental philosophy as well as corpuscularism. He does not treat these commitments as opposed or competing with one another. The work is not divided into a speculative disquisition about corpuscles, on one hand, and metaphysically neutral experimental reports, on the other. Instead, Guglielmini relies on premises and arguments that conform to the epistemological and methodological strictures of experimental philosophy in order to develop corpuscularist explanations. This work shows that at least one early modern experimental philosopher could and did consistently entertain corpuscularist views.[12] Corpuscularist explanations were not necessarily in contrast with experimental philosophy, nor did they always belong to the realm of speculation.

Natural Philosophical Principles in Late Seventeenth-Century Italy

Corpuscularism was a widely held view among late seventeenth-century Italian *novatores*. They did not provide any explicit, detailed discussions of principles; however, a survey of their texts reveals a broad agreement on two claims. The first is that natural philosophers should not stop at "experimenting, and narrating," as the proem to the Cimento's *Saggi* states,[13] but they should also search for the causes or principles of natural phenomena.[14] The second claim is that these principles are neither the traditional four elements nor the Paracelsian *tria prima* nor water, as van Helmont had claimed. They are either corpuscles or motion and matter, which, in turn, consists of corpuscles.[15] For instance, Giuseppe Valletta states that, for modern philosophers, corpuscles "are the first principles of all material things."[16] Donato Rossetti (1667: 14) claims that the "Democritean principles," that is "corpuscles and atoms," are necessary to explain natural phenomena. He divides atoms into dark and bright, and he calls both types of atoms principles (Rossetti 1671: 1). Rossetti's adversary, Geminiano Montanari, officially denies that we are already able to establish whether corpuscles are "first principles" (Montanari 1980: 547). In his view, we will conclusively identify the first principles only once we have charted all of their effects (540). Yet, it is telling that, having surveyed a variety of opinions on whether the true principles are those of the Presocratics, Democritus's atoms and vacuum, Plato's matter and ideas, or Aristotle's matter, form, and privation, Montanari ignores all those theories except Democritus's and goes on to discuss "corpuscles, that is, atoms" (Montanari 1980: 538, 544–548).[17]

Perhaps the most instructive example of how widespread the adoption of corpuscularist principles was among Italian *novatores* is provided by Francesco Redi. At first sight, he might appear to provide a nice illustration of the discontinuity between experimental philosophy and corpuscularism. He is often portrayed as the prototype of a "superficial" style of inquiry (Baldini 1980: 427) that focuses on "macroscopic and behavioral features of animal species" (Bernardi 1996 [1668]: 7) and eschews any "hypotheses on the basic structure of phenomena" (Baldini 1980: 450). There is good reason to believe that this approach was motivated, at least in part, by Redi's concern to avoid conflicts with the Aristotelians and the church. Those conflicts might have endangered not only his privileged position in the Medici court but also the freedom of teaching and research of Florentine *novatores*. Redi was instrumental in dissuading Rossetti from publishing the *Polista fedele*, a work on the compatibility of corpuscularism with the Catholic faith, whose appearance would have raised the ire of the traditionalists (Gómez López 2011: 231). Yet, in an anonymous text, Redi was quick to declare that "the truly natural principles" of the sensible world are atoms or corpuscles.[18]

Despite these endorsements, corpuscularism was far from being universally accepted, undisputed, or uncontroversial in late seventeenth-century Italy. It was a distinctive view of natural philosophical *novatores*. Traditionalist Aristotelian philosophers and Church authorities rejected it on several grounds. Their most vocal objections were theological, especially those concerning the incompatibility of corpuscularism with the dogma of transubstantiation (see e.g., Borrelli 1995b: 13–50). These objections were the ground for charges of atheism and smear campaigns. In Naples, these were followed by trials. More broadly, traditionalist Aristotelians perceived corpuscularism as subverting the entire edifice of *scientia* (Torrini 1979b: 18–20), along with the positions of cultural and political power to which they saw Aristotelianism as being subservient. As the Aristotelian Giovanni Maffei (1995 [1670]: 1327) candidly stated, Aristotelianism was "more useful than any other [doctrine] to the attainment of those ends to which the monarchs of the Earth aspire."

Although corpuscularism was a controversial natural philosophical view, the authors who defended it most vocally, Francesco D'Andrea (1995 [1685]) and Giuseppe Valletta (1975 [1691–1697]), were not primarily natural philosophers but lawyers. If we look at the most significant contributions to natural philosophy and medicine that were published by Italian authors— including Lorenzo Bellini, Domenico Guglielmini, Marcello Malpighi, Alessandro Marchetti, Geminiano Montanari, and Francesco Redi—we can easily identify corpuscularist assumptions underlying specific arguments or entire theories (see e.g., Baldini 1977: 11–12). Yet, none of those authors published any explicit, extended development of a corpuscularist matter theory or a defense of corpuscularist principles.[19]

This might be explained in two ways. In the first place, one might note that, in the light of Galileo's condemnation and given the hostility of church authorities, the most effective strategy for spreading corpuscularism was not to publish explicit defenses of its principles or replies to the attacks of the Aristotelians. It was to publish descriptions and explanations of specific natural phenomena that presupposed corpuscularist principles, sometimes even mentioned them in passing, but did not emphasize them (Vasoli 1979: 205–206). When the Aristotelians—even the most progressive ones— engaged with the *novatores* on specific empirical questions, unprejudiced readers could often see that the empirical evidence weighted on the side of the *novatores*.[20] These could hope that an increasing acceptance of their explicit empirical results would pave the way for the acceptance of their implicit corpuscularism. In the second place, one might claim that the most prominent natural-philosophical *novatores* did not provide any explicit defense of corpuscularist principles because they were incompatible with the adherence to experimental philosophy that was key to their successes. In what follows, we will see that Guglielmini provides a counterexample to the latter claim.

Before we turn to Guglielmini's views, it is worth acknowledging that not all corpuscularists were adherents of experimental philosophy. Tommaso Cornelio held that natural philosophers should not start by carrying out experiments and observations but by formulating hypotheses and axioms (Cornelio 1688: 78–81). Rossetti held that, before turning to experiments and observations, natural philosophers should develop a theory of nature as a whole (Rossetti 1669: 11). Giovanni Alfonso Borelli, a key member of the Accademia del Cimento and the author of a seminal work on biomechanics, relied on *a priori* arguments to show that certain animals cannot move in given ways because they are not sufficiently simple, economical, or conducive to the achievement of natural purposes (Borelli 1680–1681, 1: 266–267). Unlike Cornelio, Rossetti, and Borelli, several corpuscularists openly endorsed experimental philosophy. Among them is Domenico Guglielmini.

Guglielmini's *Philosophical Reflections Derived from the Figures of Salts*

Domenico Guglielmini was a Bolognese natural philosopher and physician whose most notable contributions lie in the fields of crystallography (*Reflections* and Guglielmini 1719c [1705]) and fluid mechanics (Guglielmini 1697; see Maffioli 2002). He presents his *Reflections* as a discourse that was read in Bologna, at a meeting of a "philosophical-experimental academy"[21] in the tradition of the Accademia del Cimento and the Royal Society. Despite their brevity, the *Reflections* is one of the most significant seventeenth-century works on crystals.[22] It reflects his practice with the procedure for obtaining salt crystals,[23] his familiarity with the views of the Aristotelians, Descartes (whom he often criticizes[24]), "the most famous Boyle,"[25] and others on what determines the shape of crystals, and his knowledge of Antoni van Leeuwenhoek's (1687: 119–148) microscopic observations of Cyprus vitriol and tartrate floating in water, published only one year before the *Reflections*.

In 1669, Nicholas Steno had highlighted the difference between the regular figures of crystals and the irregular figures of petrified living things. He proposed that quartz crystals "grow through the deposit of layers parallel to their surfaces" (Gohau 2002: 835). He also spelt out, with more precision than his predecessors, what the uniformity of their shapes amounts to. Their angles have always the same measure, whereas the relative length of their facets can change.[26] With this proposal, Steno anticipated the law of constancy of interfacial angles, that Jean-Baptiste Romé de l'Isle would generalize and confirm in 1783.[27]

Guglielmini's *Reflections* addresses the issue of what determines the constancy of interfacial angles. Aristotelians could explain it by appealing to substantial forms. Boyle had denied that it is necessary to appeal to substantial forms to account for the figures of "Allom, Vitriol, and other Salts, that are so curiously and Geometrically shap'd." "[T]hese bodies themselves may receive their shape from the coalition of such singly invisible corpuscles" and from the way in which they "are determin'd to stick together" (*Origin and Virtues of Gems*, Boyle 1999–2000, 7: 29).[28]

Boyle's comments on the figures of the corpuscles of specific crystals were characteristically cautious (Burke 1966: 32), unlike Descartes's and Hooke's. In the *Meteorology*, Descartes had claimed that, "since we observe salt grains to be square, they must be made up of oblong shaped particles arranged side by side, to form a square" (Hattab 2011: 73).[29] As Helen Hattab notes, "[t]his is fairly typical of Cartesian explanations," many of which are "hasty inferences from observed effects to the supposed geometrical properties and arrangements of unobservable material particles."[30] Unlike Descartes, Hooke (1665: 85–86) supposed that the particles of all crystals may be spherical and that the combination of spheres of different sizes determines the variation in the shapes of crystals.

In the *Reflections* (18), Guglielmini extends the constancy of interfacial angles from macroscopic crystals to their smallest constituent corpuscles, that is *minima*. Guglielmini argues that the *minima* of salts have the same interfacial angles as the crystals that they compose. The *minima* of common salt are cubes, those of vitriol rhombohedra, those of nitre hexagonal prisms, and those of alum tetrahedra.[31]

This claim provides the basis for Guglielmini's explanation of why all instances of the same crystal have the same interfacial angles. This is not due to their substantial forms, nor does it depend on which acids can be found in the crystals (*Reflections*: 20–21). It is due to the fact that the visible instances of any given salt are combinations of *minima* which are tightly stacked together and which have the same interfacial angles of the salt that they compose. The *Reflections* falls squarely within the tradition of corpuscular philosophy: they argue for a claim concerning corpuscles, and they employ it to account for the properties of macroscopic objects.

Four Methods

Guglielmini prefaces his corpuscularist arguments with a discussion of four natural philosophical methods and a clear-cut endorsement of the method of experimental philosophy. He begins by criticizing traditional philosophers. Instead of starting the study of nature by observing specific natural phenomena, they endorse certain general principles and derive propositions on specific phenomena from them. Yet, experience shows that their conclusions are false. This is because they rely on principles whose truth is dubious (*Reflections*: 3).

Hypothetical philosophers seek to avoid the error of traditional philosophers by starting their inquiries from experience.[32] Having observed certain phenomena,

> they formulate a hypothesis on the constitution and nature of principles, which is suitable to explain [*rendere ragione*] effects that are ordinarily observed. They claim [*pretendono*] that the agreement of the supposed principles with observations is a sufficient demonstrative proof of the hypothesis.
>
> (*Reflections*: 4)

Their assumption is mistaken because alternative hypotheses can explain the same empirical facts equally well. Guglielmini shows this by using the familiar example of alternative astronomical systems. If they are to provide persuasive explanations, philosophers should not rush to devise hypotheses for any given phenomenon. They should first gather extensive observations. Although Guglielmini does not mention any hypothetical philosophers, Descartes's argument on the figure of salt provides a good example of their way of proceeding, and experimental philosophers had often contrasted Descartes's premature reliance on hypotheses with their reliance on experience. As a consequence, Guglielmini's readers could hardly fail to read the passage as a criticism of Descartes and his disciples.[33]

Superficial philosophers make a mistake opposite to that of hypothetical philosophers. They gather a large amount of observations of natural phenomena, but they refrain from identifying their causes. Although their efforts are commendable, they are not authentic natural philosophers, because natural philosophy is a search for the causes of phenomena (*Reflections*: 6–7). Maurizio Mamiani (1987: 248) holds that superficial philosophy "is certainly the method of natural history, with an old tradition, recently renewed by Boyle." Yet, far from turning his back on causal inquiries, Boyle conceived of experimental natural history as a preliminary to the search for causes. Guglielmini was certainly aware of this. He is more likely to have identified superficial philosophers with ancient and Renaissance natural historians, like Ulisse Aldrovandi, or with the superficial style of natural philosophical inquiry of Francesco Redi and his disciples.

According to Guglielmini (*Reflections*: 8), natural philosophers should collect a large number of "replicated and well-regulated" experiments and organize them in a "natural history," that will provide "a solid and necessary foundation" for identifying the "causes" of "nature's operations" (*Reflections*: 7–8).[34] This method "provided the opportunity for the establishment of many famous academies" (*Reflections*: 7) in England, Italy, and elsewhere. It is the method of experimental philosophy, as it was described at beginning of this chapter. Guglielmini states that it is the method of his exposition in the *Reflections*.[35] This indicates that he views the *Reflections* as a work that conforms to the dictates of experimental philosophy.

Experimental Philosophy and Corpuscularism in the *Reflections*

To see if the *Reflections* really conforms to the dictates of experimental philosophy, we should focus on two issues: whether the organization of the work is in line with the methodological precepts of experimental philosophy, as Guglielmini states, and whether its contents conform to the epistemic strictures of this movement. As for the organization of the *Reflections*, it does not contain a proper experimental natural history of salts. However, they do have a two-part structure which broadly conforms to the preferred methodology of experimental philosophers: first, they provide a description of given phenomena, and then they put forward explanations. After the methodological preface (*Reflections*: 1–9), Guglielmini devotes several pages to the description of the figures of five types of salts (*Reflections*: 9–17). This is followed by an explanatory section that extends the constancy of interfacial angles to corpuscles, relies on it to account for the figures of salts, and refutes alternative accounts (*Reflections*: 17–33).

Regarding its contents, the *Reflections* will conform to the epistemic strictures of experimental philosophy if they adhere to the prescription that any firm commitment to substantive claims or theories must be justified by experiments and observations. Guglielmini establishes his claim on the figures of corpuscles both positively, by providing three arguments for it, and negatively, by responding to objections that may be raised against his view and refuting alternative accounts of the figures of salts.

The first argument establishes its conclusion by means of an inference from a feature of visible crystals to a feature of the *minima*. Guglielmini invites his readers to observe

> that all visible crystals of the same salt, whether big or small, have the same figure, so that the arrangement of their parts is independent from the greater or smaller quantity of their matter; indeed, one can observe that the efflorescences of nitre on walls . . . are arranged in very subtle rows which have the same figure that is displayed by its crystals.

He then infers that

> the salts that our senses cannot perceive, too, will have the same figure . . . by applying the same reasoning to the smallest parts, we will know that the ultimate parts of matter, that is, those that no natural agent can divide into smaller particles, have a given figure.
>
> (*Reflections*: 18)[36]

The reasoning that underpins this inference is an analogical reasoning. This raises the question of whether Guglielmini holds that analogical reasoning presupposes *a priori* principles. Analogical reasoning is a close relative of induction and one of Guglielmini's correspondents, Gottfried Wilhelm Leibniz, held that induction can only be warranted by substantive *a priori* principles.[37] If Guglielmini held that analogical reasoning presupposes any such principles, his first argument on the *minima* of salts would rely not only on observations and experiments, but also on substantive *a priori* truths. If so, one might worry that Guglielmini's reliance on *a priori* truths is in contrast with the professed reliance of experimental philosophy on experiments and observations alone.

In response, it should be noted that there is no evidence for the view that Guglielmini took analogical reasoning to rely on *a priori* assumptions. As far as I am aware, neither Guglielmini, nor those of his peers who wrote extended methodological discussions, like Giorgio Baglivi and Marcello Malpighi, discuss whether the foundations of analogical reasoning are empirical or non-empirical.

The comments of Giuseppe Antonio Barbari and Giorgio Baglivi on analogical reasoning are telling in this regard. Barbari is a little-known author who, like Guglielmini, studied in Bologna under Montanari and who published a treatise on vision. Its preface discusses analogical inferences from the macroscopic domain to the sub-microscopic domain in some detail. Barbari (1678: vii–xi) discusses their degree of reliability, their potential pitfalls, and their psychological basis, which he takes to be innate. Yet, he does not even raise the question of whether analogical inferences presuppose non-empirical assumptions. The same holds for Baglivi, who devotes an entire chapter of *The Practice of Physick* (1696: Bk. I, Ch. 6) to analogical reasoning. He praises its usefulness, sets limits to its employment, and defends the legitimacy of the analogies employed by mechanist philosophers and physicians. Yet, he does not discuss their empirical or non-empirical basis. As far as I know, this issue was not on the table for seventeenth-century Italian *novatores*. Nor is there any reason to believe that Guglielmini took his recourse to analogy to presuppose non-empirical truths.

Guglielmini's employment of analogical reasoning to establish a claim about corpuscles might raise another worry. This is whether Chalmers's objections against Boyle's analogical arguments concerning corpuscles apply to Guglielmini's argument. Chalmers notes that one of Boyle's arguments applies the law of fall to corpuscles and ascribes weight to them. Yet, corpuscles have only fundamental, mechanical qualities or affections. Boyle does not include weight among them, nor does he explain how weight might derive from the properties or collisions of corpuscles (Chalmers 1993: 549). Boyle's other arguments explain features of bodies by means of analogies between the behavior of the corpuscles that compose them and that of strands of fleece, clocks, or watches. Yet, Boyle fails to account for "the elasticity" of the fleece and "the spring and the rigidity of the gear wheels" in terms of "the shapes, size, motions and arrangement of corpuscles" (Chalmers 1993: 550).

Chalmer's objections against Boyle do not apply to Guglielmini. This is because, unlike Boyle's analogical arguments, Guglielmini's argument does not ascribe to corpuscles properties like gravity, weight, rigidity, and elasticity, which he fails to explain in terms of the fundamental properties of corpuscles. The only properties that Guglielmini's argument ascribes to corpuscles are shapes and figures. These were routinely included in early modern lists of basic, primary qualities, including Guglielmini's own list. In his view, shape and size are the only fundamental, intrinsic qualities of all material bodies (Guglielmini 1719a: 466, 468).

While Guglielmini's first argument does not raise concerns, the second argument is more problematic:

> When we separate some salt from water, the parts of salt are ordered in such a way that they form an exquisite figure. What can we imagine that could bring about such figures? Nothing else than the inclination of the planes of their smallest parts. Since they all have the same inclinations, as they gradually and orderly join each other, the size will grow, but the figure will not change.
>
> (*Reflections*: 18–19)

This argument is problematic because, rather than resembling the argumentative style of experimental philosophers, it recalls the all-too-quick flight of hypothetical philosophers from experience to theories. Guglielmini notes an empirical fact. He sketches a corpuscular story that accounts for it. He claims that the story provides the only explanation for the fact. Instead of pausing to justify this claim, he takes it as established that the explanation must be accepted.[38]

It is hard to believe that Guglielmini could have given much weight to this argument. This is because he was well aware that his corpuscular story does not provide the only explanation for that empirical fact. Just one page later (*Reflections*: 20–21), he discusses and then refutes an alternative explanation, according to which the shape of salts is due to "the spirit, or volatile acid that predominates in the salts" (*Reflections*: 20). In the light of this, the second argument is best seen as a brief rhetorical parenthesis between the first and third arguments. These carry the real argumentative weight of Guglielmini's view. They aim to establish that his explanation is not the only possible explanation of the facts but the best and most probable explanation. Like many of his Italian peers, Guglielmini holds that natural philosophical theories can attain only probability, not certainty.[39]

The experiences on which his third argument relies are the microscopic observations reported by Leeuwenhoek in his *Anatomia seu interiora rerum* (1687: 122–126). Guglielmini states that Leeuwenhoek saw the fact mentioned at the beginning of the second argument. Specifically, he saw that particles of vitriol and tartrate floating in water have the same interfacial angles of their macroscopic conglomerates (*Reflections*: 19). Leeuwenhoek also reported that he saw the particles of salts increasing in size, while maintaining the same figure. This provides further evidence for the scale invariance of the figures of salts on which the first argument relies.

Experimental philosophers stressed the importance of first-person experience and the necessity of verifying the testimony of others whenever possible. In the light of this, it may seem surprising that Guglielmini appeals to Leeuwenhoek's testimony, instead of providing first-person reports of those observations. Guglielmini was skilled in the use of the microscope. He had learned it from his teachers Geminiano Montanari and Marcello Malpighi, both accomplished microscopists.[40] Presumably, he too observed the crystals of tartrate floating in water under a microscope, as Leeuwenhoek had done.

Two remarks help explain Guglielmini's reference to Leeuwenhoek's observations. In the first place, as Steven Shapin (1994) has stressed, experimental philosophers relied on a significant extent on "borrow'd Observation[s]" of "Authors not to be distrusted" (*Certain Physiological Essays*, Boyle 1999–2000, 2: 190). Boyle approves the reliance of experimental philosophers on the testimony of "Shepherds, Plowmen, Smiths, Fowlers, &c.," who "are conversant with the Works of Nature" (*Christian Virtuoso, I*, Boyle 1999–2000, 11: 313), and the reliance of "the most rational physicians" on the testimony of their patients and earlier physicians (Boyle 1999–2000, 11: 308).[41] In the eyes of Guglielmini and his peers, Leeuwenhoek was a trustworthy source of information. Although, occasionally, accomplished experimentalists reacted with caution to some of Leeuwenhoek's observations,[42] by 1688 he had established a strong reputation. He had published no fewer than twenty-seven articles in the *Philosophical Transactions*[43] and two volumes in Latin, and he had been elected Fellow of the Royal Society. In the light of this, even though Guglielmini probably replicated Leeuwenhoek's observations, noting that Leeuwenhoek had carried them out might have helped lend plausibility to them.

In the second place, Guglielmini's use of Leeuwenhoek's observations to establish a conclusion concerning *minima* was in line with the assumptions and expectations of many *novatores*. Bacon had related the use of the microscope to the vision of *minima* as early as 1620. He stated that, "if he [Democritus] had seen" a microscope, he "would have been overjoyed and would have thought that a means of seeing atoms . . . had been discovered" (*New Organon*, II, §39, Bacon 2004: 343–345). In 1664, Henry Power claimed that microscopes allow us to see "the very Atoms and their reputed Indivisibles and least realities of Matter" (Power 1664: Preface, sig. b2).[44] Newton's *Opticks*, first published in 1704, sixteen years after Guglielmini's *Reflections*, states that microscopes will allow us to see at least the largest particles, perhaps even most of them (Newton 1730: 236–237). Giuseppe Gazola, a promoter of experimental philosophy who, like Guglielmini, studied with Montanari, states in a posthumous discourse that microscopes enable "modern physicians" to see "the figure of the smallest [*menome*] particles that make up compound bodies" (Gazola 1716: 172–173). Guglielmini does not state that the particles seen by Leeuwenhoek are themselves *minima*.[45] However, his appeal to microscopic observations to defend a conclusion about *minima* is in line with the view, widely shared by *novatores*, that there is no radical discontinuity between the *minima* and those corpuscles that can be observed with the microscope.

Guglielmini's reference to Leeuwenhoek's observations is especially interesting because some of them may appear to disprove Guglielmini's views. Guglielmini holds that the *minima* of alum are square pyramids with adjacent bases, so as to form octahedra. In his later, systematic treatise on crystals (Guglielmini 1719c: §xxi), Guglielmini refers to Leeuwenhoek's observations in order to establish this claim. Yet, as the *Reflections* notes in passing and his systematic treatise explains (Guglielmini 1719c: §xxiii), Leeuwenhoek describes "the figures of alum as being mostly hexagonal" rather than octahedral. Guglielmini confirms that, "[u]sing the microscope, one can really see them as having a hexagonal, sometimes even pentagonal shape."

Guglielmini explains the apparent hexagonal or pentagonal shape of alum by noting that, in the salts observed with a microscope, "the distance between the opposed sides is minimal" and their "transparency, even if it may be low, cancels completely the effects of distance itself." As a result, what is actually an octahedron can appear as a hexagon or a pentagon. Consider two adjacent *minima* of alum, which form an octahedron. If its depth is not perceived, then, depending on the observer's position, it may appear as a square, a rhombus or a hexagon, as can be seen from Figure 5.1. A single *minimum* of alum, which has the shape of a square pyramid, may appear as a square, a triangle, or a pentagon, as can be seen from Figure 5.2.

Figure 5.1 Octahedra: The three figures on the top row are octahedra in different positions. If their depth is not perceived, they may appear like the square, hexagon, and rhombus on the bottom row.

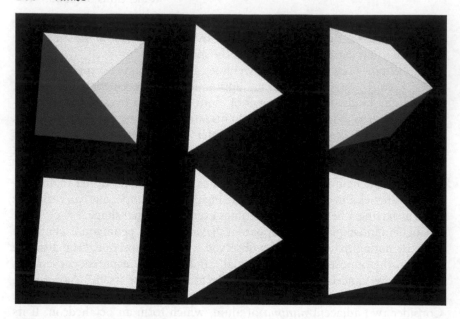

Figure 5.2 Tetrahedra: The three figures on the top row are tetrahedra in different positions. If their depth is not perceived, they may appear like the square, triangle, and rhombus on the bottom row.

Guglielmini's explanation of the observations by Leeuwenhoek that appear to disprove his view exemplifies the kind of "genuine empirical support" for corpuscularism that, according to Alan Chalmers, Boyle never provided. Chalmers writes,

> It is conceivable that genuine empirical support for the corpuscular hypothesis could be arrived at by (*a*) appealing to some phenomena to determine something about the shapes, sizes and motions of corpuscles and then (*b*) to employ those characteristics to explain or predict other phenomena, but I am not aware that Boyle achieved any successes of that kind.
>
> (Chalmers 1993: 553, letters added)

Guglielmini (*a*) appeals to phenomena concerning the macroscopic crystals of salts to determine the figures of their corpuscles. In particular, he determines that the corpuscles of alum, when combined two by two, have an octahedral figure. Leeuwenhoek observed, to employ Chalmers's words, "other phenomena" (the apparent hexagonal figures of alum) that Guglielmini's theory does not account for. Guglielmini (*b*) explains those phenomena by combining his claim that paired corpuscles of alum have an octahedral figure with the remark that microscopic observations of salts do not reveal their depth.

This pattern of argument conforms to Bacon's recommendation to ascend and then descend "a double scale or ladder":[46] "to extract . . . from works and experiments causes and axioms, and in turn from causes and axioms new works and experiments" (*New Organon*, I, §117, Bacon 2004: 175). Guglielmini ascends from phenomena concerning salts to their material cause, which is the figure of their corpuscles, and then descends from that cause to the explanation of Leeuwenhoek's observations.[47]

Having made a positive case for his claim on the figures of corpuscles, Guglielmini replies to objections and criticizes alternative accounts of what determines the figure of given salts. He thoroughly discusses a series of objections concerning alum. Some of them are empirical, whereas others are *a priori*. Guglielmini does not hesitate to provide *a priori* replies to the *a priori* objections, but he combines them with empirical remarks. Consider, for instance, the following objection. Visible crystals of alum appear to have the shape of an octahedron. Yet, it is hard to see how one could form an octahedron by combining smaller octahedra (*Reflections*: 22–23). This is an *a priori*, geometrical worry. Guglielmini replies that an octahedron can derive from the combination of two tetrahedra (square pyramids) with adjacent bases. He proposes that the *minima* of alum are not octahedra but tetrahedra and that each octahedral crystal of alum is made up of twelve tetrahedra. He explains how they must be arranged to form an octahedron.

This reply may recall Descartes's "hasty inferences" from given *explananda* "to the supposed geometrical properties and arrangements of unobservable material particles" (Hattab 2011: 73). Yet, in contrast to Descartes, Guglielmini rushes to back up his proposal that the crystals of alum are tetrahedral with observations. For instance, he observes that

in the crystals formed by solutions of tartrate, mixed with alum, one can see that the figures of alum are composed by other, similar figures, for one can see that the surface of one of the superficial triangles [i.e. tetrahedra] is composed by many other triangles [tetrahedra] of the same nature, even though, sometimes, one of those triangles protrudes a bit on the outside. This happens because the matter of tartrate entered between those parts of alum and, when it hardened, it separated them from one another.

(*Reflections*: 23–24)

Guglielmini's recourse to a geometrical model to explain the disposition of salt crystals is consistent with the tenets of experimental philosophy, even if geometrical reasoning may be seen as a form of *a priori* reasoning. This is because experimental philosophers were not averse to *a priori* reasoning as such, but to non-empirical justifications of claims on the natural world. Guglielmini establishes the correctness of his geometrical model by means of observations. He then discusses three objections. He answers each of them with arguments based on his experiments and observations (*Reflections*:

24–30). This way of proceeding is in line with his later claim that "the number and figure" of the angles of salt particles cannot be "contrived in one's mind or established *a priori*" but must be derived "from experiments and observations" (Guglielmini 1719c: §xv).

In sum, Guglielmini's *Reflections* is a corpuscularist treatise whose methodological preface explicitly endorses experimental philosophy. Guglielmini does not endorse experimental philosophy to conceal or downplay his corpuscularist commitments. He does not portray experimental philosophy as being merely descriptive, uninterested in the causes of phenomena, or independent from theoretical claims on the existence and the properties of corpuscles. On the contrary, the *Reflections* is explicitly devoted to establishing a claim about the figures of the most basic corpuscles and constituents of salts, their *minima*.[48] This claim is not relegated to a speculative section of the work, as distinct from an experimental or natural historical section. On the contrary, the structure of Guglielmini's corpuscularist treatise conforms to the two-step method that experimental philosophers favoured. With the exception of his second argument, which plays a merely rhetorical role, Guglielmini's positive arguments on the figures of corpuscles and his replies to objections conform to the *desideratum* of experimental philosophers that any substantive claim on the natural world be based on experiments or observations. Even his response to an *a priori*, geometrical objection is not limited to *a priori* considerations, but is backed up by observations. Guglielmini's *Reflections* shows that corpuscularist theories can be in line with the methodological prescriptions, epistemological strictures, and preferred argumentative styles of experimental philosophers.

Notes

1 I would like to thank Tom Sorell and the participants in the symposium "Principles in Early Modern Thought" for their helpful comments. This work was supported by the Arts and Humanities Research Council [grant number AH/L014998/1].

2 References are to the first edition of 1688. The *Reflections* was re-edited, along with a Latin translation, as part of Guglielmini's *Collected Works* (Guglielmini 1719d, 1: 65–104).

3 Boyle's views were known in Italy from the early 1660s (Pighetti 1988: 20). By the 1680s, when Guglielmini's *Reflections* was published, Italian scholars could rely on the English translations of numerous works by Boyle, including a ten-volume edition (Boyle 1677).

4 See Henry Oldenburg's introductory comments on Rooke 1665: 140.

5 As Laudan (1981: 34) has noted, in the seventeenth century "Bacon was not praised (or condemned) as an inductive philosopher so much as an experimental one."

6 It is clear, however, that what they called induction or deduction is not a matter of replacing "some" with "all" in given sentences but a matter of identifying a hypothesis that entails the evidence. It is not enumerative induction but hypothetical induction. For this distinction, see Norton 2005. For an overview of the history of experimental philosophy, see Anstey and Vanzo 2016.

7 This characterization combines (*a*) a commitment to corpuscles with (*b*) a commitment to explaining phenomena in terms of matter moving according to the laws of nature. One can endorse (*a*) without endorsing (*b*), and vice versa. Among the authors working in seventeenth-century Italy, Claude Berigard and Johann Chrysostom Magnenus endorsed (*a*) but not (*b*). It should also be noted that (*b*) expresses a commitment to a specific kind of mechanism. This is distinct from the commitment to explaining natural phenomena by analogy with the functioning of machines (which can, in turn, be understood in various ways; see Roux 2011).

8 In an earlier study, Maria Laura Soppelsa (1974: 132) stated that Montanari's empirical research "appears to rule out a philosophical reflection that may attain knowledge of the first principles."

9 Chalmers (2002: 197) claims that Boyle's experimental successes "owed nothing to his allegiance to the mechanical philosophy in the strict sense [i.e. corpuscularism as characterized in this paper] and offered no support to that philosophy." For discussion, see Anstey 2002 and Pyle 2002. In this chapter, I do not discuss whether this is true for Guglielmini's experimental successes, for example in river hydraulics.

10 Anstey and Hunter 2008: 96. According to Anstey 2011: 4–5, corpuscularism was "legitimate [. . .] in the eyes of experimental philosophers" as a "generic hypothesis," which was neutral on the question of the divisibility of matter." Yet, being a "speculative hypothesis," it belonged to the domain of speculative philosophy. On the experimental/speculative distinction in seventeenth-century England, see Anstey 2005.

11 See for example Boschiero 2007: 190. Boschiero 2009 also argues that Montanari presented his work on capillaries in an experimentalist fashion to give the reader the impression that his "actual experimental procedures were not corrupted by speculations about matter theory and physical causes injected into the experimental research process." Yet, the "process of knowledge-making" underlying that work was related to Montanari's "concerns as a mechanical and corpuscular natural philosopher" (Boschiero 2009: 204, 205, 207).

12 Arguably, the same holds for Montanari, who was one of Guglielmini's teachers. See Vanzo 2016: 58–59.

13 Magalotti 1667: Proemio a lettori, sig. +2 4. In spite of this prefatory disclaimer, "any unbiased reader" could see that many of the Cimento's experiments had "destructive implications" for "central Aristotelian doctrines" on the causes and principles of natural phenomena (Galluzzi 1981: 804).

14 For instance, according to Caramuel y Lobkovitz 1670, 1: 712, the Investiganti academicians did not stop at performing experiments but searched for their reasons (*rationes*) and discussed whether they confirm or destroy the Peripatetic philosophy. On the importance of searching for causes in natural philosophy and medicine, see, respectively, *Reflections*: 6–7; Di Capua 1681: 508.

15 See, for example, Cornelio 1688: 96–100 on matter and motion as principles and 112 on corpuscles as components of matter; D'Andrea's 1995 [1685] defence of corpuscularism and his claim that "it is unnecessary to introduce any other principles within nature than matter and motion" (D'Andrea 1995 [1673–1675?]: 151); Bianchini 1785 [c.1687]. 7–8 on body and motion as "mechanical principles," and 14 on corpuscles as components of body.

16 Valletta 1975 [1691–1697]: 49. Valletta calls them atoms and uses *atom* as a synonym of *corpuscle*. The quoted passage ascribes this view of principles to the "Platonic, Democritic, and Epicurean Philosophy," endorsed by Valletta.

17 Montanari's student, Guglielmini, does not identify "the first principles of all natural things" with corpuscles but with their figure, size, and motion. See

Guglielmini 1719b: 463. Among physicians, Baglivi calls "figure and motion" a "general principle that is common to all things and most evident" (Baglivi 1696: Book I, Ch. xii, §9, p. 117). The Tuscan physician Pirro Maria Gabbrielli, who adapted the philosophy of "Democritus and Epicurus" to "the modern custom, which is based solely on experiments," identified the "principles of all things" with "seeds [*semi*]," that is corpuscles (Crescimbeni 1708: 201–202).

18 Anon. 1698. On the authorship of this text, see Bernardi online, http://www. francescoredi.it/Database/redi/redi.nsf/b4604a8b566ce010c125684d00471e00/ db89f5de884dcab0c12569fa005f0a14 (archived at http://www.webcitation. org/6VM1oTYtQ), http://www.francescoredi.it/Database/redi/redi.nsf/pagine/ 126DDC04C8A0620EC12569F300511AF2 (archived at http://www.webcita tion.org/6VM1torcv).

19 A significant exception is Donato Rossetti, who developed and published a rather detailed corpuscularist theory. He "admitted defeat" in the bitter dispute on corpuscularism that took place in Pisa in the early 1670s by leaving the city and taking up a position in Turin as court mathematician (Bernardi online, http:// www.francescoredi.it/Database/redi/redi.nsf/b4604a8b566ce010c125684d004 71e00/805e5cf3a5f06a1cc12569f4003d1f7b, archived at http://www.webcita tion.org/6YBK2drVg). Guglielmini developed a corpuscularist matter theory in two brief essays 1719a, 1719b, but he did not publish them.

20 As Torrini 1979b: 25 noted, this was the case, for instance, for the controversy between Francesco Redi and the Jesuit Filippo Buonanni concerning spontaneous generation.

21 *Reflections*, frontispiece. On this academy, see Cavazza 1990: 51–56.

22 A more detailed development of their themes can be found in Guglielmini 1719c. For Guglielmini's definition of salt, see his 1719c: §iii–v.

23 He obtained them by incinerating or calcining a substance containing salts, boiling the ash or calx in water, filtering it repeatedly with felt, and making it evaporate slowly until, after a few days, crystals appear on the bottom and the side of the container. If the crystals are dissolved in water, the process can be repeated multiple times. Crystals appear every time, and their interfacial angles have always had the same measure. See *Reflections*: 10–11.

24 See for example *Reflections*: 11 and Guglielmini 1719c: 76 on the figure of the particles of nitre; Guglielmini 1719c: 74 on Descartes's doctrine of the three elements; Guglielmini 1719c: §vii on the infinite divisibility of matter; Guglielmini 1719a: 468 on space. By the 1680s, knowledge of Descartes's natural philosophy was common among Italian *novatores*. Cartesianism had been influential in Naples since the 1640s (see, e.g., Cornelio's letter *De cognatione aeris et aquae*, written in 1649, in Cornelio 1688: 387). Montanari read Descartes's natural philosophical works in 1657–1658 (Rotta 1971: 153n97). Redi asked a correspondent to purchase all of Descartes's works for him in 1665 (Bernardi online, http://www.francescoredi.it/Database/redi/redi.nsf/b4604a8b566ce010c12568 4d00471e00/51931bcb175458c6c12569fb0051b3d1, archived at http://www. webcitation.org/6VOGOCvac).

25 *Reflections*: 28.

26 Steno 1669: "Explicatio Figurarum," trans. in Steno 1916: 272.

27 The law, in Romé de L'Isle's formulation, states that the faces of crystals of the same species "can vary in their shape and in their relative dimensions, but the respective inclination of the same faces is constant and invariable in each species" (Maitte 2013: 6).

28 See Boyle 1999–2000, 5: 368. Eighteenth-century authors would put forward several other views on what determined the figures of crystals. According to William Homberg, "the figures belonged to the alkalis rather than to the acids."

Which figure a crystal took depended on which alkali "it had crystallized" (Burke 1966: 25). For Johann G. Wallerius, "salt itself possessed no crystalline figure before it combined with something metallic" and "the figure of a mineral crystal was due to its metallic ingredients" (26). Torbern Bergman held that "the external configuration of salts depended upon the joint combination of acid and alkali" (27).

29 See *Meteores*, AT VI 256–260, trans. in Descartes 1965: 280–283; Hattab 2009: 126–135.

30 Hattab 2011: 73. This is not to deny that some of Descartes's observations were meticulous, or that he was aware of the different roles that experiments can play. Yet, Descartes's outlook on natural philosophical explanations accords secondary roles to experiments and observations. See Roux 2013: 52–54.

31 Guglielmini holds that the *minima* of alum are typically paired two by two, so as to give rise to octahedra. Guglielmini's *De Salibus* (1719c: §xvii) identifies four basic, simple crystals. This view is not present in the earlier *Reflections*.

32 This label, like the others in this section, is mine. Guglielmini calls the method of these philosophers "philosophizing through hypotheses" (*Reflections*: 4–5).

33 Mamiani 1987: 248 noted this.

34 As was customary for experimental philosophers, Guglielmini assigns to both experiments and observations the role of identifying the matters of fact that provide the basis for theorizing. In a posthumous treatise, Guglielmini explains that natural philosophers should aim to ascend gradually from immediate to more remote causes, until they have explained all properties of bodies in terms of their essential, fundamental properties, namely, shape and size. It does not pertain to natural philosophers to explain why bodies have these essential properties and not others. See Guglielmini 1719a: 467.

35 *Reflections*: 8. His later treatise on salt, instead, follows the "synthetic" method of exposition (1719c: 81).

36 Guglielmini defends the view that there are such ultimate parts of matter against Descartes in his 1719c: §vii. Like Boyle (*Forms and Qualities*, 1999–2000, 5: 325–326), Guglielmini (1719a: 468) holds that God can divide those particles.

37 See Leibniz 1966 [1670]: 432, trans. in Leibniz 1969: 130. For Guglielmini's correspondence with Leibniz, see Cavazza 1987.

38 This pattern of reasoning was explicitly endorsed by Descartes, to whom Guglielmini's criticism of hypothetical philosophers alludes. See Descartes's *Principles of Philosophy*, Part 4, §1, AT IXB 201.

39 Guglielmini 1697, "A' benigni lettori"; see, e.g., Bianchini 1785 [*ca.*1687]: 17; Borelli 1680–1681, 2: 57, 72–73; Di Capua 1681: 164–165; Montanari 1671: 15; Rossetti 1669: 12.

40 Montanari reported microscopic observations in his works (e.g., Montanari 1667: 12). He used to build microscopes and grind lenses. Malpighi employed the microscope systematically to observe animals and plants from the 1660s.

41 See also the example of Columbus at p. 214.

42 See, for example, Redi's reaction to Leeuwenhoek's observation of spermatozoa, as reported in Bernardi online, http://www.francescoredi.it/database/redi/redi.nsf/b4604a8b566ce010c125684d00471e00/fec5577faaaa81c0c12569fb0051cad8 (archived at http://www.webcitation.org/6VM1zGVFk).

43 See Anderson online, http://lensonleeuwenhoek.net/category/bibliography/Philosophical Transactions (archived at http://www.webcitation.org/6VM26YMtz).

44 Power is more cautious at p. 155.

45 A few decades later, Jacopo Riccati (1762: 542) would emphatically deny it.

46 Bacon 2000: 50. Guglielmini's Italian peers knew and employed this image. See, for example, Lana Terzi 1977 [1670]: 52.

47 The same argumentative pattern is found in Montanari's works. They often proceed from explanations to theories, derive predictions from those theories, and then confirm those predictions empirically (e.g., Montanari 1715 [1678]: 89). For a methodological statement, see Montanari 1980: 550.

48 There is a significant difference between Guglielmini and Boyle in this regard. Guglielmini, like the Paracelsians, appears to have held that the "first," most basic components or particles of salts are themselves particles of salts (*Reflections*: 30; 1719c: §§vii, viii, xviii). Boyle (*Producibleness of Chymical Principles*, 1999–2000, 9: 33), instead, held that "the first Saline Concretions that were produc'd by Nature" are "made of Atoms, or of Particles, that before their conjunction, were not Saline." Boyle (1999–2000, 2: 105–106) took his experiment on the redintegration of nitre to show this. While Guglielmini knew Boyle's essay on nitre, I do not know how he interpreted Boyle's experiment. On a more general level, Guglielmini's matter theory distinguishes between three layers, just like Boyle's: elements, which have only basic properties like shape and size, are indivisible by natural powers, and all are composed by the same matter; their compounds, that is, molecules; and macroscopic bodies (Guglielmini 1719a: 467, 468). On Boyle's multilayered theory and its relation to Paracelsus's theory, see Newman's chapter in this book.

Bibliography

Achinstein, P., ed. (2005) *Scientific Evidence: Philosophical Theories and Applications*, Baltimore: Johns Hopkins University Press.

Altieri Biagi, M. L. and Basile, B., eds. (1980) *Scienziati del Seicento*, Milan: Ricciardi.

Anderson, D. (online) *Lens on Leeuwenhoek*, http://lensonleeuwenhoek.net (Accessed: 17 June 2016).

Anon. (1698) "A chi leggerà" in Giovanni Michele Milani, *La luce*, Amsterdam: Starck.

Anstey, P. R. (2002) "Robert Boyle and the heuristic value of mechanism," *Studies in History and Philosophy of Science*, 33: 157–170.

———. (2005) "Experimental versus speculative natural philosophy" in eds. P. R. Anstey and J. A. Schuster 2005, pp. 215–242.

———. (2011) *John Locke and Natural Philosophy*, Oxford: Oxford University Press.

———. (2014) "Philosophy of experiment in early modern England: The case of Bacon, Boyle and Hooke," *Early Science and Medicine*, 19: 103–132.

Anstey, P. R. and Hunter, M. (2008) "Robert Boyle's 'Designe about Natural History'," *Early Science and Medicine*, 13: 83–126.

Anstey, P. R. and Schuster, J. A., eds. (2005) *The Science of Nature in the Seventeenth Century: Patterns of Change in Early Modern Natural Philosophy*, Dordrecht: Springer.

Anstey, P. R. and Vanzo, A. (2016) "Early modern experimental philosophy" in eds. J. Sytsma and W. Buckwalter 2016, pp. 87–102.

Bacon, F. (2000) *The Advancement of Learning*, ed. M. Kiernan, Oxford Francis Bacon, vol. 4, Oxford: Clarendon Press; 1st edn 1605.

———. (2004) *The* Instauratio magna *Part II*: Novum organum *and Associated Texts*, ed. G. Rees, Oxford Francis Bacon, vol. 11, Oxford: Clarendon Press; 1st edn 1620.

Baglivi, G. (1696) *De praxi medica ad priscam observandi rationem revocanda libri duo*, Rome; trans. as *The Practice of Physick*, London, 1704.

Baldini, U. (1977) "Il corpuscolarismo italiano del Seicento: Problemi di metodo e prospettive di ricerca" in *Ricerche sull'atomismo del Seicento: Atti del Convegno di studio di Santa Margherita Ligure (14–16 ottobre 1976)*, Florence: La Nuova Italia, pp. 1–76.

———. (1980) "L'attività scientifica nel primo Settecento" in ed. G. Micheli, *Storia d'Italia: Annali*, vol. 3: *Scienza e tecnica nella cultura e nella società dal Rinascimento a oggi*, Turin: Einaudi, pp. 465–529.

Barbari, G. A. (1678) *L'iride: Opera fisicomatematica* [. . .] *nella quale si espone la natura dell'Arco Celeste, e si commenta il testo oscurissimo d'Aristotele* De figura iridis *nel Terzo delle Meteore*, Bologna: Manolessi.

Barbin, E. and Pisano, R., eds. (2013) *The Dialectic Relation between Physics and Mathematics in the XIXth Century*, Dordrecht: Springer.

Bartoli, D. (1677) *La tensione e la pressione disputanti qual di loro sostenga l'argento vivo ne' cannelli dopo fattone il vuoto*, Bologna: Longhi, 1677.

Bernardi, W. (1996) "Introduzione" in Redi (1996 [1668]), pp. 5–60.

———. (online) *Francesco Redi: Scienziato e poeta alla Corte dei Medici*, http://www.francescoredi.it (Accessed: 17 June 2016).

Beretta, M., Clericuzio, A., and Principe, L. M., eds. (2009) *The Accademia del Cimento and its European Context*, Sagamore Beach: Science History Publications.

Bianchini, F. (1785 [*ca.*1687]) "Dissertazione di Monsignor Francesco Bianchini tolta da' sui MSS. lasciati alla Libreria del Capitolo di Verona: Da lui recitata nella radunanza dell'Accademia degli Aletofili in Verona" in *Nuova raccolta d'opuscoli scientifici, e filologici*, vol. 41, Venice: Occhi.

Biener, Z. and Schliesser, E., eds. (2014) *Newton and Empiricism*, New York: Oxford University Press.

Borelli, G. A. (1680–1681) *De motu animalium*, 2 vols, Rome: Bernabò; trans. as *On the Movement of Animals*, ed. P. Maquet, Dordrecht: Springer, 1989.

Borrelli, A. (1995a) *D'Andrea atomista: L' "Apologia" e altri inediti nella polemica filosofica della Napoli di fine Seicento*, Naples: Liguori.

———. (1995b) "D'Andrea atomista" in ed. Borrelli 1995a, pp. 1–50.

Boschiero, L. (2007) *Experiment and Natural Philosophy in Seventeenth-Century Tuscany*, Dordrecht: Springer.

———. (2009) "Networking and experimental rhetoric in Florence, Bologna and London during the 1660s" in eds. M. Beretta, A. Clericuzio, and L. M. Principe 2009, pp. 195–210.

Boyle, R. (1677) *Opera varia*, 10 vols, Geneva: Tournes.

———. (1999–2000) *The Works of Robert Boyle*, 14 vols, eds. M. Hunter and E. B. Davis, London: Pickering and Chatto.

Burke, J. G. (1966) *Origins of the Science of Crystals*, Berkeley: University of California Press.

Caramuel y Lobkovitz, J. (1670) *Mathesis biceps: Vetus, et nova*, Naples: Officina Episcopalis.

Casellato, S. and Sitran Rea, L., eds. (2002) *Professori e scienziati a Padova nel Settecento*, Treviso: Antilia.

Cavazza, M. (1987) "La corrispondenza inedita tra Leibniz, Domenico Guglielmini, Gabriele Manfredi," *Studi e memorie per la storia dell'Università di Bologna*, new series, 6, pp. 51–79.

———. (1990) *Settecento inquieto: Alle origini dell'Istituto delle Scienze di Bologna*, Bologna: Il Mulino.

———. (1998) review of S. Gómez López, *Le passioni degli atomi, Universitas: Newsletter of the International Centre for the History of Universities and Science* (Bologna), 11 (May 1998), http://www.cis.unibo.it/NewsLetter/111998Nw/cavazza.htm (Accessed: 17 June 2016), archived at http://www.webcitation.org/6VNibjk7x.

Chalmers, A. (1993) "The lack of excellency of Boyle's mechanical philosophy," *Studies in History and Philosophy of Science*, 24: 541–564.

———. (2002) "Experiment versus mechanical philosophy in the work of Robert Boyle: A reply to Anstey and Pyle," *Studies in History and Philosophy of Science*, 33: 191–197.

———. (2012) "Intermediate causes and explanations: The key to understanding the Scientific Revolution," *Studies in History and Philosophy of Science*, 43: 551–562.

Clarke, D. and Wilson, C., eds. (2011) *The Oxford Handbook of Philosophy in Early Modern Europe*, Oxford: Oxford University Press.

Cornelio, T. (1688) *Progymnasmata physica*, 2nd edn, Naples: Raillard.

Crescimbeni, G. M. (1708) *L'Arcadia*, Rome: Rossi.

D'Andrea, F. (1995 [1673–1675]) "Lezioni" in ed. Borrelli 1995, pp. 141–169.

———. (1995 [1685]) "Apologia in difesa degli atomisti" in ed. Borrelli 1995, pp. 59–109.

Delon, M., ed. (2002) *Encyclopedia of the Enlightenment*, London: Routledge.

Descartes, R. (1965) *Discourse on Method, Optics, Geometry, and Meteorology*, ed. P. J. Olscamp, Indianapolis: Bobbs-Merrill.

———. (1996) *Œuvres de Descartes*, eds. C. Adam and P. Tannery, 11 vols, Paris: Vrin.

Di Capua, L. (1681) *Parere: Divisato in otto ragionamenti, ne quali partitamente narrandosi l'origine, e 'l progresso della medicina, chiaramente l'incertezza della medesima si fa manifesta*, 2nd ed., Naples: Bulison.

Dobre, M. and Nyden, T., eds. (2013) *Cartesian Empiricisms*, Dordrecht: Springer.

Gaillard, A., Goffi, J. Y., Roukhomovsky, B., and Roux, S., eds. (2011) *L'automate: Modèle, métaphore, machine, merveille*, Bordeaux: Presses universitaires de Bordeaux.

Galluzzi, P. (1974) "Libertà scientifica, educazione, e ragione di stato in una polemica universitaria pisana del 1670" in *Atti del XXIV Congresso Nazionale di Filosofia: L'Aquila 28 aprile—2 maggio 1973*, vol. 2, Rome: Società Filosofica Italiana, pp. 404–412.

———. (1981) "L'Accademia del Cimento: 'Gusti' del principe, filosofia e ideologia dell'esperimento," *Quaderni Storici*, 48: 788–844.

———. (1995) "La scienza davanti alla Chiesa e al Principe in una polemica universitaria del secondo Seicento" in *Studi in onore di Arnaldo d'Addario*, ed. L. Borgia, vol. 4.1, Lecce: Conte, pp. 1325–31.

Gaukroger, S. (2006) *The Emergence of a Scientific Culture: Science and the Shaping of Modernity, 1210–1685*, Oxford: Clarendon Press.

———. (2014) "Empiricism as a development of experimental natural philosophy" in eds. Z. Biener and E. Schliesser 2014, pp. 15–38.

Gazola, G. (1716) *Il mondo ingannato da falsi medici e disingannato*, Prague: Mayer.

Glanvill, J. (1668) *Plus Ultra: Or, the Progress and Advancement of Knowledge since the Days of Aristotle*, London.

Gohau, G. (2002) "Mineralogy" in ed. M. Delon 2002, pp. 835–837.

Gómez López, S. (2011) "Dopo Borelli: La scuola galileiana a Pisa" in ed. L. Pepe 2011, pp. 223–232.

Guglielmini, D. (1688) *Riflessioni filosofiche dedotte dalle figure de' sali: Espresse in un discorso recitato nell'Accademia Filosofica Esperimentale di Monsign. Arcidiacono Marsigli la sera delli 21. Marzo 1688*, Bologna.

———. (1697) *Della natura de' fiumi Trattato Fisico-Matematico*, Bologna.

———. (1719a) "De origine, & proprietatibus primarum affectionum materiæ" in Guglielmini (1719d), vol. 2, pp. 464–469.

———. (1719b) "De primis materiæ affectionibus" in Guglielmini (1719d), vol. 2, pp. 457–463.

———. (1719c [1705]) "De salibus dissertatio epistolaris physico-medico-mechanica" in Guglielmini (1719d), vol. 2, pp. 73–200.

———. (1719d) *Opera omnia mathematica, hydraulica, medica, et physica*, 2 vols, Geneva: Cramer, Perachon & Socii.

Hattab, H. (2009) *Descartes on Forms and Mechanisms*, Cambridge: Cambridge University Press.

———. (2011) "The mechanical philosophy" in eds. D. Clarke and C. Wilson 2011, pp. 71–95.

Hooke, R. (1665) *Micrographia: Some Physiological Descriptions of Minute Bodies Made by Magnifying Glasses with Observations and Inquiries Thereupon*, London.

———. (1705) *The Posthumous Works of Robert Hooke*, London.

Lana Terzi, F. (1977 [1670]) *Prodromo dell'Arte Maestra*, ed. A. Battistini, Milan: Longanesi.

Laudan, L., (1981) "The clock metaphor and hypotheses: The impact of Descartes on English methodological thought, 1650–1670" in *Science and Hypothesis: Historical Essays on Scientific Methodology*, Dordrecht: Reidel, pp. 27–58.

Leeuwenhoek, A. van (1687) *Anatomia seu interiora rerum cum animatarum tum inanimatarum*, Leiden.

Leibniz, G. W. (1966 [1670]) "Marii Nizolii de veris principiis et vera ratione philosophandi contra pseudophilosophos, libri IV" in eds. Berlin-Brandeburgische Akademie der Wissenschaften and Akademie der Wissenschaften in Göttingen, *Sämtliche Schriften und Briefe* 1923–, 6th series, vol. 2, Berlin: Akademie, pp. 398–444; partial trans. as "Preface to an edition of Nizolius" in Leibniz 1969, pp. 121–130.

———. (1969) *Philosophical Papers and Letters*, ed. L. E. Loemker, 2nd edn, Dordrecht: D. Reidel.

Maffei, G. (1995 [1670]) "Relazione di Giovanni Maffei" in P. Galluzzi, "La scienza davanti alla Chiesa e al Principe in una polemica universitaria del secondo Seicento" in ed. L. Borgia, *Studi in onore di Arnaldo d'Addario*, Lecce: Conte, vol. 4.1, pp. 1325–1331.

Maffioli, C. F. (2002) "Domenico Guglielmini" in eds. S. Casellato and L. Sitran Rea 2002, pp. 505–530.

Magalotti, L. (1667) *Saggi di naturali esperienze fatte nell'Accademia del Cimento sotto la protezione del serenissimo principe Leopoldo di Toscana e descritte dal segretario di essa accademia*, Florence; trans. as *Examples of Experiments in*

Natural Philosophy: Made in the Academy del Cimento under the Protection of the Most Serene Prince Leopold of Tuscany and Described by the Secretary of that Academy, in W. E. Knowles Middleton, *The Experimenters: A Study of the Accademia del Cimento*, Baltimore: Johns Hopkins Press, 1971, pp. 83–254.

Maitte, B. (2013) "The construction of group theory in crystallography" in eds. E. Barbin and R. Pisano 2013, pp. 1–30.

Mamiani, M. (1987) "Il metodo della filosofia naturale nelle *Riflessioni filosofiche dedotte dalle figure de' sali* di Domenico Guglielmini" in ed. S. Rossi 1987, pp. 247–252.

Montanari, G. (1667) *Pensieri fisico-matematici sopra alcune esperienze [. . .] intorno diversi effetti de' liquidi in cannuccie di vetro, & altri vasi*, Bologna: Manolessi; partial trans. as "Physico-mathematical thoughts" in *Italy in the Baroque: Selected Readings*, ed. B. Dooley 1995, New York: Garland, pp. 100–115.

———. (1671) *Sopra la sparizione d'alcune stelle et altre nouità celesti: Discorso astronomico*. Undated work, probably published in Bologna in 1671.

———. (1715 [1678]) *Discorso sopra la tromba parlante: Aggiuntovi un trattato postumo del Mare Adriatico e sua Corrente esaminata, co la naturalezza de Fiumi scoperta, e con nove forme di ripari corretta*, Venice: Albrizzi.

———. (1980) "Della natura et uso degli atomi o sia corpuscoli appresso i moderni: Trattato primo fisico-matematico del Sig. Geminiano Montanari" in eds. M. L. Altieri Biagi and B. Basile 1980, pp. 537–552.

Newton, Sir I. (1730) *Opticks: Or, a Treatise of the Reflections, Refractions, Inflections & Colours of Light*, 4th edn, London.

———. (1999) *The Principia: Mathematical Principles of Natural Philosophy*, eds. I. B. Cohen and A. Whitman, Berkeley: University of California Press; 1st edn 1687.

Norton, J. D. (2005) "A little survey of induction" in ed. P. Achinstein 2005, pp. 9–34.

Oldroyd, D. R. (1987) "Some writings of Robert Hooke on procedures for the prosecution of scientific inquiry, including his 'Lectures of things requisite to a natural history'," *Notes and Records of the Royal Society*, 41: 145–167.

Pepe, L., ed. (2011) *Galileo e la scuola galileiana nelle Università del Seicento*, Bologna: CLUEB.

Pighetti, C. (1988) *L'influsso scientifico di Robert Boyle nel tardo '600 italiano*, Milan: Franco Angeli.

Power, H. (1664) *Experimental Philosophy*, London.

Pyle, A. (2002) "Boyle on science and the mechanical philosophy: A reply to Chalmers," *Studies in History and Philosophy of Science*, 33: 175–190.

Redi, F. (1996 [1668]) *Esperienze intorno alla generazione degl'insetti*, ed. W. Bernardi, Florence: Giunti; trans. as *Experiments on the Generation of Insects*, ed. M. Bigelow 1909, Chicago: Open Court.

Riccati, J. (1762) *Opere, vol. 2: Dei principj, e dei metodi della fisica*, Lucca: Giusti.

Rooke, L. (1665) "Directions for sea-men, bound for far voyages," *Philosophical Transactions*, 1: 140–143.

Rossetti, D. (1667) *Antignome fisico-matematiche con il nuovo orbe, e sistema terrestre*, Livorno: Bonfigli.

———. (1669) *Insegnamenti fisico-matematici: Dati ad O. Finetti scolare del Dott. G. Montanari sopra la Prostasi che quegli stampò per questi*, Livorno: Bonfigli.

————. (1671) *Composizione, e passioni de' vetri, overo dimostrazioni fisico-matematiche delle Gocciole, e de' Fili del Vetro, che rotto in qualsisia parte tutto quanto si stritola*, Livorno: Bonfigli.

Rossi, S. ed. (1987) *Science and Imagination in XVIIIth-Century British Culture: Proceedings of the Conference*, Milan: UNICOPLI.

Rotta, S. (1971) "Scienza e 'pubblica felicità' in G. Montanari" in *Miscellanea Seicento*, vol. 2, Florence: Le Monnier, pp. 65–208.

Roux, S. (2011) "Quelles machines pour quels animaux? Jacques Rohault, Claude Perrault, Giovanni Alfonso Borelli" in eds. A. Gaillard, J.-Y. Goffi, B. Roukhomovsky, and S. Roux 2011, pp. 69–113.

————. (2013) "Was there a Cartesian experimentalism in 1660s France?" in eds. M. Dobre and T. Nyden 2013, pp. 47–88.

Shapin, S. (1994) *A Social History of Truth: Civility and Science in Seventeenth-Century England*, Chicago: University of Chicago Press.

Soppelsa, M. L. (1974) *Genesi del metodo galileiano e tramonto dell'aristotelismo nella scuola di Padova*, Padova: Antenore.

Sprat, T. (1667) *The History of the Royal-Society of London*, London; reprinted in 1958.

Steno, N. (1669) *De solido intra solidum naturaliter contento dissertationis prodromus*, Florence: Typographia sub signo Stellæ; trans. as *The Prodromus of Nicolaus Steno's Dissertation Concerning a Solid Body Enclosed by Process of Nature within a Solid*, ed. J. G. Winter 1916, New York: Palgrave Macmillan.

Sytsma, J. and Buckwalter, W., eds. (2016) *A Companion to Experimental Philosophy*, Malden: Wiley-Blackwell.

Torrini, M. (1979a) "Cinque lettere di Lucantonio Porzio in difesa della moderna filosofia," *Atti dell'Accademia di Scienze morali e politiche di Napoli*, 90: 143–171.

————. (1979b) *Dopo Galileo: Una polemica scientifica (1684–1711)*, Florence: Olschki.

Valletta, G. (1975 [1691–1697]) *Lettera in difesa della moderna filosofia e de' coltivatori di essa*, in *Opere filosofiche*, cd. M. Rak, Florence: Olschki, pp. 75–215. Cited with the page numbers of the original edition.

Vanzo, A. (2016) "Experiment and speculation in seventeenth-century Italy: The case of Geminiano Montanari," *Studies in History and Philosophy of Science*, 56: 52–61.

Vasoli, C. (1979) "Sulle fratture del galileismo nel mondo della Controriforma" in *La scuola galileiana: Prospettive di ricerca*, Florence: La Nuova Italia, pp. 203–213.

6   The Principles of Spinoza's Philosophy

Michael LeBuffe

Principium is not explicitly a technical term for Spinoza. That is, he does not define it. It is a rich term, which Spinoza uses in a variety of ways. However, there is a precise sense that *principium* holds for Spinoza in some cases, on which a principle is an indubitable, clear, and distinct truth that one comes to know through a method of investigation that Descartes called analysis. In a few passages, Spinoza is careful to show that the propositions he calls principles are this sort of truth. In a few other texts, it seems likely that he does not use the term just because it would not be appropriate, given this precise sense, to use it. Recognizing where Spinoza does and does not use the term in this precise sense can matter for the interpretation of his philosophy. In some difficult passages in which the meaning of the term is otherwise unclear, Spinoza may be usefully understood to be building upon this precise sense.

Analysis for Descartes and his successors, including Spinoza, can be a confusing concept and so may make the precise sense of *principium* that I find in Spinoza difficult to understand. Analysis may be confusing because, although Descartes is widely credited with the invention of analytic geometry and happily and explicitly takes over the method of analysis in geometry by that label (AT VI 20), he also explicitly criticizes the method of presentation of the ancient geometers, which he takes to be a kind of synthesis (AT VII 156). So geometry is at once the paradigm of analysis and the paradigm of what is not analysis.

I do not offer an exhaustive interpretation of the two kinds of method as Descartes understands them. Very broadly, however, we may say that, traditionally, analysis is a method of proceeding backward in the order of argument, from the result aimed at to what is most fundamental; synthesis is the opposite. The two kinds of geometrical method may be distinguished in this way, and Descartes may be thought to approve the former. The central points in the distinctively Cartesian account of analysis, however, are that this method shows the actual order of learning and that truths understood in analysis are indubitable and clear and distinct (AT VII 155–156). Readers familiar with Descartes's *Meditations* will find in this characterization a basis for Descartes's claim that they are written in the analytic method.

Indeed, although the method of *The Geometry* is analytic in the traditional sense, it also is distinctively Cartesian in this way. That is, Descartes also tends in the work to proceed by referring to confusion, on one hand, and clarity and distinctness, on the other. For example, he argues that geometry may legitimately go beyond the ancients by discussing curves that are more complex than those of conic sections by pointing out that such curves may be understood just as clearly and distinctly as the others:

> I do not know what there is to stop one from conceiving the description of this first [of these curves] as clearly and distinctly as that of a circle or at least as those of conic sections; nor to stop one from conceiving the second, the third and all the others that one might describe just as well as the first.
>
> (AT VI 392)[1]

An emphasis on certainty and the order of understanding tends to blur the distinction between analysis and synthesis. One might approach, evaluate, and learn from the arguments that Descartes takes to be synthetic hypothetically, that is, by thinking about their validity without possessing knowledge of the premises or even attending to their truth. Doing so would certainly not be engaging in distinctively Cartesian analysis so understood. However, one might think that analysis could perhaps proceed by means of the same arguments—even those written in the ancient, synthetic geometrical order that Descartes hesitates to employ—if the process of understanding these arguments could be something like an authentic order of knowledge. In such a procedure, first the premises would themselves be rendered principles in the precise sense: the investigator would understand them clearly and distinctly and would not be able to doubt them. Then, by means of following the argument in the right way, the investigator could arrive at other principles. I think that it is with such a possibility in view that Spinoza wrote *Descartes's Principles of Philosophy* (hereafter, DPP) and also the *Ethics*. Propositions in these works are not principles in the precise sense, strictly speaking, since one may approach and evaluate the arguments without the requisite knowledge. Approached in the right way by a thoughtful reader, however, they might be principles. It is perhaps because Spinoza was convinced of this possibility that he was willing to write a great deal in the geometric order; it is perhaps because Descartes was not so convinced that he was reluctant.

In the first two sections of this chapter I introduce the passages in Spinoza that suggest that he understands *principium* in this precise sense. I will then be in a position to ask whether and how Spinoza's other uses relate to the precise sense. For some, discussed in the third section, it is clear that they simply are not related. While some important texts are mentioned here, nothing about the precise sense of *principium* helps us to understand them. Others, because they arise in a text in which the narrow sense is prominent or because they hold a place in a structure of ideas similar to that occupied by indubitable truths, may be interpreted in light of the precise sense. I argue in the fourth section that discussions of principles in Part Three of the DPP and Chapter Seven of the *Theological-Political Treatise* (hereafter, TTP) are best understood in light of the precise sense of the term. The uses in Part Three of the DPP might be thought, following Descartes in a similar part of his *Principles of Philosophy*, to invoke the precise sense despite explicit suggestions that they do not. Others, those in the TTP, might be well understood to capitalize on the quirkiness of the *Principles of Philosophy* and DPP use. The similarity of these principles to those in Part Three of Descartes's *Principles of Philosophy* and in Part Three of the DPP makes them ambiguous in a way that is helpful to Spinoza's project.

Principles and Analysis: Uses in the Precise Sense

In 1663, Spinoza completed the DPP. As Spinoza's friend Meyer introduces the book (*Opera* I/129–130), it is an attempt to recast Descartes's most important arguments, all of which were written in the analytic method, in the more formal and demonstrative synthetic order of geometry. The uses of *principium* that I take to be uses in the precise sense, arise in two passages early in the work, Spinoza's "Prolegomenon" and a scholium to a rather important proposition, p4, on which, " 'I am' cannot be the first thing known except insofar as we think," and its unsurprising corollary, that mind is better known than body:[2]

> he attempted to call all things into doubt not as a Sceptic would, who sets before himself no further purpose than to doubt, but so that he might free his mind from all prejudices and at last discover firm foundations of the sciences. By this method, if there were any, they could not escape notice. For the true principles of a science must be truly clear and certain so that they need no proof, are set beyond all risk of doubt, and nothing can be demonstrated without them. And, after meticulous doubting, he did discover them. Moreover, after he had discovered these principles it was not difficult for him to distinguish the true from the false; and to reveal the cause of error; and even to guard himself against accepting anything false and doubtful as true and certain.
>
> (Prolegomenon, *Opera* I/141–142)[3]

> Now, in the sciences nothing more can be sought or desired for our greatest certainty about things than to deduce all things from the firmest principles and so to make them as clear and distinct as the principles from which they are deduced. It follows clearly that each thing that is just as evident to us; *and* that we perceive to be as clear and distinct as the principle we have already discovered; *and* that is so consistent with this principle—*and* so depends upon this principle that if we wanted to

doubt it we would also have to doubt this principle—must be held to be most true.

(P4S, *Opera* I/153)[4]

These passages seem to treat *principium* in a uniform manner: it is a clear and distinct truth that is self-evident in the sense that it requires no proof but that is itself the sort of source that makes further truths that are derived from it as certain—that is, as clear and distinct—as it is itself. There is, to be sure, a process of deduction that Spinoza describes here and so a kind of dependence. The deduction in this order of investigation is, however, one by means of which new principles are rendered as clear and distinct as those from which they are deduced. After that point, the new principles are placed beyond doubt and stand as little in need of proof as those from which they are deduced.

In both passages, Spinoza does write about principles in the plural. His foundationalist language together with the condition of self-evidence and the place of the passages in the DPP all suggest, however, that he has one particular principle in mind, which is the starting point for the deduction of others: "I am." For the "meticulous doubting" that Spinoza describes in the "Prolegomenon" is Descartes's procedure in the *Meditations*, and the first principle that, Spinoza reports, Descartes arrived at is "I doubt, I think, therefore I am" (*Opera* I/144). The second passage, from Ip4s, just is an explanation of the claim that "I exist thinking," is the only and most certain foundation of all of philosophy. Indeed, in the "Prolegomenon," Spinoza is careful to show that the *cogito* argument does not violate the description of a foundational principle that he offers there, on which it needs no proof. He argues (*Opera* I/144) that "I think; therefore I am" is not a syllogism and that, if it were, "I am" could not be the foundation of all knowledge.

The most basic principle, then, on Spinoza's presentation of it, is the "I am" produced by the cogito argument. It is this point that constitutes positive evidence for my claim that *principium* in these passages is an indubitable, self-evident truth that one comes to know through analysis. "I am" or "I exist" is the first truth that Descartes himself establishes in the *Meditations* (AT VII 25), and Descartes writes in the Second Replies that he used analysis alone in the *Meditations* (AT VII 156). Descartes's own explicitly synthetic argument in the replies to Mersenne, the reader of the *Meditations* who had asked for a presentation of the same ideas in the geometrical order, begins by contrast with the proposition that God's existence can be known by his own nature. While Spinoza does begin the formal argument of the DPP with propositions related to his own existence (p1–p4), those propositions are distinguished from what Spinoza explicitly labels as Descartes's own propositions, which begin—after the cogito and the introduction of additional axioms taken from Descartes—with proposition 5. Proposition 5, like proposition 1 in Descartes's explicitly synthetic argument in the Second Replies, is that God's existence can be known by his own nature. Moreover, although p1 through p4 of the DPP concern our knowledge of our own existence, none of them establishes straightforwardly, after the manner of the cogito argument: that "I am." This proposition does not arise in the synthetic geometric order at all but, as we have seen, in a scholium to p4. If it is the most basic truth, then, "I am" must be the most basic truth in the order of analysis for Spinoza just as it is for Descartes. In order to gain knowledge through the use of synthetic argument in the geometric method, one must start with knowledge of a premise that is not itself the result of a demonstration in the synthetic order.

Principles and Analysis: Precise Omissions

Interesting passages in which Spinoza does not use *principium* contribute further to the case on behalf of the thesis that he sometimes understands the term in the precise way. Spinoza does use the term in senses that depart from this narrow technical sense, and I discuss some notable passages of this sort in subsequent sections. The term, however, is one that it is natural for a philosopher to use more frequently than Spinoza does in the discussion of arguments.

The clearest test case for the claim that *principium* belongs only to analysis is the geometric order. Descartes in the Second Replies and Spinoza in the DPP, both explicitly associate the geometrical order of presentation with synthesis. I discuss both authors here, however, the discussion focuses on Spinoza, who used the order much more extensively.

The geometrical order and its implications for the interpretation of Spinoza have been discussed and debated by scholars at great length. Roughly speaking scholars' views may be placed on a continuum from an interpretation of the order that takes Spinoza's propositions and demonstrations at face value, as deductive arguments that represent the only basis that he offers for endorsing a given proposition, to an interpretation that takes the geometric order to be nothing more than a mask, a distraction from the actual content of the work.[5] (My own view is that Spinoza's demonstrations are meant to be what they pretend to be and so are important resources, if not the only resources in his geometrical works, for understanding his views.)

Whether it is sincere or disingenuous, the geometrical order is technical in the sense that it is a form of expression in which Spinoza pays close attention to his terminology, as Spinoza's reservation of scholia, prefaces, and other interruptions to the order for informal explanation shows. Spinoza's use of, or hesitancy to use, *principium* in the course of his geometrical works, then, can be a good indication of its meaning. In those works, terms tend to take precise senses.

The order is full of items that one might understand to be principles or that might in other contexts be known as principles: real definitions, axioms, and propositions, for example. The term *principium*, however, simply does not arise in the synthetic order in either author's presentation. Descartes uses the term once in his brief geometrical presentation in the Second Replies, at D1, where it is used not to refer to a truth at all but to a beginning in time.[6] In the much longer DPP, Spinoza uses the term frequently and in several different senses—I have discussed some of these uses and will turn to others in the following. However, except where he refers to the title of Descartes's work, he never uses *principium* in the formal apparatus of the argument, that is, in a definition, an axiom, a proposition, a corollary, a lemma, or a demonstration. The story is similar in the *Ethics*, a very long work of synthesis, in which Spinoza uses *principium* only fourteen times. Of these uses, five appear to take the uninteresting temporal sense of "a beginning," such as where Spinoza refers to what he wrote at the beginning or to people who err in thinking that substance has a beginning in time.[7] Another two refer, in the preface to Part Five, to the others' principles,[8] and one, at 1p19, occurs in the title of Spinoza's own work on Descartes. The remaining uses are interesting, and I discuss them in the following, but only one occurs in the formal presentation of argument, at 4p22c, where Spinoza refers to the foundational normative rule, the demand of reason that one should strive to preserve oneself. Principles, then, are very nearly absent in the geometrical order, that is, in those of Spinoza's arguments that he regards as synthetic, and in this practice he follows Descartes.

A second important nonuse of *principium* may be found in Spinoza's exchanges with Henry Oldenburg. Oldenburg—shortly to become founding secretary of the Royal Society, a position for which he is best known—met Spinoza in 1661 at Rijnsburg.[9] Shortly thereafter, he sent a letter asking about Spinoza's views and including for Spinoza's examination a copy of Boyle's *Certain Physiological Essays*.[10] Spinoza responded with an account of some of his positions together with a detailed assessment of Boyle's work notable for its remarks on experimentalism and the order of philosophy. In a series of exchanges about both Spinoza's theories and Boyle's, Oldenburg seems much more willing than Spinoza to talk about principles:

Oldenburg (September 1661): . . . The third is: do you hold the axioms that you shared with me to be indemonstrable principles and known by the light of nature and needing no proof? (*Opera* IV/10)[11]

Spinoza replies (October 1661): You go on, third, to object against the things that I have proposed that axioms are not to be counted among the common notions. But I do not dispute that. (*Opera* IV/13)[12]

Spinoza concerning Boyle (April 1662): Section 25: In this section, the Most Ill. man seems to want to demonstrate that the alkaline parts are carried here and there by the impulse of the saline particles but that, by means of a characteristic [*proprio*] impulse, the saline particles lift themselves into the air. (*Opera* IV/26)[13]

Oldenburg replies (April 1663): To what you note about Section 25 . . . he responds that he has used Epicurean principles, on which motion

is inherent [*connatum*] in the particles. Indeed some hypothesis was
needed to explain the phenomenon. (*Opera* IV/50)[14]

Spinoza concerning Boyle (April 1662): §13 up to §18, the Most Illustrious man tries to show that all tangible qualities depend only upon
motion, figure, and other mechanical properties. Because these demonstrations are not presented by the Most Illustrious man as mathematical, it is not necessary to consider whether they are not convincing.
However, I do not know why the Most Illustrious Man strives so anx-
iously to gather this from his experiment since it has already been
demonstrated more than satisfactorily by Verulam [Bacon] and, later,
by Descartes. (*Opera* IV/25)[15]

Oldenburg replies (April 1663): To your observations about sections 13–
18, he responds only that he had written these things in the first place
in order to show and to reinforce the usefulness of chemistry for confirming the Mechanical principles of Philosophy. (*Opera* IV/50)[16]

Spinoza replies (July, 1663): I pass on to what the Most Ill. man asserts
with respect to Sections 13–18, and I say that I freely acknowledge
it: this reconstitution of niter is certainly a very clear experiment for
investigating the very nature of niter, whenever we already know the
Mechanical principles of Philosophy . . . (*Opera* IV/66–67)[17]

In the first set of passages here, Oldenburg asks Spinoza whether his axioms are principles and Spinoza replies that they are not common notions.
It seems to me that the correspondents agree. Oldenburg is concerned that
Spinoza is assuming too much if he takes the axioms they have discussed to
be just obvious. Implicitly suggesting that "yes" is the obvious answer, he
asks Spinoza whether axioms are demonstrable and whether proof of them
may be justifiably demanded. Spinoza in responding assures Boyle that they
agree: axioms are not among the common notions, so, even if they are self-
evident in some rarer sense, one may justifiably demand proof of them. As
Spinoza writes, he does not dispute Oldenburg's position.

It also seems to me, and this is the point that matters in the present context, that Spinoza is offering Oldenburg a quiet correction here. His response suggests that Oldenburg has somehow misspoken in using the term
principium. Spinoza himself will not describe axioms as principles, as Oldenburg has. If Spinoza had regarded axioms as belonging properly to the
method of demonstration by synthesis and principles as belonging properly
to analysis, this conviction would explain his reticence.

In both the second and third sets of passages, Oldenburg freely offers a
label for what does the explanatory work in nature, even if it is merely a
hypothesis: it is a principle. Spinoza writes instead in the first instance about
what Boyle seeks to demonstrate. In Spinoza's rejoinder to Oldenburg about
sections 13 through 18 of Boyle's work, he does use the term *principium*.
There, however, I think it is clear that he is repeating Oldenburg, perhaps
to make clear the precise issue on which he takes himself to agree. It may
also be that, so long as Oldenburg agrees that we possess certain knowledge

of these truths prior to experimentation, Spinoza is then comfortable with the label. Synthesis and analysis may come together, as they appear to in the opening sections of the DPP, where the synthetic argument is undertaken in the right spirit.

Spinoza's caution about the use of the term *principium* in the second two exchanges might be explained by any sense of the term on which it is a well-established truth. Together with the first exchange, however, the sense seems likely to be more narrow. A principle is not drawn from whatever resources we have at hand to explain what we have observed. Neither, however, is it an axiom in a synthetic order of demonstration. We are left, once again, with the strong suggestion that Spinoza reserves the proper technical use of the term for an order of argument that is not synthetic.

Explanatory Work: Obvious Nontechnical Uses

There are, as the next few sections show, too many further uses of *principium* for the attribution of any one sense of the term to Spinoza. Nevertheless, I hope that the argument of the first two sections might do some useful work: it suggests that among Spinoza's notable uses of *principium* some clearly have a fairly specific, technical sense. If that is correct, then the argument has yielded a tool for the interpretation of other, less clear uses: we can ask of such uses whether they also have this sense. In this section I begin this task. I describe some notable uses of *principium* that are clearly not technical in the sense that, even if they are analogous in important ways, they could not be similar in kind to the "I exist" in the order of analysis. Two classes of uses that seem nontechnical, by this criterion, are notable: those that prescribe certain kinds of action and those that take the traditional Aristotelian sense of a primary efficient cause, "the changer of the thing changed."[18]

Spinoza uses *principium* to describe prescriptions in several prominent passages in Part Four of the *Ethics*. Two passages refer to the most basic of Spinoza's dictates of reason, that each is bound to seek what is useful to himself. The third is something of an outlier. It may refer to normative rules, as I suggest here; it might also plausibly be thought to be a use in a generic sense on which *principium* refers to a basic, certain truth. It does resemble Spinoza's references to the Stoics' principles and Descartes's principles in the preface to Part Five. I place it here among normative rule uses because Spinoza, in writing about a plan of living and claiming that his plan agrees both with principles and with common practice, makes principles similar to two kinds of practical norms:

> 4p18s: These are the dictates of reason that I set out to present briefly here before beginning to demonstrate them in a more detailed order. I have done so in order to (if it is possible) attract the attention of those who believe this principle—that everyone must seek what is useful to himself—to be the foundation of impiety, not indeed of virtue and piety. (*Opera* II/223)[19]

4p22c: Striving to preserve oneself is the first and only foundation of virtue. For no other principle can be can be conceived prior to this one (4p22) and no virtue can be conceived without it (4p21). (*Opera* II/225)[20]

4p45s: Therefore this plan of living best agrees both with our principles and also with common practice. (*Opera* II/245)[21]

If there is no technical reason to call these principles, why does Spinoza do so? It is noteworthy that Spinoza refers to such rules, seemingly indifferently, as rules (*regulae*), dictates of reason (*rationis dictamina*), and principles. Indeed, all three terms may be found at 4p18s, which is perhaps the most important passage in Spinoza for the understanding of his normative ethics. Don Garrett has suggested that Spinoza uses a wide variety of ethical terminology in order to show that he can address various conceptions of ethics.[22] Perhaps that philosophically unexciting explanation is the best that we can offer here. Certainly reason—which is the second kind of knowledge and which is introduced and defined in the *Ethics* (2p40s2)—and the dictates of reason are more prominent than principles in Spinoza's arguments.

This explanation is not wholly satisfying, however. One problem is that the corollary to 4p22 is, after all, a usage within the formal apparatus of the *Ethics*. So it is difficult to maintain that the use is nontechnical. Another problem is that Spinoza invokes foundationalism in discussing the first principle for action. That makes this use feel like something more than casual accommodation. There may be some relevant likeness between this prescriptive principle and the descriptive principles of philosophy that Spinoza is trying to invoke here. As an aside, and this is conjecture, it seems likely to me that Spinoza finds an important similarity between good epistemological and good moral practice that the label "principle" invokes in these passages. Just as one believes a descriptive principle in the right way where doubt of that principle would require doubt of the most foundational principle "I exist," so one follows a prescriptive principle in the right way where a violation of it would be a violation of most foundational principle "seek to preserve oneself." Such a reading would accord well with cases such as those described in 4p63s: one who seeks healthy food to preserve himself acts well, but one who seeks healthy food to avoid death acts badly. Belief and motivation are thus similarly dependent on foundations.

Turning now to the second class of nontechnical uses, a notable use of *principium* to describe a primary efficient cause may be found at DPP, D1, where, in offering an example of something related to thought that is not a thought, Spinoza writes that voluntary motion has thought as its principle, although it is not itself a thought. Three uses in the same sense may be found in the *Ethics*:

As he exists for no final cause, so he acts for no final cause; rather, as he has no principle or end in existing, so he also has none in acting. But what is said to be a final cause is nothing more than a human appetite insofar as it is considered to be the principle, that is, the primary cause of something.

(E4 Preface, *Opera* II/206–207)[23]

because the essence of our mind consists in cognition alone, of which God is the principle and foundation (by Ip15 and 2p47s), it is clear to us in what way and by what cause the existence and essence of our mind follows from the divine nature and continually depends upon God.

(5p36s, *Opera* II/203)[24]

Just as there is in Spinoza's prescriptions, there can be a kind of foundational order associated with causation in this sense. I do not think that it can be more than a loose analogy, however. As Descartes and Spinoza alike are sometimes accused of conflating causes and reasons, I suppose, one might try to build a case that these references to *archê* are uses similar in kind to references to self-evident truths in the order of knowledge. Spinoza does not use the term in this sense of an *archê* very frequently however, or consistently. It seems as though less grand and more boring explanations are more likely: where he wanted a term that described an efficient cause that does a great deal of explanatory work, he sometimes found *principium* useful. Descartes, it seems, follows a similar practice, for example, in his discussion in the *Passions of the Soul* of the principle (*principe*) of corporeal motion issuing in the thesis that heat or fire is the principle of the motions of our limbs.[25] Indeed, just as Spinoza does in the DPP, Descartes in the Fourth Replies (AT VII 230) makes mind a principle for some motion in human bodies.

Explanatory Work: Principles Derived from a History

Other notable uses of *principium* may turn out to be well understood not to have the precise sense that we have found in Spinoza, even if that interpretation is not immediately clear from their context alone. With this question in mind, we may turn now to some of the most interesting uses of *principium* in Spinoza. They arise in arguments in which, at least in one instance following Descartes, Spinoza derives principles from a history.

The use of a history in the investigation of nature is associated most strongly with Francis Bacon's method of induction in the study of matter. In a letter written in June 1622, Bacon distinguishes between other sorts of science, such as mathematics, which might be understood through syllogistic reasoning alone, and the study of matter, for which syllogism is insufficient and induction is required: "I do not give up syllogism entirely . . . For mathematics why should it not be used? It is the transience of matter and the inconstancy of the physical body that calls for induction, by which it can, so to speak, be fixed and then revealed by well defined notions"

(SEH XIV 375).[26] A history becomes necessary, then, where induction is required:

> First, indeed, a sufficient and good natural and experimental history is to be prepared, and this is the foundation of the thing. For we ought neither to suppose nor to invent but to discover what nature does or might do. Truly, natural and experimental history is so varied and scattered that it confounds and breaks up the understanding unless it is set up and presented in a suitable order.
>
> (*New Organon* 2.10, SEH I 236)[27]

As I understand Bacon's account, natural history should meet two requirements. First, it should be an accurate, or "sufficient and good" account of our experiences. This requirement suggests that we should collect as many observations, including experimental observations, as we can. Second, a history should be prepared in such a way that it is suited to our understanding. In Bacon's *New Organon*, he describes this preparation in terms of different tables of instances, or collections of observations relevant to issues of particular interest; for example, a table of instances in which heat is found, a table of instances in which heat is absent, and table of instances of heat of varying degrees.[28]

In Descartes's *Principles of Philosophy* and in Spinoza's DPP alike, the use of a history arises in the transition from very general metaphysical principles to the study of what they call natural phenomena. What is interesting for the purpose of understanding Spinoza's use of *principle* is the way in which Descartes and Spinoza use history. They are particularly concerned with the second of Bacon's requirements, that is, with use of the method to generate principles that help us to understand. It is not clear, however, whether these new principles have the same status as principles in the precise sense. In Descartes's use at *Principles of Philosophy* Part Three and in Spinoza's use following him, the authors seem to allow that principles are merely heuristic and may be false. In Spinoza's later use of a history, in chapter 7 of the TTP, he does not even discuss the truth of the principles. It is meaning and not truth that concerns him.

Spinoza's DPP follows Descartes closely, so we may start with Part Three of Descartes's *Principles*. It is, as I have mentioned, a transitional point in the work. It is also a politically sensitive part of the work, in which Descartes defends doctrines in tension with Ptolemaic models of the solar system and biblical accounts of creation.

Descartes begins the section by announcing that he will turn from the sorts of principles he has already discovered to a new topic: natural phenomena and the visible world:

> The principles of material things that have been found have been sought, not from the prejudices of the senses, but from the light of reason so

that we cannot doubt their truth. Now we will see whether, from these alone, we can explain all natural phenomena. Let us begin from those that are the most universal, on which the rest depend: that is, from the general construction of the whole visible world.

(*Principles of Philosophy*, 3.1, AT VIIIA 80, ll. 5–12)[29]

From this passage, one might naturally suppose that Descartes's explanation, since it will derive from principles—that is, truths that are beyond doubt—will consist of further indubitable principles.

Such a supposition would be hasty, however. Descartes goes on, a few paragraphs later, to qualify his procedure for understanding the visible world:

The principles, however, that we have discovered to this point are so vast and fertile that more things follow from them than what we might find to be held in this visible world . . . but we will now set out for our attention a brief history of particular natural phenomena, the causes of which are to be investigated; not, however, so that we may use them as bases for proving anything. For we want bases for the deduction of effects from causes, not however of causes from effects.

(*Principles of Philosophy*, 3.4, AT VIIIA 81, ll. 19–29)[30]

An account of such vast consequences requires a new technique. Descartes proposes to describe one way the world could be, a brief history, in order to investigate the causes of the phenomena introduced. Descartes stops short, moreover, of characterizing the results of this new technique as things that are proved. The passage is a bit ambiguous on this question: notably, Descartes might plausibly be interpreted as meaning that he does not intend to use any element of his history as a premise in a deductive argument even though he does intend to make deductive arguments of some kind. If this interpretation is correct, then 3.4 does not of itself show that Descartes thinks that all results of the process of using a brief history of phenomena and working backward to causes are uncertain. Notably, it does not discuss the results of the deductive arguments that Descartes does think eventually may result from this method.

I think, however, that the rest of *Principles of Philosophy* Part Three shows that Descartes does indeed advertise the main results of the rest of his argument to be less certain than what he has established in Parts One and Two. Evidence for my reading may be found later in Part Three, at 43 and 44. At *Principles* 3.43, Descartes makes a strong, general claim on behalf of the results that he does establish by means of the method of constructing a brief history. He argues—in a manner reminiscent of his proof of the existence of external things in the *Meditations*—that, having arrived at causal explanations that agree with all natural phenomena it is nearly impossible that those explanations should not be true: such a supposition would amount to thinking that God had given us a nature incapable of avoiding such mistakes. At 3.44, however, Descartes nevertheless refers to such morally certain results as something less than certain truths. They are hypotheses, claims that agree with all observation and yield the same practical benefit as truth but which may not themselves be true:

> Still, lest we seem exceedingly arrogant in philosophizing about such things if we should assert that we have discovered the truth, it would be preferable to let this remain ambiguous. So I will set out everything that I write hereafter as a hypothesis.
> (*Principles of Philosophy*, 3.44, AT VIIIA 99, ll. 15–20)[31]

In this passage, Descartes suggests that some of the propositions defended in the *Principles of Philosophy* that depend upon brief histories yield something close to indubitable truths, morally certain hypotheses. Such hypotheses, however, may be false, an important admission for a philosopher defending conclusions which might contradict religious doctrine.

The next paragraph, however, is more dramatic: "I shall even suppose here some things that all agree to be false." At *Principles* 3.45, Descartes accepts as an element of faith that creation gave certain things in the world—Adam and Eve, the Sun, the Earth, plants—fully formed. He also argues, however, that things will be best understood if their manner of generation is well understood. This account of the origins of things will therefore be useful but fictional. It will invoke principles. In introducing his finding, Descartes argues that we will be able to offer better explanations of phenomena,

> . . . if we are able to think of principles that are very simple and easily understood, from which we can demonstrate how, as if from seeds, the stars and the Earth and everything that we can discern on this visible world might have grown . . . And because it seems to me that I have found such principles, I will relate them briefly here.
> (AT VIIIA 100, ll. 11–15, 18–20)[32]

Spinoza's account adheres closely to Descartes's argument at the beginning of Part Three of the DPP. In the Preface, Spinoza notes that, to this point, he has set forth the "most universal principles of things" already.[33] Then he goes on to write that part III will proceed to explain what follows from them. In order to do this, he argues in terms that clearly follow Descartes closely that we need first to set out our for our attention (*ob oculos ponenda est*, Opera I/226 14) a brief history (*brevis historia*, Opera I/226 13). Once we have done so, we should in considering how phenomena arise devise, seek, and discover principles that are simple, easily understood, and powerful in the sense that everything observable might be deduced from them.

Although he does not use *principium* in the formal presentation of the DPP, in an informal preface to Part Three of the DPP, Spinoza follows Descartes in his use of the term to describe the causes of phenomena understood from the method of positing a brief history. Notably, Spinoza also follows Descartes in searching for principles that are hypotheses about how phenomena arise. His criteria for such principles include the requirement that the principles explain the generation of things:

> We say, then, that we seek principles of such a kind that from them we can demonstrate how the stars and the Earth, etc. might have grown.
>
> (*Opera* I/227 5–6)[34]

Other criteria include the requirement, consonant with Bacon's second criterion for a history, that the principles be simple and easily known: "I say that we seek principles that are simple and easily known" (*Opera* I/226 29).[35] Finally, Spinoza, like Descartes, admits that the principles are fictitious and that we may even know that phenomena did not in fact arise in the way supposed:

> We have said, finally, that it is permitted to us to assume a hypothesis from which, and from a cause we can deduce the phenomena of nature, even if we know that actually they did not grow in this way.
>
> (*Opera* I/227 20–23)[36]

I have already argued that in the DPP Spinoza takes *principium* to have a precise sense, on which, because synthesis does not require the duplication of an order of knowledge, there ought to be nothing in a geometrical order that is explicitly a principle. I do not think that Spinoza violates that commitment at the beginning of Part Three. The uses of *principium* here occur in an informal passage, not part of the formal order of presentation. Moreover, Spinoza is simply paraphrasing Descartes. I do think, however, that the presence of a sense of *principium* that is so different from a different sense of the same term in the same text is some indication of the rhetorical power of the term. Spinoza takes the term, strictly understood, to be too strong, epistemologically, to be included in a geometrical presentation: it is a truth, frequently the first truth that serves as the beginning of further investigation, that is indubitable. Descartes must have shared some similar understanding of the term: he also does not refer to principles in his geometric presentation and does refer to his most basic findings as principles elsewhere, including, of course, in the title of his book. Yet, when it comes to a label for hypotheses that, in order to honor religious doctrine, he must suppose to be false, Descartes uses *principium*, qualifying his use by saying that the hypotheses are fictional. Spinoza follows him. Suddenly, the interpretation of the texts becomes very difficult. Clearly, the principles that emerge from the method of a brief history are explicitly described as anything but principles in the strict sense if we are to take Descartes's account of his own use literally.

Starting from the precise sense of *principium*, on which it is a clear and distinct, indubitable truth, one might take Descartes's use of the term as a label for what is uncovered through the method of constructing a history in at least three different ways. One might take Descartes simply to be using *principium* in a generic way, without meaning to refer specifically to the kind of truth represented paradigmatically by the "I am" of the cogito. Alternatively, one might take him to be using the term in that precise sense and with the intention of invoking the high degree of knowledge of the, "I am." On this second interpretation, Descartes's talk about hypothesis and fictions and falsehoods is only so much whitewash, intended to assuage or to fool those readers whose religious convictions require such treatment. Finally, one might take the use to be one which is precise, but which qualifies the principles of Part Three together with, retrospectively, any principles which precede them: all of these propositions are certainly true so long as revelation does not contradict them. Revelation may contradict them, however, and just as it has the authority to contradict principles discovered in Part Three, so too, does it have the authority to contradict even a principle of the status of the "I am."

It is not clear to me whether one of these interpretations is the correct one, and it is not clear to me whether readers of the *Principles of Philosophy* should settle on any particular interpretation. Descartes's discussion of the term is puzzling, and Spinoza's adherence to Descartes in a work where he is otherwise very cautious about using the term *principium* is likewise puzzling, a problem that is exacerbated by the fact that Spinoza abruptly cuts the DPP short, finding it wanting, just after the opening propositions of part III. It is not clear that he goes far beyond paraphrasing Descartes in these passages.

Regardless of whether we find anything original in Spinoza's use of *principium* at DPP 3, it is clear that principles deriving from a history are among the most interesting of Spinoza's own uses. He returns to the method in his account of the method for understanding scripture in the TTP. In that work, which is full of complex interpretive problems, the puzzle about authorial intention and the relation between religious doctrine and human understanding in the use of *principium* becomes still more complicated.

In chapter 7, Spinoza makes novel use of the same method that he and Descartes use in moving from their more general principles to an account of phenomena. He suggests that this is the method for interpreting scripture:

> I say that the method for interpreting scripture differs not at all from the method for interpreting nature but entirely agrees with it. For, just as the method for interpreting nature consists primarily in assembling a history of nature from which we infer the definitions of natural things from what is given with certainty, so for interpreting scripture it is

necessary to prepare a sound [*sinceram*] history and from this to infer the judgment [*mentem*] of Scripture's authors by legitimate reasoning [*consequentiis*] from what is given with certainty and principles. Indeed, in this way anyone—so long as he has admitted neither principles nor data for interpreting scripture and explaining the things that are met with in it beyond those that are drawn from Scripture itself and its history—will always proceed without any danger of error and he will be able to explain those things that surpass our grasp just as surely as those we know by the light of nature. But in order to establish clearly not only that this way is certain but that is the only way, and that it accords with the method for interpreting nature, it should be noted that scripture most often deals with things that cannot be deduced from principles known by the light of nature.

(TTP Ch. 7, *Opera* III/98–99)[37]

Spinoza notes at the beginning of this passage that the method of constructing history for the interpretation of scripture is continuous with his method of interpreting nature. Because Spinoza follows Deacartes's distinctive use of history so closely in his account of nature in the DPP, this emphasis suggests that his understanding of a history in the TTP is also distinctively Cartesian. One way in which it might be thought to be so is in its account of the relation between more basic *a priori* principles and history. As we have seen, in the *Principles* and the DPP, principles known *a priori* do have particular states of nature as their implications—a conception of the relation between the *a priori* and nature that Bacon does not admit. Those principles, however are so vast and fertile that we cannot deduce those implications except by the construction of a brief history. At the end of this passage Spinoza contends that history as a method for interpreting scripture accords with his method for interpreting nature in that it "deals with things that cannot be deduced from principles known by the light of nature." This sentence invokes a similarity between the two kinds of history that is to be found in their purpose: they help us to deduce what we could not otherwise know.

As Spinoza understands scripture, however, this insistence on the equal status of a history in the study of nature and scripture may be misleading. For, if Spinoza does indeed follow Descartes with respect to nature, in the case of nature the problem is one of our human limitations: *a priori* principles, which do imply the detailed truths about nature, have such vast implications that we can only get to those truths by means of a history. On Spinoza's account of scripture, however, its independence from any prior principles is stronger and different in kind. There are some tenuous connections between natural truths and scriptural ones. For example, the moral teachings which can be derived from the common notions are also the moral teaching of scripture (III/99). Generally, however, scripture is highly insulated from any natural truths outside of scripture. The difficulty of deducing historical narratives of scripture from principles known to the natural light is *not* a function of our inability to deduce the latter from the former. It arises instead from a lack of any relation at all between philosophy and theology, the central theme of the TTP. The natural light implies truths of nature which nevertheless we are unable straightforwardly to deduce. The natural light does not imply the historical narratives of theology, which after all include accounts of miracles, at all. The principles of a history of scripture, therefore, are basic in a way that the principles of a history of nature are not.

Taking scripture to be the sole source of evidence for claims about scripture isolates it as a subject of study in a way that gives the method of history great importance and so calls for a history to be detailed and thorough in a way that uses in natural philosophy do not. We have seen that Descartes in the *Principles* emphasizes the use of a "brief history" for the sake of understanding and that Spinoza echoes these views in the DPP. By contrast, Spinoza's emphasis on a sound history for the understanding of scripture in the TTP rules out a brief history. Because it is the sole source of our understanding of scripture, Spinoza requires the same kinds of detail and scrutiny in a history of scripture that Bacon demands in a history of nature. History requires a thorough knowledge of the language of the books of scripture as its authors understood it (*Opera* III/ 99–100); it should collect all of the verses that treat of the same subject together under main headings, while noting all puzzles of consistency or obscurity that they raise (*Opera* III/100); and it should include a complete account of the histories of each book, including an account of the life and character of each author (*Opera* III/101). For Spinoza, the principles of a history have a foundational importance for the understanding of scripture. Spinoza's account of our epistemic position with respect to scripture fits closely Bacon's account of the epistemic position that prompts us to construct a history in the study of nature or in other areas.[38]

Just as Descartes's use of *principium* at Part Three of the *Principles of Philosophy* is ambiguous and exploits an association with the precise sense of the term, so the TTP benefits from the powerful association of the term with clear self-evident truths. It may be correct that Spinoza is founding a kind of science of textual interpretation in his method for interpreting scripture and that the principles described in chapter 7 will be genuine principles in the sense that they are fundamental truths about meaning. Setting that question to one side, the suggestion in this passage that scripture cannot be understood from natural principles but that it has principles of its own, lends scripture a kind of status. Rhetorically, it elevates scripture to the same level as nature and so serves the purpose of reinforcing the message that philosophy and theology, rightly understood, are equally valuable sources of wisdom that are not informed by one another.[39]

Conclusion

For our understanding of principles in the history of early modern philosophy, Spinoza's use of the term *principium* shows the variety of its meanings and the danger of taking what looks like a technical term to have the same meaning in different uses. Even where it can be distinguished from its other ordinary senses, such as that of a beginning in time, *principium* can refer to a foundational truth, a cause, a well-known truth, a basis for argument, or the result of argument. This is a lesson, perhaps, that might be drawn from many authors.

More specifically, the precise sense of *principium* in Spinoza, who follows Descartes in this use, should move us to consider carefully the identification of principles with premises in demonstrative arguments, and especially arguments in the geometrical order. For Spinoza, and perhaps also Descartes, a principle precisely understood is not merely a truth. It is indubitable knowledge. I think that, for both authors, the epistemological dimension of principles comes first. For example, supposing that all things depend upon God but that I need to know my own existence before I know God's existence, "I am" will be the most basic principle, and "God exists" will only become a principle once it is understood in the proper way. If premises in arguments in the geometrical order are, for other early modern writers, paradigm principles, Spinoza's precise sense and Descartes's similar use that precedes it, suggest that there is a notable tradition that uses the term in a different way.

Notes

1 AT VI 392: "Mais je ne voy pas ce qui peut empescher qu'on ne conçoive aussy nettement & aussy distinctement la description de cete premiere, que du cercle ou, du moins, que des sections coniques; ny ce qui peut empescher qu'on ne conçoive la second, & la troisiesme, & toutes les autres qu'on peut descrire, aussy bien que la premiere. . . ." Translations in this chapter are my own. Because details of authors' use are frequently at issue, I normally include the original language in footnotes.

2 "Ego sum non potest esse primum cognitum, nisi quatenus cogitamus."

3 Prolegomenon, *Opera* I/141–142: "omnia in dubium revocare aggreditur, non quidem ut Scepticus, qui sibi nullum alium praefigit finem, quàm dubitare: Sed ut animum ab omnibus praejudiciis liberaret, quò tandem firma, atque inconcussa scientiarum fundamenta, quae hoc modo ipsum, siquae essent, effugere non possent, inveniret. Vera enim scientiarum principia adeò clara, ac certa esse debent, ut nullâ indigeant probatione, extra omnem dubitationis aleam sint posita, & sine ipsis nihil demonstrari possit. Atque haec, post longam dubitationem reperit. Postquam autem haec principia invenisset, non ipsi difficile fuit, verum à falso dignoscere, ac causam erroris detegere; atque adeò sibi cavere, ne aliquid falsum, & dubium pro vero, ac certo assumeret."

4 P4S, *Opera* I/153: "Et cùm in scientiis nihil aliud quaeri, neque desiderari possit, ut de rebus certissimi simus, quàm omnia ex firmissimis principiis deducere, eaque aequè clara, & distincta reddere, ac principia, ex quibus deducuntur: clarè sequitur, omne, quod nobis aequè evidens est, quodque aequè clarè, & distinctè,

atque nostrum jam inventum principium percipimus, omneque, quod cum hoc principio ita convenit, & ab hoc principio ita dependet, ut si de eo dubitare velimus, etiam de hoc principio esset dubitandum, pro verissimo habendum esse."

5 Garrett 2003: 99–103 offers a detailed interpretation of the method and a useful survey of recent interpretations.

6 It is possible that the use is more philosophically substantive: Descartes might refer to the principle of voluntary motion in the Aristotelian sense.

7 I refer here to uses at 1d7, 1p8s2, 1p33s2 (*Opera* II/75), 5p20s and 5p33s.

8 Spinoza refers to others' principles in criticizing the Stoics (*Opera* II/277) and Descartes (*Opera* II/279).

9 Letter 1 of Spinoza's correspondence (*Opera* IV/5) refers to this visit.

10 I refer again to letter 1 here. Curley (Spinoza 1985: 173 note 14) argues that Oldenburg sent a Latin translation of Boyle's work published in the same year as the English original, 1661.

11 *Opera* IV/10: "Tertia est, an axoimata illa, quae mihi communicasti, habeas pro Principiis indemonstrabilibus, & Naturae luce cognitis, nullaque probatione egentibus?"

12 *Opera* IV/13: "tertiò in ea, quae proposui, objicere, quòd Axiomata non sunt inter Notiones communes numeranda. Sed de hâc re non disputo."

13 *Opera* IV/26: "Sect. 25. in hac sect. videtur vir cl. velle demonstrare partes alcalisatas per impulsum particularum salinarum huc illuc ferri particulas vero salinas proprio impulsu se ipsas in aerem tollere."

14 *Opera* IV/50: "Quae ad Sect. 25. notas . . . iis respondet usum se fuisse principiis Epicuraeis, quae volunt, motum particulis inesse connatum; opus enim fuisse aliquâ uti Hypothesi ad Phaenomeni."

15 *Opera* IV/25: "§13. Usque ad 18. conatur Vir Clarissimus ostendere, omnes tactiles qualitates pendêre à solo motu, figurâ, & caeteris mechanicis affectionibus, quas demonstrationes, quandoquidem à Clarissimo viro non tanquam Mathematicae proferuntur, non opus est examinare, an prorsùs convincant. Sed interim nescio, cur Clarissimus Vir hoc adeò sollicitè conetur colligere ex hoc suo experimento; cùm jam hoc à Verulamio, & postea à Cartesio satis superque demonstratum sit."

16 *Opera* IV/50: "Ad ea, quae in Sect. 13.—— — 18 animadvertis, hoc tantùm reponit, se haec scripsisse imprimis, ut Chymiae usum ad confirmanda principia Philosophiae Mechanica ostenderet, assereretque . . ."

17 *Opera* IV/66–67: "ad id, quod Cl. Vir ad ea, quae in Sectione 13 —— — 18. ponit, transeo, atque dico, me libenter fateri, hanc Nitri redintegrationem praeclarum quidem experimentum esse ad ipsam Nitri naturam investigandam, nempe ubi priùs principia Philosophiae Mechanica noverimus . . ."

18 Aristotle's *Physics* 2, 3 194b23–25 is the classic source of this definition of *archê*.

19 *Opera* II/223: "Haec illa rationis dictamina sunt, quae hic paucis ostendere proposueram, antequam eadem prolixiori ordine demonstrare inciperem, quod ea de causa feci, ut, si fieri posset, eorum attentionem mihi conciliarem, qui credunt hoc principium, quod scilicet unusquisque suum utile quaerere tenetur, impietatis, non autem virtutis et pietatis esse fundamentum."

20 *Opera* II/225: "Conatus sese conservandi primum et unicum virtutis est fundamentum. Nam hoc principio nullum aliud potest prius concipi (per prop. praec.), et absque ipso (per prop. 21. hujus) nulla virtus potest concipi."

21 *Opera* II/245: "Hoc itaque vivendi institutum et cum nostris principiis et cum communi praxi optime convenit . . . "

22 See Garrett 1996b: 288. I agree with Garrett's assessment.

23 E4 Preface, *Opera* II/206–207: "Ut ergo nullius finis causa existit, nullius etiam finis causa agit; sed ut existendi, sic et agendi principium vel finem habet nullum.

Causa autem, quae finalis dicitur, nihil est praeter ipsum humanum appetitum, quatenus is alicujus rei veluti principium seu causa primaria consideratur." Here I think that Shirley errs in taking *principium* in a purely temporal sense. Spinoza is certainly discussing the push and pull of causation.

24 5p36s, *Opera* II/203: ". . . quia nostrae mentis essentia in sola cognitione consistit, cujus principium et fundamentum Deus est (per prop. 15. p. 1. et schol. prop. 47. p. 2.), hinc perspicuum nobis fit, quomodo et qua ratione mens nostra secundum essentiam et existentiam ex natura divina sequatur et continuo a Deo pendeat." Here I agree with Shirley and disagree with Curley in translating *principium* by "principle" rather than "beginning." Spinoza does use the term in the latter sense, as we have seen. Here, however, Spinoza can not plausibly be thought to refer to God's causal influence at a particular point in time. Indeed the remark that follows—the effect, Spinoza writes, continually depends on God—suggests, as the references to Ip15 and 2p47s do, that God's influence is not properly characterized as a beginning.

25 The discussion in Descartes may be found in *Passions of the Soul*, art. 6–8, AT XI 330–333.

26 SEH XIV 375: "Non est meum abdicare in totum syllogismum . . . Ad Mathematica quidni adhibeatur? Cum fluxus materiae et inconstantia corporis physici illud sit, quod inductionem desideret; ut per eam veluti figatur, atque inde eruantur notiones bene terminatae."

27 SEH I 236: "Primo enim paranda est historia naturalis et experimentalis, sufficiens et bona; quod fundamentum rei est: neque enim fingendum, aut excogitandum, sed inveniendum, quid natura faciat aut ferat. Historia vero naturalis et experimentalis tam varia est et sparsa, ut intellectum confundat et disgreget, nisi sistatur et compareat ordine idoneo."

28 For Bacon's most-thorough account of a natural history, see his *New Organon* II.10–20, SEH I 235–238. My account of Bacon is informed by Jardine1974; Malherbe 1996; Garber 2001: 296–328; Anstey 2012; and Jalobeanu 2012. Both of the passages I quote here may be found in Jardine (86–88).

29 *Principles of Philosophy*, 3.1, AT VIIIA 80, ll. 5–12: "Inventis iam quibusdam principiis rerum materialium, quae non a praejudiciis sensuum, sed a lumine rationis ita petita sunt, ut de ipsorum veritate dubitare nequeamus, examinandum est, an ex iis solis omnia naturae phaenomena possimus explicare. Incipiendumque ab iis quae maxime universalia sunt, et a quibus reliqua dependent: nempe a generali totius huius mundi adspectabilis constructione."

30 *Principles of Philosophy*, 3.4, AT VIIIA 81, ll. 19–29: "Principia autem quae iam invenimus, tam vasta sunt et tam Foecunda, ut multo plura ex iis sequantur, quam in hoc mundo aspectabili contineri videamus . . . Sed iam brevem historiam praecipuorum naturae phaenomenon (quorum causae hic sunt investigandae), nobis ob oculos proponemus; non quidem ut ipsis tanquam rationibus utamur ad aliquid probandum: cupimus enim rationes effectuum a causis, non autem e contra causarum ab affectibus deducere . . ."

31 *Principles of Philosophy*, 3.44, AT VIIIA 99, ll. 15–20: "Verumtamen, ne etiam nimis arrogantes esse videamur, si de tantis rebus philosophando, genuinam earum veritatem a nobis inventam esse affirmemus, malim hoc in medio relinquere, atque omnia quae deinceps sum scripturus tanquam hypothesin proponere."

32 AT VIIIA 100, ll. 11–15, 18–20: "si quae principia possimus excogitare, valde simplicia et cognitu facilia, ex quibus tanquam ex seminibus quibusdam, et sidera et terram, et denique omnia quae in hoc mundo aspectabili deprehendimus, oriri potuisse demonstremus . . . Et quia talia principia mihi videor invenisse, ipsa breviter hic exponam."

33 "Principiis rerum naturalium universalissimis sic expositis . . ."

34 *Opera* I/227 5–6: "Dicimus deinde, nos talia principia quaerere, ex quibus et sidera, et terram etc. oriri potuisse deomstremus."
35 *Opera* I/226 29: "Dico, nos quaerere principia simplicia, et cognitu facilia."
36 *Opera* I/227 20–23: "Diximus denique, nobis licere hypothesis assumere, ex qua, tanquam ex causa, naturae phaenomena deducere queamus; quamvis ipsa sic orta non fuisse, probe sciamus."
37 TTP, Ch. 7, *Opera* III/98–99: "dico methodum interpretandi Scripturam haud differre a methodo interpretandi naturam, sed cum ea prorsus convenire. Nam sicuti methodus interpretandi naturam in hoc potissimum consistit, in concinnanda scilicet historia naturae, ex qua, utpote ex certis datis, rerum naturalium definitiones concludimus: sic etiam ad Scripturam interpretandam necesse est ejus sinceram historiam adornare, & ex ea tanquam ex certis datis & principiis mentem authorum Scripturae legitimis consequentiis concludere: sic enim unusquisque (si nimirum nulla alia principia, neque data ad interpretandam Scripturam & de rebus, quae in eadem continentur, disserendum, admiserit, nisi ea tantummodo, quae ex ipsa Scriptura ejusque historia depromuntur) sine ullo periculo errandi semper procedet, & de iis, quae nostrum captum superant, aequè securè disserere poterit, ac de iis, quae lumine naturali cognoscimus. Sed ut clare constet, hanc viam non tantum certam, sed etiam unicam esse, eamque cum methodo interpretandi naturam convenire, notandum, quod Scriptura de rebus saepissime agit, quae ex principiis lumine naturali notis deduci nequeunt . . ."
38 My thanks to Daniel Garber and Peter Anstey for pressing me on the nature of Spinoza's use of a history in the TTP. Both were rightly suspicious of an earlier version of this chapter, in which I interpreted Spinoza's use of a history there as largely continuous with the brief history of DPP.
39 In her account of Spinoza's views on scripture (James 2012: 146–151), Susan James discusses some of the concepts and issues I discuss here, including the relation between the DPP and the TTP and the relation of Spinoza's views about method to the Cartesian distinction between analysis and synthesis.

Bibliography

Anstey, P. R. (2012) "Francis Bacon and the classification of natural history," *Early Science and Medicine*, 17: 11–31.
Aristotle. (1984) *The Complete Works of Aristotle*, 2 vols, ed. J. Barnes, Princeton: Princeton University Press. References here are to Bekker pagination.
Bacon, F. (1861–1879) *The Works of Francis Bacon*, 14 vols, eds. J. Spedding, R. L. Ellis, and D. D. Heath, London. Cited by "SEH," volume number, and page number.
Descartes, R. (1996) *Oeuvres de Descartes*, 11 vols, eds. C. Adam and P. Tannery, Paris: Vrin.
Garber, D. (2001) *Descartes Embodied: Reading Cartesian Philosophy through Cartesian Science*, Cambridge: Cambridge University Press.
Garrett, A. (2003) *Meaning in Spinoza's Method*, Cambridge: Cambridge University Press.
Garrett, D., ed. (1996a) *The Cambridge Companion to Spinoza*, Cambridge: Cambridge University Press.
———. (1996b) "Spinoza's ethical theory" in ed. D. Garrett 1996a, pp. 267–314.
Jalobeanu, D. (2012) "Francis Bacon's natural history and the Senecan natural histories of early modern Europe," *Early Science and Medicine*, 17: 197–229.

James, S. (2012) *Spinoza on Philosophy, Religion, and Politics*, Oxford: Oxford University Press.

Jardine, L. (1974) *Francis Bacon: Discovery and the Art of Discourse*, Cambridge: Cambridge University Press.

Malherbe, M. (1996) "Bacon's method of science" in ed. M. Peltonen 1996, pp. 75–98.

Peltonen, M., ed. (1996) *The Cambridge Companion to Bacon*, Cambridge: Cambridge University Press.

Spinoza, B. (1925) *Spinoza Opera*, 4 vols, ed. C. Gebhardt, Heidelberg: Carl Winter.

——. (1985) *The Collected Works of Spinoza*, vol. 1, ed. E. Curley, Princeton: Princeton University Press.

——. (2002) *Spinoza: Complete Works*, ed. S. Shirley, Indianapolis: Hackett.

7 Principles in Newton's Natural Philosophy*

Kirsten Walsh

> Natural philosophy should be founded not on metaphysical opinions, but on its own principles.
>
> —An unpublished preface to the *Principia*[1]

As this volume attests, principles *mattered* in the early modern period. Calling a proposition a principle signaled its importance. It told you that the proposition was, for example, foundational, universal, essential, self-evident, or demonstrable. Principles played a central role in the philosophies of Descartes, Leibniz, Spinoza, and Hume, to name just a few of the figures featured in this volume.

Newton's great work of "rational mechanics" is supposedly about principles. The term is even in the title: *Philosophiae naturalis principia mathematica* (*Mathematical Principles of Natural Philosophy*). And in his preface to the first edition (1687), Newton expressed his hope that "the principles set down here will shed some light on either this mode of philosophizing or some truer one" (Newton 1999: 383). Statements such as this one, and the choice of title, suggest that principles had an important role to play in Newton's natural philosophy. In view of this, one might have several expectations. First, given the huge amount of Newton scholarship conducted over the last three hundred years, one might expect some of Newton's commentators to have carried out conceptual analysis of Newton's use of the term *principle*. Second, given his emphasis on principles, one might expect that Newton used the term in a systematic and transparent way.

One might be disappointed.

Newton's commentators have spent a lot of time analyzing key terms, such as *force* (e.g., Janiak 2007; Westfall 1971), *hypothesis* (e.g., Cohen 1962, 1969; Walsh 2012b), *cause* (e.g., Janiak 2013; Schliesser 2013), *experimental philosophy* (e.g., Shapiro 2004), *query* (e.g., Anstey 2004), *explanation* (e.g., Ducheyne 2012: 47–49) and *experimentum crucis* (e.g., Bechler 1974; Jalobeanu 2014)—to name just a few! But there has been surprisingly little analysis of his usage of the term *principle*.[2] I redress this oversight by exploring the notion of a principle in Newton's natural philosophy.

Moreover, Newton's use of the term *principle* appears to be *unsystematic* and *opaque*. As we shall see, on first appearances, Newton's use of the term is rough and messy. Was *principle*, then, just a buzzword for Newton? Or is there something important underlying his use of the term? I support the latter. While the term *principle* did not play a central role in Newton's methodology (indeed, I suggest that *hypotheses* and *theories* are the important concepts for Newton), once we disambiguate his usage, something systematic arises. What it takes to be a principle, by Newton's lights, is to *play a certain kind of role*.

I proceed as follows: I start by examining the principles in Newton's published work, both his *Principia* and his *Opticks*. I show, first, that Newton used the term in two ways, so I distinguish between *propositional-principles* and *ontic-principles*. Then, focusing more closely on the propositional-principles, I show, secondly, that Newton applied the label "principle" to propositions that (a) are deduced from phenomena and (b) function as premises in his inferences. Next, I give an account of Newton's propositional-principles in their broader methodological context. Drawing on Newton's epistemic distinction between theories and hypotheses, I argue that Newton's propositional-principles are a kind of theory. I then show that what differentiates Newton's principles from other kinds of theories is the function they serve—Newton's principles support his mathematico-experimental method in a crucial way. Finally, I test my account by turning to a draft manuscript in which Newton enumerated four "Principles of Philosophy." Here, my account of principles is illuminating: of those so-called principles, only the ones that (a) are deduced from phenomena and (b) function as premises, became principles in Newton's published work. From this discussion, a particular feature of Newton's use of the term *principle* emerges: labelling or referring to a proposition as a principle tells us about the *function*, rather than the *content*, of the proposition. I argue that this highlights a more general lesson, namely, that when we study Newton's methodology, we should emphasize functions and distinctions over content. And so, I digress to study a case that illustrates this broader point. I conclude that Newton's application of the label "principle" is exactly what we should expect, given his methodology.

Two Kinds of Principles

Newton's use of the term *principle* appears to be *opaque* and *unsystematic*. First, it is not at all clear what kind of thing the term *principle* is supposed to pick out. Neither the *Principia* nor the *Opticks* contains any propositions explicitly labelled principle. Instead, the term is found in the discussions following the introduction of new propositions. That is, while no propositions are *labelled* principle, many propositions are *referred* to as principles in the *scholia*. So a careful reading is required to figure out what the principles are. Second, once we identify the principles in Newton's work, it still seems that his use of the term is neither predictable nor consistent. One might assume, for example, that the *principia mathematica* referred to in the title of Newton's work are the propositions labelled laws. However, when the term *principle* appears in the *Principia*, it refers variously to the laws, lemmas, other mathematical propositions, and philosophical propositions.[3] Moreover, in the *Opticks*, the term refers to the axioms of optics, universal gravitation, and the forces, powers, and dispositions relating to the behavior of light. So, at first glance, Newton's use of the term *principle* appears to be unsystematic.

Newton's usage, however, becomes clearer once we disambiguate between two kinds of principles: *propositional-principles* and *ontic-principles*.[4] Table 7.1 summarizes the key features of these two kinds of principles.

In this section, I consider some examples of both kinds of principles. But before we begin, there are several things to notice about this distinction. First, these are related notions. Very broadly speaking, principles are *foundational* in both senses. Propositional-principles are foundational in that they are the *premises* from which other propositions are inferred. Ontic-principles are foundational in that they are the *causes* of phenomena. Second, these two notions are not well differentiated in Newton's work. In particular, we shall see that Newton's laws of motion have an ambiguous status in this classification.

Since we are interested in principles, it seems fitting to begin with the *Principia*. Here, we will find propositional-principles but not ontic-principles.[5]

In the *Principia*, Newton started with a set of axioms—his *laws of motion*. These provided the fundamental mathematical conditions from which his system would be built. Armed with his mathematical machinery—a geometrical form of infinitesimal calculus introduced in a series of lemmas in Book One, section 1—Newton proceeded, in Books One and Two, to explore the mathematical consequences of his laws. These increasingly complex, mathematical consequences were stated as a series of propositions, and further labelled as either "theorems" or "problems." These propositions

Table 7.1 Two kinds of principles

Propositional-principle	Ontic-principle
A truth or proposition on which others depend.	A power, force or disposition.
1. A claim about a thing	1. The thing itself
2. Functions as a premise	2. Functions as a cause
3. Known from the phenomena	3. Known from the phenomena
4. Truth-apt	4. Not truth-apt
E.g., laws of motion, mathematical propositions, theory of universal gravitation, the *Opticks* axioms	E.g., the cause by which light is reflected and refracted, forces of attraction, passive and active forces

addressed topics such as the laws and effects of centrally directed forces, the three-body problem, the motion of minimally small bodies, the effects of air resistance on pendulums, wave motion and the motion of sound, and the physics of vortices.[6] In Book Three, Newton shifted from considering an abstract mathematical system to a concrete physical system: the system of the world. Armed with a set of "rules for the study of natural philosophy" and a list of phenomena (detailing the motions of the planets and the Moon), Newton employed the mathematical propositions from Books One and Two to infer his theory of universal gravitation, and then to apply it to other phenomena such as the shape of the Earth, the precession of the equinoxes, and the motions of comets.

As I noted earlier, in the *Principia*, Newton did not *label* any proposition a principle, but he did *refer* to various propositions as principles. For instance, in the scholium to the laws of motion, where he provided justification for the laws, he referred to them as principles. He wrote, "The principles I have set forth are accepted by mathematicians and confirmed by experiments of many kinds" (Newton 1999: 424). However, the term *principle* did not just apply to the axioms or laws of motion; it had a broader application. For example, in the scholium at the end of section 1, in which Newton provided a mathematical system in the form of a series of lemmas, Newton referred to the lemmas as principles. He explained that he had included the lemmas to provide preliminary proofs of his mathematical tools so that he would not have to present them in detail later. For,

> we shall be on safer ground using principles that have been proved.
> . . . [Moreover,] the force of such proofs always rests on the method of the preceding lemmas.
>
> (Newton 1999: 441–442)

Furthermore, at the beginning of Book Three, Newton referred to the (mathematical) propositions in Books One and Two as principles:

> In the preceding books I have presented principles of philosophy that are not, however, philosophical but strictly mathematical—that is, those on which the study of philosophy can be based. These principles are the laws and conditions of motions and of forces, which especially relate to philosophy . . . It still remains for us to exhibit the system of the world from these same principles.
>
> (Newton 1999: 793)[7]

Among other things, these principles include the laws of Keplerian motion.[8] Some of the philosophical propositions of Book Three were also referred to as principles. For example, in Book Three, proposition 22, Newton wrote, "*All the motions of the moon and all the inequalities in its motions follow from the principles that have been set forth*" (Newton

1999: 832). In this context, the principles include Newton's theory of universal gravitation.

So far we have seen that, in the *Principia*, Newton's use of the term *principle* was sporadic and obscure. He did not use the term often, but when he did, he used it to refer to a range of propositions—certainly not just the laws of motion. It might appear that Newton was using the term *principle* haphazardly as a generic term for any kind of claim or statement. But there are two constraints on Newton's use of the term. The first constraint is that each of the propositions referred to as a principle had high epistemic status (at least according to Newton). We have seen that Newton considered his laws of motion to have been confirmed by experiment and the lemmas proved mathematically. The rest of the mathematical propositions followed deductively from the laws and lemmas, and so were considered certain too. Finally, the philosophical propositions were deduced from mathematical propositions and the phenomena, which, in Newton's mind, made them certain too. As we shall see, Newton considered all these propositions, including the laws, to have been "deduced from phenomena."[9] Moreover, Newton did not use the term *principle* to refer to hypotheses or queries—that is, those propositions that were not deduced from phenomena and so were uncertain.[10] The second constraint is that each time Newton referred to a proposition as a principle, it was functioning as a premise. This tells us that, for Newton, the term *principle* was context-specific: the proposition was playing a foundational role in the context of a particular argument or inference. This gives us two conditions for calling a proposition a principle:

1. It is deduced from the phenomena, and
2. It functions as a premise.

Now let us turn to the *Opticks*, in which Newton employed both kinds of principles.

The *Opticks* begins in a similar way to the *Principia*—with a list of axioms. However, these axioms are of a different sort to Newton's laws of motion. Where the laws of motion are about forces or causes of motion, the axioms of the *Opticks* describe generalized regularities or correlations.[11] For example,

> Axiom I *The Angles of Reflexion and Refraction, lie in one and the same Plane with the Angle of Incidence.*
> Axiom II *The Angle of Reflexion is equal to the Angle of Incidence.*
> Axiom III *If the refracted Ray be returned directly back to the Point of Incidence, it shall be refracted into the Line before described by the incident Ray.*

> (Newton 1952: 5)

These are statements about the geometrical properties of light. They had been established by experiment and so they meet condition (a), they are deduced from phenomena. Following these axioms, Newton wrote,

> I have now given in Axioms and their Explications the sum of what hath hitherto been treated of in Opticks. For what hath been generally agreed on I content my self to assume under the notion of Principles, in order to what I have farther to write. And this may suffice for an Introduction to Readers of quick Wit and good Understanding not yet versed in Opticks.
>
> (Newton 1952: 19–20)

This passage indicates that, as far as Newton was concerned, these axioms were uncontroversial. They were supposed to provide a summary of the current state of optics; the basic mathematics required for geometrical optics. Moreover, the passage states that these axioms would feature as basic assumptions in Newton's treatise. Thus, they would provide a foundational role in the inferences in the *Opticks*,[12] and so, because they also meet condition (b), they function as premises.

As with the *Principia*, in the *Opticks* the term *principle* appears repeatedly. Whereas the first usage looks similar to those we saw in the *Principia*, other usages look very different: they are ontic-principles. First, the term refers to the cause by which light is reflected and refracted. For example, in Book Three, query 4, Newton wrote,

> Do not the Rays of Light which fall upon Bodies, and are reflected or refracted, begin to bend before they arrive at the Bodies; and are they not reflected, refracted, and inflected, by one and the same Principle, acting variously in various Circumstances?
>
> (Newton 1952: 339)

In this passage, the term *principle* applies to some sort of mechanism or power which causes light to bend. There are similar usages of the term in Books One and Two.

Second, the term is used to describe forces of attraction. For example, in Book Three, query 31, Newton invoked principles in his consideration of the following question:

> Have not the small Particles of Bodies certain Powers, Virtues, or Forces, by which they act at a distance, not only upon the Rays of Light for reflecting, refracting, and inflecting them, but also upon one another for producing a great Part of the Phænomena of Nature?
>
> (Newton 1952: 375–376)

Arguing that we can only learn about such powers, virtues or forces by studying their effects, he went on to discuss gravitational attraction, magnetic attraction, electrical attraction, and finally chemical reactions, eventually asking,

> is it not for want of an attractive virtue between the Parts of Water and Oil, or Quick-silver and Antimony, of Lead and Iron, that these Substances do not mix; and by a weak Attraction, that Quick-silver and Copper mix difficultly; and from a strong one, that Quick-silver and Tin, Antimony and Iron, Water and Salts, mix readily? And in general, is it not from the same Principle that Heat congregates homogeneal Bodies, and separates heterogeneal ones?
>
> (Newton 1952: 383)

Here, the term *principle* applies to the forces, powers or dispositions that cause bodies to interact in certain ways.

Finally, in Book Three, query 31, Newton distinguished between *active* and *passive* principles. For example, in his discussion of the force of inertia, Newton wrote,

> The *Vis intertiæ* is a passive Principle by which Bodies persist in their Motion or Rest, receive Motion in proportion to the Force impressing it, and resist as much as they are resisted. By this Principle alone there never could have been any Motion in the World. Some other Principle was necessary for putting Bodies into Motion; and now they are in Motion, some other Principle is necessary for conserving the Motion.
>
> (Newton 1952: 397)

Here, Newton tells us that a material body on its own is passive—that is, brute and inanimate. If at rest, it cannot begin to move, and if in motion, it can neither stop completely nor change speed or direction. This disposition to remain at rest or in motion is what Newton called the *vis intertiæ*. Newton conceived of this force of inertia as a passive principle. For a material body to change its motion, it requires an *impressed force*—that is, an external force. Newton conceived of such forces as active principles. He argued that active principles are necessary, if there is to be any kind of motion in the world.[13]

So in the *Opticks, principle* referred to two kinds of things: truths or propositions and forces, dispositions, or powers. I have already identified two conditions for calling a proposition a principle in that former sense, and we have seen that Newton's axioms meet both conditions. However, the majority of references to principles in the *Opticks* are references to ontic-principles. Ontic-principles are powers, forces, or dispositions that function as causes of phenomena. Thus, the term was used to refer to some unknown cause of some particular effect. It is significant that Newton relied heavily on ontic-principles in the queries to the *Opticks*. These queries, particularly query 31, are well known for their speculative content. They explore the nature of light, whether it can act on bodies, its relationship to heat, its role in vision, and the nature of luciferous æther. These topics concern a lot of unknown causes. And the main function of the queries was to set out a future research program in order to learn about these causes. These causes could only be discovered by investigating their effects—they could only be known from the phenomena. Thus, the term *principle* seems to be a generic term, denoting some unknown cause of some particular effect.

To summarize, we have seen that in Newton's published work the term *principle* had two different kinds of referent: propositions and things in the world. In the *Principia*, Newton referred to propositions (e.g., axioms, lemmas, mathematical propositions, and philosophical propositions) as principles. Similarly, in the *Opticks* he used the term to refer to the propositions labelled as axioms. We found, however, that propositions are not *labelled* as principles, but only *referred* to as principles in specific contexts. Thus, in this first sense, principles are propositions that (a) are deduced from phenomena and (b) function as premises. And so *principle* indicates both the *epistemic status* and the *function* of a proposition. In the *Opticks*, we also found a second, nonpropositional, usage of the term. Newton used the term to refer to the powers, forces, or dispositions of objects. In this second sense, a principle does not refer to *a claim* about a thing in the world, but to *the thing itself*. Thus, we can distinguish between *propositional-principles* and *ontic-principles*. In the *Principia*, we found only propositional-principles; in the *Opticks* we found both.

The two kinds of principle were not well differentiated in Newton's work. For example, in the *Opticks*, Book Three, query 31, Newton was apparently talking about the laws of motion and universal gravitation as ontic-principles:

> It seems to me farther, that these Particles have not only a *Vis intertiæ*, accompanied with such passive Laws of Motion as naturally result from that Force, but also that they are moved by certain active Principles, such as is that of Gravity, and that which causes Fermentation, and the Cohesion of Bodies. These Principles I consider, not as occult Qualities, supposed to result from the specifick Forms of Things, but as general Laws of Nature, by which the Things themselves are form'd; their Truth appearing to us by Phænomena, though their Causes be not yet discover'd.
> (Newton 1952: 401)

However, in his letter to Roger Cotes (28 March 1713), Newton was treating the laws of motion as propositional-principles:

> These Principles [in this context, the laws of motion] are deduced from Phænomena & made general by Induction: which is the highest evidence that a Proposition can have in this Philosophy.
> (Newton 1959–1977, 5: 397)

I do not think there is any contradiction here. In some situations, Newton was thinking about the laws as causes in the world; in other situations, Newton was thinking about the laws as axioms in a mathematical system.

Neither of these notions is a typical usage of the term *principle* in early modern philosophy. Consider, for example, the entry for "Principle" in John Harris's *Lexicon Technicum* (1708). Harris described *principle* as "a Word very commonly and very variously used" (Harris 1708). He provided six different uses of the word:

1. "a Maxim, an Axiom, or a good Practical Rule of Action";
2. "a Thing Self-evident, and as it were Naturally known, and then 'tis usually called, a *First Principle*";
3. "Rudiments or Elements; as when we say, the *Principles of Geometry, Astronomy, Algebra*; we mean the Doctrine or Rules of those Sciences";
4. "in Chymistry particularly, 'tis taken for first Constituent and Component Particles of all Bodies";
5. In "modern chymistry," "there are five kinds, or different Sorts of Bodies, which may by Fire be drawn from many mix'd Natural Bodies, and therefore which may in a large sense be called *Principles*"; and
6. In a general sense, "the first cause of any Things *Existence*, or *Production*, or of its becoming *Known* to us," such as the Aristotelian elements of earth, water, air and fire, the Epicurean principles of magnitude, figure and weight, and Boyle's mechanical principles of matter, motion and rest.

The first usage is intended in a moral or religious context and does not capture the usage of Newton's propositional-principles. Nor does (2), for Newton's principles were not self-evident but deduced from phenomena. Newton's propositional-principles might, to some extent, be characterized by (3). However, Newton's usage of the term was both broader and more nuanced than this suggests. Usages (4) and (5) are ontological, as opposed to propositional. However, where Newton's ontic-principles refer to forces, powers or dispositions, these chymical principles refer to entities. As Newman's chapter in this volume attests, the latter is a more typical application of the term. Finally, Newton's principles cannot be characterized by (6), since, as we shall see, Newton did not conceive of his principles as the first causes. Rather, he thought there were further causes to discover. And so, Newton's principles do not fit any of the particular uses identified by Harris.

In the *Opticks*, Book Three, query 31, Newton made the following statement about principles:

> But to derive two or three general Principles of Motion from Phænomena, and afterwards to tell us how the Properties and Actions of all corporeal Things follow from those manifest Principles, would be a very

great step in Philosophy, though the Causes of those Principles were not yet discover'd: And therefore I scruple not to propose the Principles of Motion above-mention'd, they being of very general Extent, and leave their Causes to be found out.

(Newton 1952: 401–402)

This passage tells us several things about Newton's notion of propositional-principles. First, principles should be derived from observation and experiment, that is, "from Phænomena." Second, principles are foundational in one sense but not in another. On one hand, the object is to understand "how the Properties and Actions of all corporeal Things follow from those manifest Principles." So principles are premises from which other propositions are inferred. That is, they are foundational in that, once obtained, they provide the foundation for other propositions. But on the other hand, "the Causes of those Principles were not yet discover'd." So principles are not foundational, since they do not necessarily identify or stipulate first causes (or, indeed, any kind of cause!). For, as Newton said in the *Opticks*, query 28,

[a]nd though every true Step made in this Philosophy brings us not immediately to the Knowledge of the first Cause, yet it brings us nearer to it, and on that account is to be highly valued.

(Newton 1952: 370)

Nevertheless, Newton's contemporaries appeared to be comfortable about his usage of *principle*. So, while his usage was unusual, I take it that it was broadly consonant with the usage at the time. In other words, Newton's principles fit the spirit, if not the particulars, of Harris's definition. Cotes certainly recognized it as appropriate to call Newton's theory of universal gravitation a principle. For example, in his editor's preface to the second edition, he wrote,

I know indeed that some men, even of great reputation, unduly influenced by certain prejudices, have found it difficult to accept this new principle [of gravity] and have repeatedly preferred uncertainties to certainties.

(Newton 1999: 386)

The Method of Principles

So far, we have distinguished between propositional-principles and ontic-principles in Newton's methodology. In this section, I take a closer look at Newton's use of propositional-principles, in order to understand where principles fit in Newton's methodological framework. I begin by introducing Newton's distinction between theories and hypotheses, which, I argue, is the central epistemic distinction in Newton's methodology. I then show that, in Newton's framework, propositional-principles are a kind of theory. I argue that what differentiates principles from other kinds of theories is neither their epistemic status nor their content but, rather, their function.[14]

A well-known feature of Newton's methodology is his distinction between certainty and uncertainty.[15] Newton contrasted the certainty of his own natural philosophical claims with the mere hypotheses and speculations which other philosophers found appealing. Consider, for example, the following methodological statement from Newton's earliest publication, "A New Theory of Light and Colors," from 1672:

> A naturalist would scearce expect to see the science of [colors] become mathematicall, & yet I dare affirm that there is as much certainty in it as in any other part of Opticks. For what I shall tell concerning them is not an Hypothesis but most rigid consequence, not conjectured by barely inferring 'tis thus because not otherwise or because it satisfies all phænomena (the Philosophers universall Topick,) but evinced by the mediation of experiments concluding directly & without any suspicion of doubt.
>
> (Newton 1959–1977, 1: 96–97)

When it was written, this statement was quite scandalous. At a time when the Royal Society valued epistemic responsibility, never claiming certainty when the evidence only supported high probability,[16] Newton was making strong claims to certainty—and apparently without any special warrant!

In fact, Newton thought he was warranted in making such claims, because he had a reliable methodology. Newton's approach was based on the idea that mathematics is a bearer of certainty—if one begins with certain axioms, one can reason deductively to certain theorems without epistemic loss. He thought it was possible to apply this method of reasoning to natural philosophy: one can reason deductively from laws and principles to propositions in natural philosophy. So, if one can establish *certain* natural philosophical laws or principles, it is possible to reason mathematically to *certain* propositions. By reasoning in this way, Newton thought he could achieve a mathematical science. The challenge, then, was to identify laws or first principles that met this requirement of certainty—via *deduction from phenomena*.

Newton's distinction between certainty and uncertainty is best characterized as a distinction between "theories" and "hypotheses" (outlined in Table 7.2).[17] In Newton's methodology, theories and hypotheses deal with different subject matter, have different epistemic statuses, and perform different roles in theorizing. Theories systematize the observable, measurable properties of things; hypotheses describe the (unobservable) nature of things. Theories are inferred from observation and experiment; hypotheses are speculative. For example, Newton saw universal gravitation—that is,

Table 7.2 Definitions of *theory* and *hypothesis*

Theory	Hypothesis
A proposition is a "theory" iff it meets the following conditions:	A proposition is a "hypothesis" iff it meets one or more of the following conditions:
T1. It is certainly true, because it is reliably inferred from experiment;	H1. It is, at best, only highly probable; or
T2. It is experimental—something that has empirically *testable* consequences; and	H2. It is a conjecture or speculation—something not based on empirical evidence; or
T3. It is concerned with the *observable, measurable properties* of the thing rather than its nature.	H3. It is concerned with the nature of the thing rather than its observable, measurable properties.

the proposition that any two bodies in the universe attract each other with a force that is directly proportional to the product of their masses and inversely proportional to the square of the distance between them—was a theory, since it was inferred from celestial and terrestrial observations, had empirically testable consequences, and was used to systematize those observations. However, an explanation of the nature and cause of gravity would be a hypothesis, since it concerns the unobservable nature of things, and is speculative, rather than inferred from experiment—and thus, any account Newton could give would be, at best, only probable.[18]

The distinction between theories and hypotheses is central to Newton's methodology. For Newton, theories were on epistemically surer footing than hypotheses because they were grounded on phenomena, whereas the latter were grounded in speculations. And so hypotheses could never trump theories. When faced with a disagreement between a hypothesis and a theory (e.g., suppose our theory seems to imply action-at-a-distance, but the most plausible hypothesis about the nature of motion tells us that action-at-a-distance is impossible), we should modify the hypothesis to fit our theory, and not *vice versa*. The distinction is nicely captured in a draft letter from Newton to Cotes (28 March 1713):

> And therefore as I regard not hypotheses in explaining the phenomena of nature, so I regard them not in opposition to arguments founded upon phenomena by induction or to principles settled upon such arguments. In arguing for any principle or proposition from phenomena by induction, hypotheses are not to be considered. The argument holds good till some phenomenon can be produced against it.
>
> (Newton 2004: 120)

While Newton railed against hypotheses—determined to preserve the certainty of his propositions and to avoid epistemic loss by keeping speculative

conjectures apart—hypotheses played an important role in Newton's nego-
tiations between certainty and speculation.[19]

I now argue that Newton's (propositional-)principles were a kind of
theory but performed a specific role. We have already seen that Newton conceived of principles as propositions that had been deduced from
phenomena and therefore certain. Thus, they fit the preceding definition
of a "theory." Moreover, in statements such as this one from his letter to
Cotes (28 March 1713), Newton explicitly contrasted his principles with
hypotheses:[20]

> These Principles [in this case, the laws of motion] are deduced from
> Phænomena & made general by Induction: which is the highest evi-
> dence that a Proposition can have in this philosophy. And the word
> Hypothesis is here used by me to signify only such a Proposition as is
> not a Phænomenon nor deduced from any Phænomena but assumed or
> supposed without any experimental proof.
>
> (Newton 1959–1977, 5: 397)

In the previous section, we saw that Newton used the term *principle* to refer
to several different kinds of proposition: laws, axioms, lemmas, mathematical propositions, and philosophical propositions. Where the term *theory*
unifies all of these propositions on the basis of their epistemic status, *principle* picks out a certain kind of theory, namely, a theory that functions as a
premise. Let us examine the function of a principle more closely by considering the inference structure of the *Principia*.

The inference structure of the *Principia* is roughly as follows: Newton
started with his laws of motion. He claimed that these laws are supported
by experiment. From these laws, with the help of his mathematical tools
stated in a series of lemmas (i.e. his geometrical form of infinitesimal calculus), Newton inferred the propositions in Books One and Two of the
Principia (the "mathematical propositions"). From the mathematical propositions and supported by celestial phenomena, Newton inferred his theory
of universal gravitation. From his theory of universal gravitation and some
planetary observations (more "phenomena"), Newton inferred other features of the system, such as his theories of the comets and the shape and
motion of the Moon.[21] The tiers of this extended inference are summarized
in Figure 7.1.

We have seen that Newton applied the term *principle* to propositions at
every level of this diagram except the top one. When Newton used the laws
of motion to infer his mathematical propositions, he referred to his laws as
"principles." Newton referred to his lemmas as "principles" in a similar context. When he used his mathematical propositions to infer his theory of universal gravitation, he referred to them as "principles." And finally, when he
used his theory of universal gravitation to develop his propositions concerning comets and the moon, he referred to his theory of universal gravitation

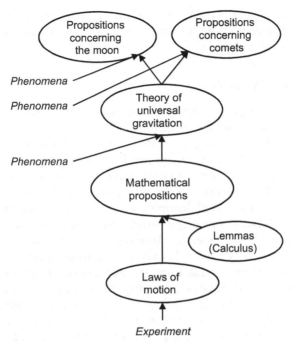

Figure 7.1 The structure of inference in the *Principia*

as a "principle." In fact, the only propositions listed in Figure 7.1 that *were not* referred to as principles in the *Principia* were the propositions concerning the Moon and the propositions concerning the comets. In the context of the *Principia* this is not surprising, since these were the final propositions in the *Principia*—they were not premises for any further arguments.

It should not be surprising, therefore, that these laws and propositions were not labelled as principles but only referred to as principles in specific contexts. The label was bestowed only circumstantially—to indicate a specific temporary function. It was, nevertheless, an important function. As we saw earlier, Newton's mathematico-experimental approach had two crucial components: (1) establishing certain principles by deducing them from phenomena and (2) reasoning deductively from those principles to other propositions (maintaining certainty). The *Principia* gave every appearance of following Newton's prescriptions for a mathematico-experimental science. Starting with laws and a strong mathematical framework of lemmas, Newton inferred mathematical propositions and, eventually, philosophical propositions. Moreover, in the scholium to the laws, Newton asserted that they were certainly true because they had been confirmed by experiment. Thus, he felt warranted in using them to deduce his theory of universal gravitation. And so Newton conceived of principles as a *kind* of theory, namely, a theory that functioned as a premise in an argument or inference.

Principles are important then, because they are theories from which other theories are deduced. I now turn to a test case: a draft manuscript intended as a preface for the *Opticks*.

Some Puzzling Principles

In a draft manuscript, originally intended as the preface to the *Opticks*, Newton wrote an account of principles.[22] The manuscript, which J. E. McGuire has called the "Principles of Philosophy" (McGuire 1970), provides the most extensive discussion of principles that can be found in Newton's writings. In this manuscript, Newton gave an account of his method of principles and then enumerated the four principles that he regarded as the foundation for his natural philosophy. The manuscript presents a puzzle for my account of Newton's principles since this is the only place we find principles explicitly labelled as such. Moreover, of the four principles identified in this manuscript, only one of them is referred to as a principle in Newton's published work. Here, we shall find my account of principles to be illuminating: of those so-called principles, only the ones that meet my two conditions for "principle" appear as principles in Newton's published work. My discussion highlights a particular feature of Newton's method of principles, namely, that labelling or referring to a proposition as a principle tells us about the *function*, rather than the *content*, of the proposition. I then suggest that Newton's method of principles is an example of a more general feature of Newton's methodology. I digress to study a case that illustrates this broader point, namely, that when we study Newton's methodology, we should emphasize functions and distinctions over content.

The manuscript opens with a discussion of methodology, which supports the framework I have outlined. For example, the discussion ends with the following passage:[23]

> Tis much better to do a little with certainty &[24] leave the rest for others that come after, than to explain all things by conjecture without[25] making sure of any thing. And there is no <other> way of doing anything with certainty then by drawing[26] conclusions from experiments & phænomena untill you come at general[27] Principles &[28] then from those Principles giving an account of Nature. Whatever[29] is certain in Philosophy is owing to this method & nothing can be done without it. <I will instance in some particulars.>

(MS Add. 3970, fol. 479r)

The themes in this passage are by now familiar to us. Newton's goal was to produce determinate natural philosophical claims. That is, he wanted to achieve certainty. Thus, in the trade-off between explanatory scope and epistemic strength, Newton chose the latter. As far as Newton was concerned, the only way to achieve certainty in natural philosophy was to start by making observations and experiments and then to infer general principles from the phenomena. Such general principles would form the basis for an account of nature.

Following these methodological preliminaries, Newton proceeded to enumerate four general principles of philosophy, providing evidence or support for them. I shall discuss them in detail shortly. But for now I simply summarize them:

1. There exists an Intelligent Creator;
2. Matter is impenetrable;
3. All bodies in the Universe gravitate towards all other bodies in the Universe, in accordance with Newton's theory of universal gravitation; and
4. Sensible bodies are composed of tiny particles of matter with gaps between them.

These "principles of philosophy" appear to present a counterexample to my account of Newton's method of principles, since they do not function as premises. However, I now argue that the manuscript supports my account. My argument rests on the recognition that Newton's "Principles of Philosophy" is a draft that was never published. Only one of the propositions was ever referred to as a "principle" in Newton's published work. Significantly, it was *also* the only proposition that fit both conditions for calling a proposition a "principle." While we can trace versions of the other principles in the *Principia* and the *Opticks*, they are never referred to as "principles." I therefore argue that, while Newton may have *hoped* to present these four propositions as principles, he changed his mind.

Let us start by considering the proposition that *was* referred to as a "principle" in print: principle 3, Newton's theory of universal gravitation. That is, that "all bodies[30] in the Universe have a tendency toward one another proportional to the quantity of matter contained in them & that this tendency in receding from the body decreases & is reciprocally proportional to the square of the distance from the body."[31] In support, Newton summarized the evidence he provided for universal gravitation in the *Principia* Book Three. For example, the pendulum experiments that he used in proposition 4 to demonstrate that the force that keeps the Moon in orbit is the same force that causes heavy bodies fall to the ground, namely, gravity. Also the evidence for proposition 5: that the moons of Jupiter and Saturn gravitate toward Jupiter and Saturn, respectively, and the primary planets gravitate toward the Sun, and this force of gravity is responsible for keeping the moons and planets in orbit. And finally, he noted that many comets seem to gravitate toward the Sun as well. In the *Principia*, principle 3 met both conditions for "principle." First, it was deduced from celestial and terrestrial phenomena such as the motions of the planets. Second, it functioned as a premise from which the motions of celestial bodies such as comets were deduced.

Now consider principle 1, that there exists an intelligent creator: "a God or Spirit infinite eternal omniscient omnipotent." To support this principle, Newton put forward an argument from design, explaining that "the best argument for such a being is the frame of nature & chiefly the contrivance of the bodies of living creatures." To this end, he pointed, first, to isomorphic traits among land animals, such as the observation that they all have two eyes, a nose, and a mouth and, second, to the ingenuity of the functions of such traits. Asking, for example, "What [is] more difficult then to fly? &[32] yet was it by chance that all creatures can fly which have wings?" Newton evidently found these arguments from the phenomena extremely compelling:

> These & such like considerations are the most convincing arguments for such a being & have convinced mankind in all ages that the world & all the species of things therein were[33] originally framed by his power & wisdom. And to lay aside this argument is unphilosophical.[34]
>
> (MS Add. 3970, fol. 479r)

The ideas expressed in this passage were clearly very important to Newton, since traces of them can be found in both the *Opticks* and the *Principia*. For example, in the General Scholium to the *Principia*, there is an extended discussion of God:

> This most elegant system of the sun, planets, and comets could not have arisen without the design and dominion of an intelligent and powerful being . . . It is agreed that the supreme God necessarily exists, and by the same necessity he is *always* and *everywhere* . . . This concludes the discussion of God, and to treat of God from phenomena is certainly a part of natural philosophy.
>
> (Newton 1999: 940–943)

The similarities between this passage and the draft passage above are striking. Here, Newton was arguing from design for an intelligent creator. Since arguments for intelligent design rest on observable phenomena of the natural world, Newton argued that understanding God and his creation are legitimate topics for natural philosophy. A similar passage appears in the *Opticks* Book Three, query 31:

> Now by the help of these Principles [in this context, the laws of motion], all material Things seem to have been composed of the hard and solid Particles above-mention'd, variously associated in the first Creation by the Counsel of an intelligent Agent. For it became him who created them to set them in order. And if he did so, it's unphilosophical to seek for any other Origin of the World, or to pretend that it might

arise out of a Chaos by the mere Laws of Nature; though being once form'd, it may continue by those Laws for many Ages.

(Newton 1952: 402)

Principle 1 was not referred to as a principle in the *Principia* or the *Opticks*. And not because it was not deduced from the phenomena—indeed, Newton thought that it had been deduced from the phenomena and was certainly true. Rather, principle 1 did not appear as a principle because it never functioned as a premise and, thus, did not perform the role of a principle. That is, it meets condition (a) but not condition (b).

Now consider principle 2, that matter is impenetrable. Newton explained, "This is usually looked upon as a maxim known to us by the light of nature, altho we know nothing of bodies but by sense." That is, while others have taken this principle to be self-evident, Newton thought that compelling support for this principle would come only from sensory experience and the absence of counter-instances. He explained,

And such observations occurring every day to every man this property of bodies is acknowledged by all men[35] without any dispute & looked upon as an Axiom.[36]

(MS Add. 3970, fol. 479v)

Newton was clearly committed to principle 2. He thought that it was well-established and even deduced from phenomena. It never appeared as a principle in the *Principia* or the *Opticks* and yet traces of this principle are found in both works. For example, in the *Opticks* Book Three, query 31, Newton wrote,

this seems to be as evident as the universal Impenetrability of Matter. For all Bodies, so far as Experience reaches, are either hard, or may be harden'd; and we have no other Evidence of universal Impenetrability, besides a large Experience without an experimental Exception.

(Newton 1952: 389)

Again, the similarities between this published passage and the draft passage are striking. The impenetrability of matter is established from experience without exception—that is, from the phenomena. The same idea is expressed in the *Principia* Book Three, where Newton used the impenetrability of matter as his main example of the application of rule three.[37] He wrote,

That all bodies are impenetrable we gather not by reason but by our senses. We find those bodies that we handle to be impenetrable, and hence we conclude that impenetrability is a property of all bodies universally.

(Newton 1999: 795)

Newton argued that the proposition that matter is impenetrable was deduced from the phenomena, so it met condition (a). However, the impenetrability of matter never functioned as an explicit premise and, thus, did not perform the role of a principle.[38] That is, it met condition (a) but not condition (b).

Finally, consider principle 4, that sensible bodies are composed of tiny particles of matter with gaps between them. That is, "all sensible bodies are aggregated of particles laid together with many interstices or pores between them." To support this principle, Newton described phenomena such as the absorption of water and other liquids by various substances, the ability of acids to dissolve metals, and the transmission of light through various substances (e.g., air, water, oil, and crystals). Evidently, this principle was an important one for Newton's theory of optics:

> As by the <third>[39] Principle we gave an account <heretofore> of the motions of the[40] Planets & of the flux & reflux of the sea, so by this Principle we shall in[41] the following treatise give an acct of the permanent colors of natural bodies, nothing further being requisite for the production of those colors then that the colored bodies abound with pellucid particles of a certain size & density.[42]
>
> (MS Add. 3970, fol. 479v)

And yet, it was not referred to as a "principle" in the *Opticks*. In fact, in query 31, Newton argued that the proposition that all sensible bodies are composed of hard particles "seems to be as evident as the universal Impenetrability of Matter" (Newton 1952: 389). But in query 31 he was more interested in how the particles of matter stick together, than in what follows from this idea. He concluded,

> There are therefore Agents in Nature able to make the Particles of Bodies stick together by very strong Attractions. And it is the Business of experimental Philosophy to find them out.
>
> (Newton 1952: 394)

And so, Newton thought that the proposition that all sensible bodies are composed of tiny particles of matter with gaps in them was true and presumably deduced from the phenomena. But when it was introduced in query 31, Newton's focus was different. Instead of using this proposition to deduce the properties of colored bodies, Newton wanted to understand how the particles cohere. So this proposition might have met condition (a) but certainly not condition (b).

To summarize, while all four of Newton's "principles of philosophy" can be traced in Newton's published work, only one of them was referred to as a "principle." Moreover, that principle was the only one that met the two conditions for being a principle. This case highlights something important about how we should view Newton's principles. If we focus on content, they look unsystematic. But when we focus on function, Newton's use of the term starts to look systematic and important. Thus, if we want to understand how Newton used the term *principle*, we should focus on the functions of principles rather than on their content.

This lesson can be generalized. Newton's approach to the term "principle" was similar to his approach to other methodological terms such as *hypothesis*, *rule*, and *query*. Namely, the label emphasized the function of the proposition rather than its content. In fact, in Newton's work, it was entirely possible for the same proposition or idea to appear in one context with one label and in another context with a different label. I think this is a fairly important lesson about how to understand Newton's methodology. So now I briefly digress to illustrate this general point by examining another draft manuscript: a plan for the fourth book of the *Opticks*.[43]

Several drafts of the *Opticks* Book Four exist in various stages of development.[44] The topic of Book Four was to be "the nature of light & the power of bodies to refract and reflect it." One draft begins with a list of three items under the heading "Observations":

1. Of the 3 faciae of colors made by inflected light.
2. Of bent of several rays which passe at several differences from the edge of the knife.
3. Conclusions concerning the vernicular motions of the inflected rays.

(MS. Add. 3970, fols 342r–343r)

These are clearly topics to be covered by the observations rather than the observations themselves.[45]

The draft that I shall focus on here contains a list of propositions.[46] But since, as Alan Shapiro has pointed out, this is a draft of an outline or plan of the book, rather than a draft of the book itself, it contains little in the way of argument and virtually no discussion of experimental evidence (Shapiro 1993: 141–143). Thus, the propositions are things that Newton *hoped* to prove. For example,

> Prop. 1. The refracting power of bodies in vacuo is proportional to their specific gravities.
> Prop. 2. The refracting power of two contiguous bodies is the difference of their refracting powers in vacuo.

Newton listed about twenty propositions before he made a structural change: he went back to proposition 17 and relabeled it "hypothesis 1." He also added a heading: "The Conclusion." Then he began to renumber the propositions that followed. By the time Newton had finished with it, the draft contained eighteen propositions and a conclusion consisting of five hypotheses.

Let us focus on "Hypothesis 2," which I shall quote *in extenso*:[47]

> <Hypoth 2>.[48] As all the great motions in the world depend upon a certain kind of force (which[49] in this earth we call gravity) whereby great bodies attract one another at great distances: so all the <little>[50] motions in the world depend upon certain kinds of forces whereby minute bodies attract or dispell one another at little distances.
>
> How <the great bodies of the earth Sun <moon> & Planets gravitate towards one another what>[51] are the laws &[52] quantities of their gravitating forces at all distance from them & how all[53] the motions of those bodies are regulated by those their gravities I shewed in my Mathematical[54] Principles of Philosophy <to the satisfaction of my readers>: And if Nature be most simple & <fully> consonant to her self she observes the same method in regulating the motions of smaller bodies which she doth in regulating those of the greater. This[55] principle of nature being very remote from the conceptions of Philosophers[56] I forbore to describe it in <that Book[57]>[58] least I should be accounted an extravagant freak & so prejudice my Readers against all those things which were the main designe of the[59] Book: but &[60] yet I hinted at it both in the Preface <& in>[61] the book it self where I speak of the inflection of light & of the elastic[62] power of the Air: but the design of the book being secured by the approbation of Mathematicians, I had not <scrupled>[63] to propose this Principle in plane[64] words. The truth of this Hypothesis I assert not because I cannot prove it, but I think it very probable because a great part of the phaenomena of nature do easily flow from it which seem otherways inexplicable.
>
> (MS. Add. 3970, fol. 338v)

I. Bernard Cohen has described this as "a 'whale' of an hypothesis" (Cohen 1969: 320)—and he is right! When Newton started writing out this statement, he intended for it to be "Proposition 18"—a theory. But at some point, he scratched out "Prop 18" and re—branded it as "Hypoth 2." As we have seen, for Newton there was an important epistemic shift here. Theories were things that he was able to assert as true. Hypotheses were things that he was unable to assert because he did not have the evidence. Newton probably hoped to assert Proposition 18. But as he started to explicate it, he must have realized that he could not prove it. Thus, he relabelled it as a hypothesis—something speculative and, thus, uncertain. Newton also made it clear that he wanted to use this as a principle, but he was not certain of it, so he could not.

When Newton abandoned the idea of a fourth book, and restructured his *Opticks*, this Hypothesis 2 appears to have been re-worked to become part of Query 31 in the *Opticks*:[65]

> Have not the small Particles of Bodies certain Powers, Virtues, or Forces, by which they act at a distance, not only upon the Rays of Light

for reflecting, refracting, and inflecting them, but also upon one another for producing a great Part of the Phænomena of Nature? For it's well known, that Bodies act one upon another by the Attractions of Gravity, Magnetism, and Electricity; and these Instances shew the Tenor and Course of Nature, and make it not improbable but that there may be more attractive Powers than these. For Nature is very consonant and conformable to her self.

(Newton 1952: 375–376)

Moreover, similar themes appear in the *Principia* Book One. This is a much more technical discussion in which Newton aimed to solve "*[t]he motion of minimally small bodies that are acted on by centripetal forces tending toward each of the individual parts of some great body*" (Newton 1999: 622). In the scholium to propositions 94, 95, and 96, Newton noted that

[t]hese attractions are very similar to the reflections and refractions of light made according to a given ratio of the secants, as Snel discovered, and consequently according to a given ratio of the sines, as Descartes set forth. . . . Moreover, the rays of light that are in the air . . . in their passing near the edges of bodies, whether opaque or transparent . . . are inflected around the bodies, as if attracted toward them; and those of the rays that in such passing approach closer to the bodies are inflected the more, as if more attracted . . . And those that pass at greater distances are less inflected . . . Therefore because of the analogy that exists between the propagation of rays of light and the motion of bodies, I have decided to subjoin the following propositions for optical uses.

(Newton 1999: 625–626)

And in Newton's discussion of Rule 3 in the *Principia*, he wrote, "nor should we depart from the analogy of nature, since nature is always simple and ever consonant with itself" (Newton 1999: 795). And finally, in Newton's discussion of proposition 6 in the *Opticks* Book One, part I, he wrote, "That it should be so is very reasonable, Nature being ever conformable to her self; but an experimental Proof is desired" (Newton 1952: 76).

The lesson is simple. Here we have seen that similar ideas emerged in different contexts throughout Newton's work. The same content was variously referred to as a "hypothesis," a "rule," a "proposition," a "principle," and even a "query," depending on what role it was playing. This tells us that, if we want to understand Newton's usage of a methodological term such as "principle" or "hypothesis," we should look at what work the idea is doing rather than its content. And we cannot expect the same idea always to bear the same label.

I want to close this section with a final comment about Newton's "Principles of Philosophy." We should interpret this manuscript very much as a work in progress. Newton was attempting something very ambitious, but it obviously did not work out. It is striking that these principles do not form a coherent system of philosophy. That is, there is no framework, mathematical or philosophical, that ties the principles together. This suggests that Newton had a different goal in mind when writing this. He was trying to give an account of his natural philosophy in broad brush strokes—as befits a preface.[66] In his commentary on this manuscript, McGuire sheds some light on this intended preface. He dates the draft to late 1703, not long before the *Opticks* was published (McGuire 1970: 179). At that time, Newton was focused on the various conceptual problems arising from his theories of universal gravitation and the composition of light, such as extending the application of the laws of motion to the smallest units of matter: "In general he wished to make clearer the significance of his natural philosophy" (McGuire 1970: 178). One interesting thing about this manuscript is that we get the sense that Newton conceives of his two great works as two lines of inquiry in a unified natural philosophical project. They are unified by their methodology and by these four principles.

Conclusion

In a draft methodological fragment, related to the "Principles of Philosophy," Newton wrote,

> Thus in the Mathematical Principles of Philosophy I first shewed <from Phænomena> that all[67] bodies endeavour[68] by a certain force proportional to their matter to approach one another, that this force in receding from the body grows less & less in reciprocal proportion to the square of the distance from it & that it <is>[69] equal to gravity & therefore <is>[70] one & the same force with gravity. Then using this force as a Principle of <Philosophy>[71] I derived from it all the motions[72] of the heavenly bodies & the flux & reflux of the sea, shewing by mathematical demonstrations that this force alone was sufficient to produce all those Phænomena, & deriving from it (a priori) some <new> motions which Astronomers had not then observed but since appear to be true, as that Saturn & Jupiter draw one another, that the variation of the Moon is bigger in winter then in summer, that there is an equation of the Moons meane motion amounting to almost 5 minutes which depends upon the position of her Apoge[73] [sic.] to the Sun.
>
> (MS. Add. 3970. fol. 480v)

I have argued that, when considering Newton's principles, we should focus on function rather than content. And so it is significant that in this passage, Newton focused on how his theory of universal gravitation was used. Here Newton claimed that his theory of universal gravitation (a) had been deduced from phenomena (i.e. "I first shewed from Phænomena") and (b) had functioned as a premise (i.e. "I derived from it . . ."). It is significant for my interpretation of Newton's principles, that Newton described the latter step as "using this force as a Principle." I have argued that *principle* was a context-specific term. Any kind of theory could be a principle, provided it served the right kind of function. Therefore, it seems appropriate for Newton to talk of using some concept as a principle.

An important lesson emerged from my discussion of Newton's method of principles: labelling or referring to a proposition as a principle tells us about the *function*, rather than the *content*, of the proposition. I have argued that this is an example of a more general feature of Newton's methodology, namely, that when we study Newton's methodology we should emphasize functions and distinctions over content. Newton's use of the term *principle* fits what I call his "rhetorical style." Newton took the already-familiar term and stretched it to fit his methodology. It is well known that Newton did this with many of his innovative philosophical terms, such as *force* and *mass*. However, this is also a feature of many of Newton's methodological concepts: he "borrowed" familiar terms and "massaged" them to fit his own needs. Steffen Ducheyne has argued that Newton did this with his dual methods of analysis and synthesis (Ducheyne 2012: 5). Because Newton bent methodological terms to fit his needs, it is a mistake to focus too closely on the content of propositions. We should instead understand his methodology in terms of the *roles* which concepts play.

I shall close with a final comment about the title of Newton's *Principia*. The origin of the title is contentious. Many have taken it as an open expression of Newton's hostility toward Cartesian science. The title, *Philosophiæ naturalis principia mathematica*, is usually recognized as a simple alteration of Descartes's *Principia philosophiæ* and thus been taken as a bold declaration of the anti-Cartesian bias of the work. However, D. T. Whiteside has presented an alternative view. In a letter to Edmond Halley (20 June 1686), Newton discussed the proposed title of the work:

> The two first books without the third will not so well beare the title of *Philosophiæ naturalis Principia Mathematica* & therefore I had altered it to this *De motu corporum libri duo*: but upon second thoughts I retain the former title. Twill help the sale of the book which I ought not to diminish now tis yours.
>
> (Newton 1959–1977, 2: 437)

Thus, the title may have been chosen for its marketing potential, in the hopes of yielding a profit (Whiteside 1991: 34). Whatever its origin, the title captures something important about Newton's work. The concept of a "principle" came to be closely tied up with Newton's mathematico-experimental method. And thus, that Newton used his method of principles to lay down the foundational premises for his new natural philosophy is significant. The inclusion of the word *principles* in the title of Newton's great work might have been accidental, convenient or even vindictive. But it turned out to be entirely appropriate.

Notes

* For valuable comments on earlier versions of this paper, I thank Peter Anstey, Zvi Biener, Adrian Currie, and Ian James Kidd.
1 Translation quoted in Newton 1999: 53–54. I. Bernard Cohen has dated this draft to the years following the publication of the second edition of the *Principia* (1713), Newton 1999: 49.
2 The only obvious exception to this is J. E. McGuire's brief discussion of principles in McGuire 1970. However, the scope of McGuire's conceptual analysis does not include Newton's published work.
3 Philosophical propositions, or *Natural* philosophical propositions, are propositions concerning natural bodies and motions as opposed to abstract mathematical ones.
4 Here I follow Peter Anstey's distinction between propositional and ontological principles (see the Introduction to this book).
5 We shall see that in the *Opticks* some of these are treated as ontic-principles.
6 While these topics suggest that Books One and Two deal with physical problems, these are treated as abstract mathematical problems. For, as Newton explained in the scholium at the end of Book One, section 11:

> Mathematics requires an investigation of those quantities of forces and their proportions that follow from any conditions that may be supposed. Then, coming down to physics [i.e. natural philosophy], these proportions must be compared with the phenomena, so that it may be found out which conditions of forces apply to each kind of attracting bodies. And then, finally, it will be possible to argue more securely concerning the physical species, physical causes, and physical proportions of these forces.
>
> (Newton 1999: 588–589)

7 In this passage Newton made a distinction between *philosophical* and *mathematical* principles. Philosophical principles are those that describe natural phenomena, and are more properly called natural philosophical principles. Mathematical principles are those that concern abstract mathematics or geometry. Newton's distinction between mathematical and philosophical propositions is important for the status of the *Principia* as a work of natural philosophy. But it is not important for the status of these propositions as principles. As we shall see, they are called principles because they perform the same general methodological function.
8 Keplerian motion can be defined by three rules now known as Kepler's laws:

1   The orbit of a planet is an ellipse, with the sun at one of the two foci;
2   A line segment joining a planet and the Sun sweeps out equal areas in equal times (this is often called the area rule); and
3   The square of the orbital period of a planet is proportional to the cube of the semi-major axis of its orbit (this is often called the harmonic rule or the 3/2 power rule). See Wilson 2000 for an account of how these propositions came to be regarded as laws.

9 Newton frequently used the phrase "deduced from the phenomena," for example, Newton 1999: 943. However, he did not always use the terms *deduction* and *induction* in their modern technical senses. See, for example, Shapiro 2004: 211–215. Therefore, in keeping with Newton's style of usage, I tend to use *deduced* interchangeably with the less technical *inferred*. For accounts of Newton's use of the terms *deduction* and *induction*, see, for example, Davies 2003, Ducheyne 2005, Fox 1999, and Worrall 2000.

10 I address Newton's distinction between certainty and uncertainty in the next section.

11 For a discussion of these differences, see Ducheyne 2012: 219–222.

12 In the *Opticks*, Newton did not make explicit reference to the axioms in his proofs. However, Peter Achinstein has pointed out that we can only make sense of the proofs once we recognize that the axioms are implicitly assumed; Achinstein 1991: 44, note 28. Thus, it seems plausible to regard Newton's axioms as hidden premises in his arguments.

13 For more detailed accounts of active and passive principles in Newton's natural philosophy, see, for example, Dobbs 1991: 24–57 & 94–96, Westfall 1980: 299–310.

14 Note that, in this context, *theory* and *hypothesis* refer to singular propositional statements as opposed to systems of propositions.

15 This distinction has been discussed by, for example, Guicciardini 2011, Shapiro 1993, and Walsh 2012a.

16 See, for example, Robert Hooke's "To the Royal Society" in his *Micrographia*, Hooke 1665.

17 Here, I am not simply juxtaposing two terms of reference but, rather, making an epistemic distinction. I use *theory* and *hypothesis* as generic terms that are intended to capture a distinction between two kinds of propositions. (By "proposition" here, I am referring to the meaning of a declarative sentence that is the primary bearer of some truth-value.) Newton rarely used the term *theory* in his publications. My definition of *theory* fits his usage of the terms *law*, *lemma*, and *proposition* (Newton frequently divided his propositions into theorems and problems). Moreover, my definition of *hypothesis* fits Newton's usage of both *hypothesis* and *query* (assuming we read the latter as assertions rather than questions). In Newton's usage, where hypotheses and queries come apart is in the role they play in his natural philosophy, see, for example, Anstey 2004. For a discussion of the distinction between theories and hypotheses in early modern philosophy more generally, see Ducheyne 2013.

18 It is worth noting that hypotheses are closely related to ontic-principles. "Principle" refers to an unknown cause of a particular effect. It is typically a place holder for a force or power to which we do not have epistemic access. A proposition describing the nature or cause of an ontic-principle would be a hypothesis.

19 For an extended discussion of the respective roles of hypotheses and queries in Newton's natural philosophy, see Walsh 2014.

20 This passage highlights a difference between Newton's and John Locke's notions of principle. Where Locke viewed principles as hypotheses (mere probabilities or conjectures), Newton saw principles as theories (i.e. deduced from phenomena and certainly true), McGuire 1970: 181. For, unlike Locke, Newton believed that universal statements could be established from phenomena. For an account of how Locke's notion of a natural philosophical principle developed in light of Newton's achievements in the *Principia*, see Anstey 2011.

21 This inference structure has been interpreted as hypothetico-deductivism. For two compelling critiques of this interpretation, see Harper 2011: Ch. 9 and Smith 2014.

22 MS. Add. 3970, fols 479r v.

23 Unless otherwise stated, all quotes in this section are from MS. Add. 3970, fols 479r–v.

24 Following "then to" deleted.

25 Following "& leave" deleted.

26 Following "drawing conclusions" deleted.

27 Following "such" deleted.

28 Following "as are" deleted.
29 Following "This is the only" deleted.
30 Following "the great" deleted.
31 McGuire notes that this is probably the earliest nonmathematical statement of universal gravitation in Newton's own words, McGuire 1970: 184, note 21.
32 Following "& that had flying creatures their wings by chance? or has any creature wings without being able to fly?" deleted.
33 Following "&" deleted.
34 Following "very" deleted.
35 Following "mankind" deleted.
36 Newton often mentioned agreement among philosophers or mathematicians in discussions of empirical support. Elsewhere I have argued that we should understand Newton's notion of certainty as "compelled assent" (Walsh 2017): the evidence compelled him undeniably to his conclusion. I take it that agreement among scholars supported this undeniability, the thought being that no rational person, having carried out the experiment, could deny the conclusion.
37 Newton 1999: 795: "Rule 3. *Those qualities of bodies that cannot be intended and remitted and that belong to all bodies on which experiments can be made should be taken as qualities of all bodies universally.*"
38 Although the concept of impenetrability almost certainly played a tacit role in Newton's thinking about collision.
39 Following "former" deleted.
40 Following "all" deleted.
41 Following "give an acct" deleted.
42 That "the following treatise" deals with colour tells us that this draft is intended for the *Opticks* rather than the *Principia*.
43 In the late 1680s, Newton conceived of the *Opticks* as a four-volume work. However, by the early 1700s, the *Opticks* had been rewritten in three volumes. Much of the material from Book Four was incorporated into Book Three of the *Opticks*. For a discussion of the delay in publishing the *Opticks*, see Shapiro 2001.
44 MS. Add. 3970, fols 335r–336v, 337r–340v, 342r–343r.
45 And these are the topics covered by the observations that were eventually published in Book Three of the *Opticks*.
46 MS. Add. 3970, fols 337r–338v.
47 The draft has been heavily edited, and many of the words are illegible.
48 Following "Prop 18" deleted.
49 Following "vulgarly called gravity" deleted.
50 Following "minute" deleted.
51 Following "all the great motions {illeg.} are regulated by the gravity of <which> great bodies <have> towards one another I shewed at large in my <Philosophiae> Principia mathematica {illeg.}" deleted.
52 Following "of" deleted.
53 Following "{illeg.}" deleted.
54 Following "Philosophiae naturalis Principia mathematica by such a convincing <mathematical> way of arguing as has given satisfaction procured the assent of {illeg.} as many {illeg.} all the <ablest> mathematicians {illeg.} have {illeg.} <who have had leasure to> had leasure & skill to examine the book" deleted.
55 Following "<But what>" deleted
56 Following "{illeg.}" deleted.
57 Following "said" deleted.
58 Following "my Principles" deleted.
59 Following "only that" deleted.
60 Following "{illeg.} those things being received by Mathematicians" deleted.
61 Following "of" deleted.

62 Following "a refraction of &" deleted.
63 Following "doubted" deleted.
64 Following "{illeg.}" deleted.
65 While there is an obvious semantic shift between hypothesis and query—the query is stated as a question—this difference is often ignored. Queries are often interpreted as assertions. Some scholars have argued that this is the *only* difference between hypotheses and queries: in the *Opticks*, queries were simply Newton's way of getting around his self-imposed ban on hypotheses. There is something to this suggestion: there is a sense in which the queries employed in the *Opticks* may be considered *de facto* hypotheses. However, there is more to the shift than this. Newton was using the semantic structure of the query to explore a possible future research program. So, while Newton could not prove it himself, he felt that it was the kind of thing that could be dealt with in his natural philosophy.
66 McGuire has suggested that we should understand these principles in the Boylean sense: they are general claims intended to help to "explicate the already known phenomena of nature," Boyle 1999–2000, 2: 12, quoted in McGuire 1970: 180–181. But, where Boyle in this context understands principles *solely* as heuristic aids to the progress of knowledge, Newton construes them as constitutive statements about the natural world.
67 Following "the" deleted.
68 Ending "ed" deleted.
69 Following "was" deleted.
70 Following "was" deleted.
71 Following "gravity, I shewed how" deleted.
72 Following "Phænomena of nature" deleted.
73 Following "Aphelion" deleted.

Bibliography

Manuscript Sources

1. MS. Add. 3970

Papers on hydrostatics, optics, sound and heat (c.1670—c.1710)
Physical Location: Cambridge University Library, Portsmouth collection.
Accessed online: Cambridge Digital Library
URL: http://cudl.lib.cam.ac.uk/collections/newton

2. MS. EL/N1/55

Isaac Newton, dated at Cambridge, to Edmund Halley (20 June 1686)
Physical Location: Royal Society Library
Accessed online: The Newton Project
URL: http://www.newtonproject.sussex.ac.uk/

Printed Works

Achinstein, P. (1991) *Particles and Waves: Historical Essays in the Philosophy of Science*, New York: Oxford University Press.
Anstey, P. R. (2004) "The methodological origins of Newton's queries," *Studies in History and Philosophy of Science*, 35: 247–269.
———. (2011) *John Locke and Natural Philosophy*, Oxford: Oxford University Press.

————, ed. (2013) *The Oxford Handbook of British Philosophy in the Seventeenth Century*, Oxford: Oxford University Press.

Bechler, Z. (1974) "Newton's 1672 optical controversies: A study in the grammar of scientific dissent" in ed. Y. Elkana 1974, pp. 115–142.

Biener, Z. and Schliesser, E., eds. (2014) *Newton and Empiricism*, Oxford: Oxford University Press.

Boyle, R. (1999–2000) *The Works of Robert Boyle*, 14 vols, eds. M. Hunter and E. B. Davis, London: Pickering and Chatto.

Buchwald, J. Z. and Cohen, I. B., eds. (2001) *Isaac Newton's Natural Philosophy*, Cambridge, MA: MIT Press.

Cohen, I. B. (1962) "The first English version of Newton's *Hypotheses non fingo*," *Isis*, 53: 379–388.

————. (1969) "Hypotheses in Newton's philosophy," *Boston Studies in the Philosophy of Science*, V: 304–326.

Dalitz, R. H. and Nauenberg, M., eds. (2000) *The Foundations of Newtonian Scholarship*, Singapore: World Scientific.

Davies, E. B. (2003) "The Newtonian myth," *Studies in History and Philosophy of Science*, 34: 763–780.

Dobbs, B. J. T. (1991) *The Janus Faces of Genius: The Role of Alchemy in Newton's Thought*, Cambridge: Cambridge University Press.

Ducheyne, S. (2005) "Bacon's idea and Newton's practice of induction," *Philosophica*, 76: 115–128.

————. (2012) *The Main Business of Natural Philosophy: Isaac Newton's Natural-Philosophical Methodology*, Dordrecht: Springer.

————. (2013) "The status of theory and hypotheses" in ed. P. R. Anstey 2013, pp. 169–191.

Elkana, Y., ed. (1974) *The Interaction Between Science and Philosophy*, Atlantic Highlands: Humanities Press.

Fox, J. (1999) "Deductivism surpassed," *Australasian Journal of Philosophy*, 77: 447–464.

Guicciardini, N. (2011) *Isaac Newton on Mathematical Certainty and Method*, Cambridge, MA: MIT Press.

Harper, W. L. (2011) *Isaac Newton's Scientific Method: Turning Data into Evidence about Gravity and Cosmology*, Oxford: Oxford University Press.

Harris, J. (1708) *Lexicon Technicum: Or, an Universal English Dictionary of Arts and Sciences: Explaining Not Only the Terms of Art, but the Arts Themselves, vol. 1*, London.

Harris, J. A., ed. (2013) *The Oxford Handbook of British Philosophy in the Eighteenth Century*, Oxford: Oxford University Press.

Hooke, R. (1665) *Micrographia: Or, Some Physiological Descriptions of Minute Bodies Made by Magnifying Glasses, with Observations and Inquiries Thereupon*, London.

Jalobeanu, D. (2014) "Constructing natural historical facts: Baconian natural history in Newton's first paper on light and colors" in eds. Z. Biener and E. Schliesser 2014, pp. 39–65.

Janiak, A. (2007) "Newton and the reality of force," *Journal of the History of Philosophy*, 45: 127–147.

————. (2013) "Three concepts of causation in Newton," *Studies in History and Philosophy of Science*, 44: 396–407.

Maclaurin, J., ed. (2012) *Rationis Defensor: Essays in Honour of Colin Cheyne*, Dordrecht: Springer.

McGuire, J. E. (1970) "Newton's 'Principles of Philosophy': An intended Preface for the 1704 'Opticks' and a related draft fragment," *British Journal for the Philosophy of Science*, 5: 178–186.

Newton, Sir I. (1952 [1730]) *Opticks: Or a Treatise of the Reflections, Refractions, Inflections & Colours of Light*, 4th edn, New York: Dover Publications.

———. (1959–1977) *The Correspondence of Isaac Newton*, 7 vols, eds. H. W. Turnbull, J. F. Scott, A. R. Hall and L. Tilling, Cambridge: Cambridge University Press.

———. (1999 [1726]) *The Principia: Mathematical Principles of Natural Philosophy*, eds. I. B. Cohen and A. M. Whitman, Berkeley: University of California Press; 1st edn 1687.

———. (2004) *Isaac Newton: Philosophical Writings*, ed. A. Janiak, Cambridge: Cambridge University Press.

Schliesser, E. (2013) "Newton and Newtonianism in eighteenth-century British thought" in ed. J. A. Harris 2013, pp. 41–64.

Shapiro, A. E. (1993) *Fits, Passions and Paroxysms: Physics, Method and Chemistry and Newton's Theories of Colored Bodies and Fits of Easy Reflection*, Cambridge: Cambridge University Press.

———. (2001) "Newton's experiments on diffraction and the delayed publication of the *Opticks*" in eds. J. Z. Buchwald and I. B. Cohen 2001, pp. 47–76.

———. (2004) "Newton's 'Experimental Philosophy'," *Early Science and Medicine*, 9: 185–217.

Smith, G. E. (2014) "Closing the loop: Testing Newtonian gravity, then and now" in eds. Z. Biener and E. Schliesser 2014, pp. 262–351.

Walsh, K. (2012a) "Did Newton adopt hypothetico-deductivism?" in *Early Modern Experimental Philosophy*. https://blogs.otago.ac.nz/emxphi/2012/12/did-newton-adopt-hypothetico-deductivism/ (Accessed: 2 December 2014).

———. (2012b) "Did Newton feign the corpuscular hypothesis?" in ed. J. Maclaurin 2012, pp. 97–110.

———. (2014) *Newton's Epistemic Triad*, PhD Thesis, University of Otago.

———. (2017) "Newton: from certainty to probability?" *Philosophy of Science*, 85.

Westfall, R. S. (1971) *Force in Newton's Physics: The Science of Dynamics in the Seventeenth Century*, New York: Elsevier Publishing Company.

———. (1980) *Never at Rest: A Biography of Isaac Newton*, Cambridge: Cambridge University Press.

Whiteside, D. T. (1991) "The prehistory of the *Principia* from 1664 to 1686," *Notes and Records of the Royal Society of London*, 45: 11–61.

Wilson, C. (2000) "From Kepler to Newton: Telling the tale" in eds. R. H. Dalitz and M. Nauenberg 2000, pp. 223–242.

Worrall, J. (2000) "The scope, limits, and distinctiveness of the method of 'Deduction from the Phenomena': Some lessons from Newton's 'demonstrations' in optics," *British Journal for the Philosophy of Science*, 51: 45–80.

8 Leibniz on Principles in Natural Philosophy

The Principle of the Equality of Cause and Effect

Daniel Garber

Leibniz was a man of principle. In fact, he was a man of many principles. Everyone knows the famous line from the *Monadology*:

> Our reasonings are based on *two great principles, that of contradiction,* in virtue of which we judge that which involves a contradiction to be false, and that which is opposed or contradictory to the false to be true. And *that of sufficient reason,* by virtue of which we consider that we can find no true or existent fact, no true assertion, without there being a sufficient reason why it is thus and not otherwise.
>
> (*Monadology*, §§31–32 (AG 217))[1]

Indeed, the Principle of Sufficient Reason (PSR) is virtually identified with Leibniz, so often does he appeal to it and use it in everything from his logic to his metaphysics to his physics to his theodicy. But these are not the only principles that Leibniz uses in his philosophy. He also uses the Principle of Perfection, the Principle of Plenitude, Principle of the Identity of Indiscernibles, and the Principle of Continuity, among others. There is an enormous literature on these principles, disputes about what exactly they are, whether they are necessary or contingent, how they are related to one another, the various roles that they play in his philosophy, and so on and so on and so on.

Leibniz appeals to principles in virtually every area of his philosophy. In this chapter, though, I would like to focus on the way in which he uses principles in one particular area, his broadly mechanistic natural philosophy. This is a large domain, one that evolved over a number of years. What I would like to do in this chapter is look especially at Leibniz's thought in this domain in the period in which Leibniz's program for dynamics is just beginning to come together, starting in the mid- and late-1670s. At that moment Leibniz is thinking quite self-consciously about framing new principles that will allow him to systematize a new account of motion and its laws that, he argues, has important consequences for his entire natural philosophy.

Central here is the Principle of the Equality of Cause and Effect. In one statement of the principle, it reads as follows:

Hence it is necessary that the cause be able to do as much as the effect and vice versa. And thus any full effect, if the opportunity offers itself, can perfectly reproduce its cause, that is, it has forces enough to bring something back into the same state that it was in previously, or into an equivalent state.

(A8.2.136)

This principle is basic to Leibniz's developing dynamics. As he wrote in an early outline of a book on natural philosophy that he was never to complete,

[c]ertain things take place in body which cannot be explained from the necessity of matter alone. Such are the laws of motion, which depend upon the metaphysical principle of the equality of cause and effect.

(A6.4.1988)

Though it may not be as well known as the PSR, say, it is clear that the Principle of the Equality of Cause and Effect, the PECE for short, was of great importance to Leibniz.

I begin by discussing an interesting text from that period where Leibniz reflects on the idea of principles, in general, and the idea of principles of mechanics, in particular. I then turn to the PECE and to some of the texts and concerns that led up to Leibniz's articulating the PECE in a series of texts written in 1675 and 1676. I then look at Leibniz's reflections on the metaphysical status of the principle: though at first presented as a necessary truth, grounded in the meaning of its terms, Leibniz later came to see it as contingent, and grounded in divine wisdom. I end by discussing briefly some of the other principles that Leibniz uses in his natural philosophy and tracing the way in which the early vision of a physics grounded in principles evolves into something rather different.

Among Leibniz's early writings, dating from the Paris period or shortly thereafter, is a piece that the Akademie Edition editors have entitled "Principia Mechanica" (A6.3.101f). While the Akademie edition dates it as having been written somewhere between 1673 and 1676, recently Richard Arthur has suggested a date of late summer 1676 to February 1677 for this text.[2] (The significance of this dating will become clearer later.) Much of the essay is concerned with motion and the relativity of motion. But it begins with some general considerations about principles. Leibniz begins as follows: "We have resolved to set down the principles of the science of motion called Mechanics" (A6.3.101 [Arthur 2013: 107]). But immediately he turns to some more general considerations about principles:

Principles should be clear, certain, few and sufficient for the explanation of the rest. When these have been constituted in each science, the rest can be accomplished almost by a kind of calculus, and reduced to a superior and simpler science. For just as problems in geometry are reduced to arithmetic by having recourse to just a few propositions

> from Euclid, so what we need to bring about is that when our principles
> are understood the difficulties of mechanics should be reduced to pure
> geometry.
>
> (A6.3.101–102 [Arthur 2013: 107, trans. slightly altered])

Leibniz continues:

> Nor should we hope that natural science can be greatly advanced before
> [this is done]. For just as mechanical explanations [*rationes*] are bor-
> rowed from geometry, so physical explanations [*rationes*] are borrowed
> from mechanics.
>
> (A6.3.102 [Arthur 2013: 107])

Further on in this initial discussion, Leibniz refers to "the Principles of Mechanics or [*sive*] laws of motion" (A6.3.102 [Arthur 2013: 107]), suggesting
that the two are identical.

The discussion is very interesting. Some of the remarks tell us in quite general terms what Leibniz thinks principles are, at least at this moment. There
are relatively few of them, but they suffice for explaining the rest of things.
Furthermore, they allow us to reduce one science to another, "superior and
simpler science."

But the most interesting remarks are specifically directed at mechanical
principles. These principles will also underlie physics, Leibniz claims, insofar as "physical explanations are borrowed from mechanics." What does
Leibniz mean by mechanics here? Traditionally, mechanics had been the science of machines, how to construct them, and how they work. This is what
it meant in connection with *Mechanica* (incorrectly) attributed to Aristotle,
a work that was very important and influential in the sixteenth and seventeenth centuries. The author of the *Mechanica* writes,

> Nature often operates contrary to human interest; for she always fol-
> lows the same course without deviation, whereas human interest is al-
> ways changing. When, therefore, we have to do something contrary to
> nature, the difficulty of it causes us perplexity and art has to be called
> to our aid. The kind of art which helps us in such perplexities we call
> Mechanical Skill.
>
> (*Mechanica* 847a, Aristotle 1984, 2: 1299)

The devices that this art provides are, of course, machines. It is this conception of mechanics that the *Académie Française* dictionary of 1694 has in
mind when it defines mechanics (*méchanique*) as "cette partie des Mathematiques qui a pour objet les machines."

But at just the moment Leibniz wrote the passage about mechanical principles that is at issue here, the meaning of the term *mechanics* was undergo-
ing an evolution. In 1670–1671 John Wallis published his *Mechanica: sive,*

de motu, tractatus geometricus. In this work, mechanics is identified more generally with the theory of motion. In the first definition Wallis writes, "Mechanicen, appello, Geometriam de Motu" (Wallis 1670–1671: 1).[3] It is more probable that Wallis is just a witness to an emerging new meaning of the term rather than its inventor, but there is no record I know of earlier than this for this new conception of mechanics. Leibniz knew Wallis's book; indeed, a number of his comments on it survive, one set dated at 1675, around the time of the "Principia Mechanica."[4] It is mechanics in this sense that Leibniz seems to have in mind: "We have resolved to set down the principles of the science of motion called Mechanics." One can see Wallis's definition of mechanics as the geometry of motion reflected in the claim that for Leibniz, mechanics is, in some sense, to be reduced to (pure) geometry.

What Leibniz says about the principles of mechanics, understood in this way, is very precise: the hope is that they will allow us to reduce the problems of mechanics to problems of geometry. That is to say that the principles of mechanics are intended to give us a way of treating mechanics in a rigorously mathematical way. Later, after I examine the PECE and its development in Leibniz's texts, I return to these general remarks and try to relate them to what Leibniz has actually done with the PECE.

In July or August 1676, most likely, Leibniz wrote a remarkable document, an essay titled "De Arcanis Motus, et Mechanica ad puram Geometriam reducenda" (On the secrets of motion and on reducing mechanics to pure geometry).[5] The "De arcanis motus" is the first essay in which Leibniz attempts to set out and defend his PECE. (Indeed, it is one of the very few essays I can think of in the Leibniz corpus devoted to the explication of one of his principles.)

In that essay, Leibniz set his goal as the perfection of mechanics, something that will not happen until it is mathematized, that is, "when from sufficient givens the effect can be predited, with the help of calculation and geometry" (A8.2.133). For this to be done, he says, "it is necessary that the laws of motion, which up until now, seem to be diverse, must be reduced to a certain single principle, with whose help we can arrive at certain analytic equations" (A8.2.133). After certain reflections on the history of the problem, Leibniz turns to his candidate for this single principle, what I called the PECE:

> Just as in geometry, the principle of reasoning usually cited is the equality between the whole and all of its parts, so in mechanics everything depends on the equality of the whole cause and the entire effect. Hence just as the primary axiom in geometry is that the whole is equal to all its parts, so the primary mechanical axiom is that the whole cause and the entire effect have the same power [*potentia*]. . . . Hence it is necessary that the cause be able to do as much as the effect and vice versa. And thus any full effect, if the opportunity offers itself, can perfectly

> reproduce it cause, that is, it has forces enough to bring something back into the same state that it was in previously, or into an equivalent state.
>
> (A8.2.135–136)

Or, as he summarizes the principle later in the essay,

> [t]he full cause and the complete effect have the same power [*potentia*]. (Power [*potentia*] is the state from which in given circumstances, there follows an effect of determinate magnitude.) Hence the full cause can reproduce the complete effect. The effect can reproduce itself. The effect cannot produce something more powerful [*potentius*] than it. If the effect is weaker than the cause, it is not complete. If the causes are similar, then the effects must also be similar.
>
> (A8.2.136–137)

Leibniz illustrates the principle with some examples:

> Hence it happens that a stone which falls from some height constrained by a pendulum can climb back to the same height, but no higher, if nothing interferes and it acts perfectly, and if nothing of the forces are removed, no lower either. And any bow, tensed and restored extends just as much in the other direction.
>
> (A8.2.136)

In this way, for example, the speed of the pendulum bob at the bottom of its fall constitutes the cause, and the height to which it restores itself is its effect. Here the assumption is that there is a measure of this power:

> For being able to measure equivalent things, it is therefore useful that a measure be assumed, such as the force necessary to raise some heavy thing to some height.
>
> (A.8.136)

In this way, the power can be understood to be something like the ability to do some work, to bring about some definite—and measurable—effect.

This, in fact, is key to the way that Leibniz uses the PECE to mathematize mechanics. For the PECE to be applicable we need to know what the power of a cause is, so that we can determine what the power of its effect must be. And for it to be capable of being mathematized, this power must be amenable to being measured in some unambiguous way. This is how he characterizes this early insight in the later *Specimen dynamicum* (1695):

> Next, I arrived at the true way of measuring forces . . . By *effect* here I understand not any arbitrary effect, but one for which the force has to be expended, or one in which the force has to be consumed, an effect

which one can therefore call *violent*. . . . Moreover, I have chosen from among violent effects the one which is most conducive to homogeneous division, that is, the one most capable of being divided into similar and equal parts, as in the ascent of a body endowed with heaviness. For the elevation of a heavy body by two or three feet is precisely double or triple the elevation of the same body by one foot, and the elevation of a heavy body, double in size, by one foot is exactly double the elevation of a single heavy body to a height of one foot.

(GM VI 243–244 (AG 127))

With a principle that tells us that the cause and the effect must have the same power, and with a conception of power interpreted in terms of distance fallen and velocity, measurable quantities, the stage is set for the mathematization of mechanics.

Though promised in the "De arcanis motus," Leibniz did not fully deliver on the mathematical transformation of mechanics until a few years after he wrote this essay. Now, a text that the Akademie editors have named "Axioma de potentia et effectu" suggests that at the time when Leibniz first framed the PECE in the "De arcanis motus," he assumed that it supported the Cartesian law of the conservation of $m|v|$, size times speed.[6] But by two years later, he realized his mistake. In a series of manuscripts, "De corporum concursu," (On the Collision of Bodies) Leibniz attempted to use that assumption to solve the problem of collision. It did not work. It was at that moment, in January 1678 that he realized that it was not Cartesian quantity of motion that was conserved, but mv^2. This is the argument as it is given for the very first time in Leibniz's notes:

> Force is the quantity of the effect. Hence the force of a body in a state of motion should be reckoned by the height to which it could raise itself. On the inclined plane AB

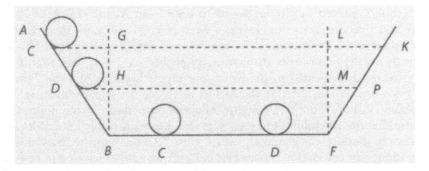

Figure 8.1 Leibniz, "De corporum concursu" (1678), LH 37.5.86r, A8.3.136

let two balls C and D, of the same material and size descend, and let them pass without being reflected onto the horizontal plane BF and there conserve their speed. It is obvious that the bodies will proceed on the plane with the speed acquired at B. Now, the speeds in question are proportional to the square roots of the heights GB, HB, or CB, DB, as established from demonstrations due to Galileo. But the forces are as CB is to DB or GB is to HB, [the elevations] from which the bodies descended, or as NF to PF or LF to MF, [the elevations] to which it can raise itself afterward, which we can here consider the same for the sake of brevity. Thus the forces of two bodies like C and D also moved on a horizontal plane BM will be proportional to the square of the speed. And thus the force remains the same not when the same quantity of motion remains the same, that is, the sum of the products of speeds and [the size of the] body, but the sum of the product of the squares of the speeds and [the size of the] body.

(Leibniz 1994: 152–153)[7]

This argument, in essence, is repeated numerous times in Leibniz's later writings, and gives rise to what came to be called the *vis viva* controversy, since the force identified with bodies in actual motion was what Leibniz came to call living force, or *vis viva*.[8] It is first published in 1686 in his "Brevis demonstratio erroris memorabilis Cartesii et aliorum circa legem naturalem . . ." (A brief demonstration of a notable error made by Descartes and others about the law of nature . . .). It was summarized that same year in §17 of the *Discourse on Metaphysics* and appears in different versions in many later publications, including the massive *Dynamica*, an attempt at a systematic treatise on dynamics, and the shorter *Specimen dynamicum*.[9] In all those presentations, the PECE is central to Leibniz's argument: through the PECE, Leibniz was able to give mathematical content to the notion of the force associated with the motion of a body. But the PECE also transforms the metaphysics of body, Leibniz argues. Here, though, the story is even more complicated.

Leibniz's interest in physics seems to date from August 1669, when he was first introduced to the Huygens/Wren laws of collision, then recently published in the *Philosophical Transactions of the Royal Society*.[10] This led directly to his first works in physics, published in 1671, the *Hypothesis physica nova*, dedicated to the Royal Society of London, and the *Theoria motus abstracti*, dedicated to the Académie royale des sciences in Paris. The Hobbes of *De corpore* was a major influence on Leibniz at that point. One central feature of Hobbes's physics was that bodies are only extension, and that only motion was genuinely efficacious: "Rest cannot be the Cause of any thing; nor can Action proceed from it, seeing neither Motion nor Mutation can be caused by it" (Hobbes 1655, Ch. 9 §9).[11] From this, it would follow, a body in motion however small could set a body at rest however large into motion, without losing any motion of its own. On Hobbes's view, there is nothing in a body at rest from which resistance to motion could derive. This is exactly what Leibniz held in his abstract theory of motion. But it is obvious that this is in conflict with our everyday experience of bodies. As we know from experience, size matters and plays a role in the collision of bodies. So a body at rest will oppose motion and will do so in proportion to its size.

The young Leibniz tried a number of ways of making his abstract theory of motion consistent with experience. The new physical hypothesis of the *Hypothesis physica nova* is exactly such an attempt to bridge the gap between the rational theory of motion and experience. The hypothesis is quite complicated, and involved a creation story, followed by light from the Sun that hits a pristine Earth causing bubbles of matter to form. It is the division of matter into parts that does the work on Leibniz's model. Very roughly, a larger body is thought to be made up of more discrete particles than a smaller body. A direct collision between the two is then analyzed as a sequence of collisions of the parts that make it up. In each collision, the end result would be that both particles would move together with a speed determined by the vector sum of the two speeds. So, insofar as the larger body has more parts, it will have a greater ability to impose its speed on the smaller body than the smaller body has to impose its speed on the larger and will appear to resist the imposition of speed on it by the smaller body.[12] The problem of how the size of a body could possibly enter into the laws of collision given that the only activity of body is motion is something that seems to have continued to worry Leibniz during his years in Paris, and he came back to this issue a number of times. In one manuscript, dated December 1675, Leibniz writes,

> That the same quantity of motion is conserved, i.e. that if the magnitude of a moving body is increased, its speed is diminished, has been observed by Galileo, Descartes, and Hobbes, and even by Archimedes. This fact has been derived from the phenomena, but no one has shown its origin in nature itself.
>
> (A6.3.466 (Leibniz 2001: 31))

Even so, Leibniz continues to hold that there is nothing in body that could give rise to resistance:

> We have assumed by a kind of prejudice that a greater body is harder to move, as if matter itself resisted motion. But this is unreasonable, for matter is indifferent to any place whatever, and thus also to change of place, or motion.
>
> (6.3.466 [Leibniz 2001; 31])

Leibniz then goes on with an attempt to derive a kind of conservation principle from the fact that bodies moving faster occupy more space than slower moving bodies do.[13] In another passage, this one dated April 1676, Leibniz tries another strategy. He writes:

> The nature of body or matter . . . contains a secret marveled at until now: namely, that magnitude compensates for speed, as if they were homogeneous things. And this is an indication that matter itself is resolved into something into which motion is also resolved, namely, a certain universal intellect. For when two bodies collide, it is clear that it is not the mind of each one that makes it follow the law of compensation, but rather the universal mind assisting both, or rather all, equally.
>
> (A6.3.493 (Leibniz 2001: 77))

On this new strategy it is God who adjusts speed to mass, and makes it more difficult to set a larger body in motion than a smaller one.

But one consequence that Leibniz draws from his PECE is that such strategies will not do: for the PECE to hold in nature, he argues, we must adopt a new conception of what body is. And so he writes in the "De arcanis motus,"

> It has been established through experience that the cause why a larger body is moved with difficulty even on a horizontal plane is not therefore heaviness, but solidity. Unless body were to resist, perpetual motion would follow, since a body resists in proportion to its mass [*moles*], since there is no other factor that would limit it [*nulla alia ratio determinandi*]. That is to say, since there is no other factor [*ratio*] which would hinder it from rebounding less than to its [original] height, since in itself, without an extrinsic impediment through the impulse of [another] body, it would give [the other body] its whole motion, and retain it as well.
>
> (A8.2.138)

All the various attempts to save appearances in collision and make it appear as if the size of a body contributed to the outcome of a collision had this one feature: at the deepest level, there are violations of conservation. If all collision is simply the vector sums of the velocities of the two bodies in collision, then when a body in motion collides with a body at rest, there is a greater volume of matter in motion than before and when two bodies in motion collide directly, then it is possible for both bodies to come to rest, and there is a smaller volume of matter in motion. In the one case, force, that is, the ability to do work and accomplish an effect, would be increased, and in the second, it would be lost, in violation of the PECE. For the PECE to hold at the deepest level, bodies must offer resistance to acquiring motion in proportion to their size. And with this, we have the beginning of Leibniz's new philosophy, where body has something more than extension, an inherent passive force that resists the acquisition of motion.

This, then, is the PECE and at least two of its main consequences. But how did Leibniz get to the PECE? Some have suggested that it may have come from his reading of John Wallis's *Mechanica*.[14] Indeed, proposition 7 of chapter 1 of the *Mechanica* reads: "Effects are proportional to their adequate causes" (Wallis 1670–1671: 15). But it is interesting to note, the causes are *proportional*, and not *equal*, as they are in Leibniz. Furthermore, Wallis's proposition is quite thin: he in no way specifies how it is that we are to measure a cause, or how we are to establish that it is proportional to the effect. Finally, although Leibniz was known to have read Wallis's *Mechanica* in 1675, at the very moment when he is formulating his PECE, there is no reason to think that he paid any special attention to this particular proposition. There are eight sheets, front and back, of Leibniz's reading notes on Wallis.[15] Though he goes into great detail on many parts of Wallis's text, that particular proposition is not even mentioned. If we are to find the roots of the PECE, I think that the place to look is in Leibniz's own reflections on questions of motion and force.

Some recently published manuscripts, dating from Leibniz's stay in Paris in 1674 show that the young Leibniz was interested in the possibility of perpetual motion at that moment.[16] These manuscripts propose two elaborate machines, one purely mechanical and one involving a configuration of magnets that would result in perpetual motion.[17] Such machines would, of course, violate a conservation principle such as the one Descartes proposed in his *Principia philosophiae*, in accordance with which the total quantity of motion would remain constant. But this would not disturb Leibniz. In abstract theory of motion in the physics of 1671, Descartes's laws are clearly violated: when two equal bodies with the same speed collide directly, Leibniz held, they both come to a halt, and both of their motions are lost completely.[18] Nor did it bother Leibniz that Descartes's conservation principle could be violated. Writing to Oldenburg in 1671, Leibniz noted,

> Nor did Descartes demonstrate that the same quantity of motion is always conserved by God in the world, for his reasoning from the immutability of God is quite feeble.

(A2.1².272)

Thus, it seems, at least in the early 1670s Leibniz would have seen no major obstacles to producing perpetual motion.

At the same time as he was thinking about perpetual motion, Leibniz was also thinking about how to measure the force of a falling body. There is a sheet, very tentatively dated from 1673, before the manuscripts on perpetual motion, which Leibniz titled "Reasoning about the force of bodies acquired through continuous natural motion" (A8.2.205ff). There, Leibniz is looking to find a measure for the force of a falling body. The measure that he plays with here is the number of spheres of a given size that a given falling body can raise to a given height. At the end of the piece, Leibniz ends with a series of questions:

> The following needs to be established by experiment: (1) from what height a given falling body can raise another given body equal to it, or unequal to it, or of the same weight, or not of the same weight, to what height. For it is certain that a body falling from a height however small can raise a body of the same weight at rest, but to a small height.
>
>   (2) It must be found by experiment what height is needed for a small falling body to raise a large body sensibly in some small height.
>
> (A8.2.206–207)

Leibniz continues along the same lines in some later pieces. In a manuscript dated May 1675, Leibniz considers a thought experiment where a pendulum bob transfers its force to raise a sphere of a given weight through a pulley, again attempting to figure out the equivalence between the power of a body in motion and the height to which a given weight can be raised (see Figure 8.2).[19]

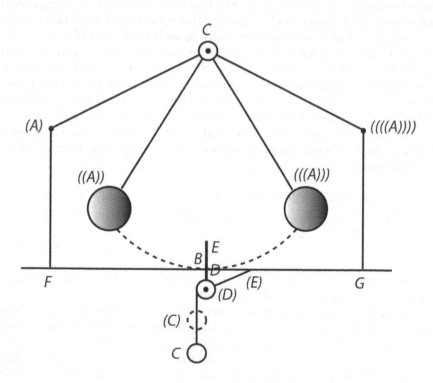

Figure 8.2 Leibniz, "De ictuum quantitate" (May 1675), A8.2.447

At roughly the same time as this last piece, he returns in another manuscript to the idea of using a column of balls to be raised as a measure of the force of a body in motion. As he states the problem here, it is one of comparing dead force, the force of weight, with living force, the force in a body in motion.[20] But in this manuscript, something new enters. Leibniz begins by positing the following situation:

> Suppose that there were a chain of contiguous globes AB of which the highest, C, falling onto pan D which, with the help of the rope DE around the pulley E raises the chain and succeeding into the place of the last one, the highest [globe] in turn falls and succeeds into its place [see Figure 8.3]. It is obvious that perpetual motion follows from this.
>
> (A8.2.163)

But, interestingly enough, at this point Leibniz reacts as follows:

> This is absurd. It is therefore obvious that the force of the blow [of C on pan D] cannot be sufficient to raise the chain [of globes] on the left so that the blows can continue. Indeed, it is certain that by falling alone, a falling body or anything equipollent to it cannot be raised higher than the height from which it fell.
>
> (A8.2.163)

This is very significant: this suggests strongly that Leibniz has now changed his mind about perpetual motion and now holds the principle that the ability to do work, understood in terms of the height to which it can raise itself, is conserved. This is a principle, Leibniz suggests, that can be confirmed

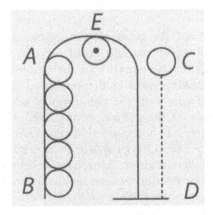

Figure 8.3 "De motu et effectu" (April 1675), A8.2.162

both by reason and by experience: "by reason since nature does not seem to act against itself, and by experience, since a pendulum never allows itself to re-ascend to its prior height" (A8.2.163).

This suggests a path to the PECE. Leibniz had been mulling perpetual motion for a few years. And he had also been mulling the way to measure the power of a body in motion and thinking of it in terms of the height to which a body in motion can raise other bodies. Now, it is not clear to me why Leibniz came to give up the possibility of perpetual motion. But when he finally did in 1675, his other work on the measure of force gave him a ready-made measure of power in terms of falling bodies and their ability to raise other bodies to a definite height. It is a short distance from this to the concerns that Leibniz discusses in the "De arcanis motus" and a statement of the PECE.

At this point I would like to turn to the metaphysical and epistemological status of the new PECE. Now, when the young Leibniz first turned to physics in the late 1660s and early 1670s, his model was Hobbes of the *De corpore*, as I mentioned earlier. And like Hobbes, the young Leibniz thought that physics was to be grounded in definitions, the definitions of motion, body, and so on. This is the project of the abstract theory of motion, an *a priori* account of motion and collision, based on the definitions of terms, which, supplemented with an *a posteriori* hypothesis about the state of the world, the new physical hypothesis of the *Hypothesis physica nova*, in principle yields an account of the phenomena we observe. Now, the discovery of the PECE as articulated in the "De arcanis motus" marks an important rupture in Leibniz's physics, the adoption of a decidedly non-Hobbesian physics. But even so, Leibniz continues for some time to think that physics should be grounded in *a priori* reasoning. In particular, he argues that the new PECE is, itself, both metaphysically necessary and knowable *a priori*. This is how he puts it in the "De arcanis motus":

> Just as in geometry, the principle of reasoning usually cited is the equality between the whole and all of its parts, so in mechanics everything depends on the equality of the whole cause and the entire effect. . . . Both axioms must be demonstrated from metaphysics, and just as the one depends on the definition of whole, part, and equal, the other depends on the definition of cause, effect, and power. . . . Hence, at any rate, it is necessary that this connection can be demonstrated, for every necessary proposition is demonstrable, at least by someone who understands it. Moreover, every demonstration takes place through their resolution into identical propositions. Therefore, it is necessary that in the end, "cause" and "effect" wind up perfectly resolved into the same thing.
>
> (A8.2.135)

This was not an idle claim for Leibniz. In the manuscripts of the period one can find a number of serious attempts to do just that, to provide an *a priori* derivation of the PECE. This is not the place to go through the variety of attempts Leibniz made to derive PECE *a priori*, but try as hard as he could, it is fairly obvious that none of them were very convincing, even to Leibniz himself.[21]

But by early 1678, Leibniz had changed his mind. Leibniz wrote to Herman Conring on 19/29 March 1678:

> *Everything happens mechanically* in nature, that is, by certain mathematical laws *prescribed by God.*
>
> (A2.1².604)

Similarly he writes in a letter to Christian Philipp sent in December 1679:

> For my part, I believe that the laws of mechanics which serve as foundation for the whole system depend on final causes, that is to say, on the will of God determined to do what is most perfect, and that matter takes on not all possible forms but only the most perfect ones. . . .
>
> (A2.1².767 (Leibniz 1969: 272))

This view is echoed in a number of other pieces that are less securely datable but which seem to have been written in the same period.[22]

Why did Leibniz change his mind? In part, I suspect, it is because of the difficulties he had finding a satisfactory *a priori* proof for the PECE. But, in addition, I suspect that there was another motivation. This is the moment when Leibniz was confronting Spinoza. And one of the things that he worried most about in Spinoza was his particular brand of necessitarianism, *blind* necessitarianism, necessitarianism without divine wisdom. This is how Leibniz characterized Spinoza in his essay "Two Sects of Naturalists," dated from 1678–1680:

> The sect of the new Stoics believes that . . . God is the soul of the world, or, if you wish, the primary power of the world, . . . but that a blind necessity determines him to act; for this reason, he will be to the world what the spring or the weight is to a clock. They further believe that there is a mechanical necessity in all things, that things really act because of his power and not due to a rational choice of this divinity, since, properly speaking, God has neither understanding nor will, which are attributes of men.
>
> (A6.4.1385 (AG 282))

Blind necessity, a necessity without purpose, without final causes, is exactly what Leibniz came to find most poisonous in Spinoza and was what he sought to avoid in his own philosophy. Leibniz came to see final causes as central to his own thought. This, combined with the difficulty of making sense of the PECE as *a priori*, I suggest, led Leibniz to see it as a contingent principle, one that derives not from the meaning of its terms but from divine wisdom.

In the beginning of this chapter, I noted some features of principles, in general, and mechanical principles, in particular, that Leibniz articulated at about the time that he first advanced the PECE. To what extent does the PECE actually satisfy them? Let us return to the essay "Principia Mechanica" that I began with, where Leibniz sets out some general thoughts about principles. We can now appreciate the fact that it is dated at 1676 or 1677, which puts it at just the moment when he has articulated the PECE. Let me review what Leibniz said:

> Principles should be clear, certain, few and sufficient for the explana-
> tion of the rest. When these have been constituted in each science, the
> rest can be accomplished almost by a kind of calculus, and reduced
> to a superior and simpler science. For just as problems in geometry
> are reduced to arithmetic by having recourse to just a few propositions
> from Euclid, so what we need to bring about is that when our principles
> are understood the difficulties of mechanics should be reduced to pure
> geometry. . . Nor should we hope that natural science can be greatly
> advanced before [this is done]. For just as mechanical explanations [*ra-
> tiones*] are borrowed from geometry, so physical explanations [*rationes*]
> are borrowed from mechanics. . .
>
> (A6.3.101–102 [Arthur 2013: 107])

How does this general characterization fit the PECE?

Well, Leibniz would certainly say that the PECE is clear and certain. But it is only as clear as the idea of power is. That is, we can understand what it means to say that the effect has exactly the same power as the cause only if we understand what power means. In the "De arcanis motus" (and in other relevant texts) this is exactly what Leibniz tried to do: find a way of specify-ing in precise terms what exactly power means. The way he chose to do it was in terms of the ability a body in motion has to raise itself or another body to a particular distance, as I pointed out earlier. In this way, Leib-niz pins down what the effect of a given cause means. But importantly, he sought to do so in a way that enables it to be quantified. It is this that allows him to go from a general metaphysical principle to a precise mathematical law, from the idea that the cause and the effect have the same power, to the claim that in interactions among bodies, the mathematical quantity mv^2 is conserved.

It is interesting to note here that, in a way, we have a double application of the PECE in the derivation of this conservation law. It is obvious that

Leibniz is appealing to the law when he articulates the conservation of mv^2: mv^2 is a measure of the force a moving body has by virtue of its motion, its ability to accomplish an effect, and if force understood in this sense is to be conserved, as the PECE demands, then mv^2 must be conserved. But the PECE seems to be assumed in the very argument that Leibniz uses to establish that mv^2 is the proper measure of force. In the argument given in the "De corporum concursu" and later texts, the assumption is made that the height to which a body in motion can raise itself by virtue of its motion is equipollent with the force in the moving body. That is to say that the assumption made is the PECE, the claim that the force in the full cause is equal to the force in the whole effect.

The idea that the PECE gives us a way of quantifying force is what I think Leibniz means when he says, both in the "Principia mechanica" and in the "De arcanis motus" that his principle allows him to reduce mechanics to pure geometry. Because of his interpretation of what power and work mean, Leibniz is able to go from the general statement about the conservation of power to a precise statement in terms of size and speed, broadly geometric concepts. Does this reduce mechanics to geometry? Obviously not, if by geometry one understands Euclidean geometry, or even Cartesian geometry. But with the conservation of mv^2, one can say that many problems in mechanics taken broadly, as Wallis did, as the geometry of motion, can be rendered soluble in terms of broadly geometrical notions. It is important to note here that the PECE is not doing this work all by itself: it is the PECE together with a particular conception of what power is and what constitutes an effect. It is these two together that get us the conservation of mv^2. In this way one must say that the PECE by itself does not exactly reduce mechanics to geometry, as Leibniz claimed. But if one reads Leibniz's claim with some (deflationary) charity, one can certainly credit him with an advance in the subject.

It is, perhaps, unsurprising that the text of the "Principia mechanica" fits the PECE so well: given its probable date, it is plausible to suppose that it was, in fact, inspired by the discovery of the new principle. Indeed, the opening of the text reads like an attempt to sketch out what a physics based on principles might look like, using the PECE as a model of what a principle might be. But after the paragraphs on principles in general in mechanics, Leibniz drifts off into a discussion that focuses largely on motion and relativity, not irrelevant but not really focused on the question of principles either.

At roughly the time he was writing the "Principia mechanica," Leibniz was also beginning to think about another principle relevant to his physics, a kind of least action principle that was particularly relevant to his work in optics at that moment. In a fragment dated to 1677 or 1678, Leibniz wrote,

> Nature does nothing in vain, nature acts through the shortest paths, as long as they are regular. Hence the shortest paths should be sought not in the refracting surfaces themselves, but in tangents.
>
> (A6.4.1367)[23]

This comment comes from a short essay on how the soul acts on the body and is, as Leibniz puts it, "just in passing." But this principle, what we might call the Shortest Path Principle, will be the subject of an important essay that Leibniz published in the *Acta eruditorum* in July 1682, "Unicum opticae, catoptricae et dioptricae principium" (A unique principle of optics, catoptrics, and dioptrics). There, using the principle that "light travels from the radiating point to the point illuminated by the easiest of all paths" (Leibniz 1682: 185),[24] Leibniz proves the laws of reflection and refraction. He writes with enthusiasm about his new principle:

> Therefore we have reduced all laws concerning [light] rays, justified through experience, to pure geometry and calculation, having made use of this singular proposition, obtained through a final cause, if you consider the matter correctly.
>
> (Leibniz 1682: 186)

As with the PECE, Leibniz thinks that this new principle also allows us to reduce problems in physics to geometry, in a certain sense. In this way, it seems to fit Leibniz's characterization of principles in the "Principia mechanica" and seems to be a principle of mechanics on the order of the PECE.

In later years, though, Leibniz added further principles to his physics. At the time he discovered the PECE and wrote the "Principia mechanica," Leibniz certainly had his famous PSR. The proposition that "nothing exists without a reason" is endorsed both by the philosopher and by the theologian in the *Confessio philosophi* of 1672–1673; indeed, in a text written shortly before, Leibniz offers an *a priori* proof of the principle![25] Furthermore, the PSR makes an appearance among the "*praedemonstrabilia*" in the *Theoria motus abstract*, Leibniz's 1671 account of the motion of bodies, where he claims that it is significant not only in natural philosophy but in "civil science" as well (A6.2.268 (Leibniz 1969: 138)). Though it entered into his earlier physics, there is no reason to believe that he thought it played any role in the new dynamics that began to emerge after the articulation of the PECE. But later, of course, the PSR would have a major role to play in Leibniz's response to Newton's absolutist account of space in the Leibniz–Clarke exchange.[26]

Other principles relevant to natural philosophy would emerge later as Leibniz develops his physics in more and more detail. The Principle of Continuity is first announced in public in an essay published in 1687, titled

"Lettre de M. L. sur un principe general utile à l'explication des loix de la nature par consideration de la sagesse divine. . ."[27] The "general principle" is as follows:

> When two instances or data approach each other continuously, so that one at last passes over into the other, it is necessary for their consequences or results (or the unknown) to do so also.
>
> (G III 52 (Leibniz 1969: 351))

Or, as he puts it elsewhere, "no change happens through a leap" (*Specimen dynamicum*, part II, GM VI 248 (AG 131)). Leibniz uses this principle in a number of ways in his physics. When first introduced, it is used in an argument against Descartes's laws of impact. He argues as we change the size of the bodies colliding with one another, the outcomes of the collision should vary continuously. But, he shows that there is a radical discontinuity as one passes from the case of the collision of equal bodies to that of unequal bodies.[28] Another use of the principle occurs in his argument that every body, no matter how small, must be elastic. If there could be perfectly hard bodies, then, Leibniz reasons, in collision their speeds and directions would change instantaneously, and there would be a discontinuity in their motion.[29] As a consequence, every body must be elastic; in collision, then, the motion of each body would be slowed gradually, and there would be no leap in nature. A consequence of this would be that there are no atoms in nature, since elasticity entails that a body is made up of smaller parts. So, if every body, no matter how small, must be elastic, every body, no matter how small, must be made up of smaller parts to infinity. And thus, there can be no atoms.[30]

Leibniz also attacks the existence of atoms using his Principle of the Identity of Indiscernibles. Here the argument is that if there were atoms, identical in every way, then there would be no way of distinguishing them, and they would not be different from one another.[31] The Principle of the Identity of Indiscernibles will also later be used in an argument for why there has to be something in bodies over and above extension: for if bodies were just extension, then, in principle, there could be nothing to distinguish one body from another. Cartesians, of course, appeal to motion in this context. But Leibniz shows by way of an ingenious argument that not even motion could be used to distinguish bodies from one another if all they were was extension made real.[32] Though these later principles seem to change the game, they do not really fit the model of principles that he is describing in the "Principia mechanica." In the "Principia mechanica," Leibniz seems to have a rather precise conception of what it is to be a principle: principles are supposed to allow us to reduce a branch of natural philosophy to a "superior and simpler science." In particular, "when our principles are understood the difficulties of mechanics should be reduced to pure geometry."

242   *Garber*

This, in a sense, is what Leibniz thought that the PECE could do, when first proposed, as well as the Shortest Path Principle. But these other principles are of a different sort: while they all have roles to play in his natural philosophy, they do not obviously lead to the mathematization of nature in the way in which the principles he first contemplated do. This goes even more so with the variety of principles that shape the other parts of Leibniz's philosophy.

Leibniz begins his investigation of nature with a rather precise and narrow sense of what a principle is and what it is supposed to do. As he builds his natural philosophy, though, this conception expands considerably: the later principles that he adds, such as the Principle of Continuity and the Principle of the Identity of Indiscernibles, do not fit the conception of principle that he articulated in 1677 or 1678 in the "Principia mechanica," when he was first thinking about a natural philosophy grounded in principles, principles that allow us to reduce natural philosophy to mechanics, and mechanics to "pure geometry." This leads naturally to other questions: What exactly is the new conception of principle that will replace this earlier conception, and how is natural philosophy now thought of as grounded in principles? But understanding the larger dimensions of Leibniz's principle-driven physics is a project for another day.

Notes

1 When not my own, the reference to the translation is given in parentheses following the original language reference.
2 See Arthur 2013: 101.
3 A century later, this will be the principal definition of mechanics. In the *Académie Française* dictionary of 1772, mechanics (*mécanique*) is defined as "La partie des Mathématiques, qui a pour objet les lois du mouvement, celles de l'équilibre, les forces mouvantes, &c."
4 See Rivaud 1914–1924, items 941 and 944 and A8.2.64–125.
5 Dating this text is a somewhat delicate. In a letter to Henry Oldenburg, 17/27 August 1676, Leibniz talks about a "very pretty metaphysical axiom concerning motion, which is of no less importance for motion than the axiom that the whole is greater than the part is with respect to magnitude" in a way that suggests that this is something he recently discovered. (A3.1.586) This, of course, echoes the language of the "De arcanis." A8.2 gives a date of February to September 1676, based on the watermark, which is consistent with the Oldenburg letter.
6 See A8.2.235 (Vorausedition, downloaded 10/16; since the text is otherwise unavailable, I cite it in this preliminary edition.). Fichant refers to this text in his commentary on the *De corporum consursu*, Leibniz 1994: 284. Rivaud 1914–1924, item 1502 dates this as August 1676. On the basis of watermark evidence, the Akademie editors date it at February to September 1676, identical to their dating of the "De arcanis motus."
7 See also Fichant's French translation, Leibniz 1994: 308–309. This text will appear in A8.3, which is currently available as a Vorausedition.
8 On the controversy, see the classic article on the subject, Iltis 1971. For Leibniz's mature catalogue of forces, where he defines *vis viva*, see the account he gives in the *Specimen dynamicum* (1695), GM VI 236–239 (AG 119–122).

9 For the *Dynamica*, see GM VI 287–292 (AG 105–111); for the *Specimen dynamicum*, see GM VI 243–246 (AG 127–130). For the place of this argument in Leibniz's larger dynamical project, see Garber and Tho forthcoming.

10 See Garber 2009: 14. For the Huygens and Wren texts, with commentary, see Murray et al. 2011.

11 The translation is from Hobbes 1656.

12 See Garber 2009: 20–21.

13 For details, see Garber 2009: 100–101.

14 See Garber 2009: 103.

15 LH 35.14.117–124, A8.2.64–106.

16 Leibniz's interest in perpetual motion predates the trip to Paris. There is a manuscript dated June 1671 in which he discusses the question. See A8.1.554ff. But I focus on texts that more directly lead up to the "De arcanis motus." For somewhat broader reflections on perpetual motion in Leibniz's thought, see Freudenthal 2002 and Stan 2016.

17 A8.2.731–733, 735–737. The date and location, "1674. Parisiis." is on the first manuscript itself. The first manuscript also contains a later addition (A8.2.733–734), dated on the manuscript as May 1678, noting the paralogism in the argument.

18 See Garber 2009: 17–18.

19 A8.2.446–448. (Vorausedition, downloaded 10/16) This manuscript has at the top "Maii 1675."

20 A8.2.162–163. The piece is tentatively dated at April 1675.

21 See Garber 2009: 237–244 for a discussion of some of those arguments.

22 For example, "For the general laws of mechanics are decrees of the divine will . . ." [A6.4.1367], dated by the Akademie editors at early 1677 to early 1678; preface to the book outlined in the "Conspectus libelli," quoted earlier, dated by the Akademie editors at summer 1678 to winter 1678/9, A6.4.2009–2010 (Leibniz 1969: 289); "Principium mechanicae universae novum," LH 35.10.5.3r in Leibniz 1994: 287 note 1, dated by Fichant at 1679–1680.

23 The principle is also announced in an optical sketch dated 1677; see Leibniz 1906: 71. Thanks to Jeffrey McDonough for the reference. For further discussion of the principle and its use in Leibniz's optics, see McDonough 2009 and McDonough forthcoming.

24 On the difference between shortest path and simplest path, see Duchesneau 1993: 263–264.

25 For the *Confessio*, see A6.3.118 (Leibniz 2005: 33). For the *a priori* proof, see A6.2.483.

26 Garber 1995: 302f.

27 G III 51–55 (Leibniz 1969: 351–353).

28 G III 53 (Leibniz 1969: 352). See also the extended discussion of Descartes's account of impact in Leibniz's commentary on Descartes's *Principia philosophiae*, *Animadversiones in partem generaleem Principiorum Cartesianorum*, G IV 375–384 (Leibniz 1969: 397–403 and 412).

29 See *Specimen dynamicum*, part II, GM VI 248–249 (AG 131–133).

30 GM IV 249 (AG 132–133).

31 See Leibniz to de Volder, 20 June 1703, G II 250 (AG 175).

32 *De ipsa natura* §13, G IV 514 (AG 164).

Bibliography

Antognazza, M. R., ed. (forthcoming) *The Oxford Handbook of Leibniz*, Oxford: Oxford University Press.

Aristotle. (1984) *The Complete Works of Aristotle*, 2 vols, ed. J. Barnes, Princeton: Princeton University Press.

Arthur, R. T. W. (2013) "Leibniz's *Mechanical Principles* (c. 1676): Commentary and translation," *Leibniz Review*, 23: 101–116.

Duchesneau, F. (1993) *Leibniz et la méthode de la science*, Paris: Presses Universitaires de France.

Freudenthal, G. (2002) "*Perpetuum mobile*: The Leibniz–Papin controversy," *Studies in History and Philosophy of Science*, 33: 573–637.

Garber, D. (1995) "Leibniz: Physics and philosophy" in ed. N. Jolley 1995, pp. 270–352.

———. (2009) *Leibniz: Body, Substance, Monad*, Oxford: Oxford University Press.

Garber, D. and Tho, T. (forthcoming) "Force and dynamics" in ed. M. R. Antognazza forthcoming.

Hess, H. J. (1978) "Die unveröffentlichten naturwissenschaftlichen und technischen Arbeiten von G. W. Leibniz aus der Zeit seines Parisaufenthaltes. Eine Kurzcharakteristik" in eds. Gottfried-Wilhelm-Leibniz-Gesellschaft and Centre national de la recherche scientifique (France), *Leibniz à Paris: 1672–1676*, vol. 1, Wiesbaden: Steiner, pp. 183–217.

Hobbes, T. (1655) *Elementorum philosophiae sectio prima de corpore*, London.

———. (1656) *Elements of Philosophy the First Section concerning Body*, London.

Iltis, C. (1971) "Leibniz and the Vis Viva controversy," *Isis*, 62: 21–35.

Jalobeanu, D. and Anstey, P. R., eds. (2011) *Vanishing Matter and the Laws of Motion: Descartes and Beyond*, New York: Routledge.

Jolley, N., ed. (1995) *The Cambridge Companion to Leibniz*, Cambridge: Cambridge University Press.

Leibniz, G. W. (1682) "Unicum opticae, catoptricae et dioptricae principium," *Acta eruditorum*, June: 185–190.

———. (1845–1863) *Leibnizens mathematische Schriften*, ed. C. I. Gerhardt, 7 vols, Berlin: A. Asher.

———. (1875–1890) *Die philosophischen Schriften*, ed. C. I. Gerhardt, 7 vols, Berlin: Weidmann.

———. (1906) *Leibnizens nachgelassene Schriften physikalischen, mechanischen und technischen Inhalts*, ed. E. Gerland, Leipzig: Teubner.

———. (1923–) eds. *Sämtliche Schriften und Briefe*, Deutsche Akademie der Wissenschaften zu Berlin, Berlin: Akademie-Verlag.

———. (1969) *Philosophical Papers and Letters*, ed. and trans. L. E. Loemker, 2nd edn, Dordrecht: D. Reidel.

———. (1989) *Philosophical Essays*, eds. and trans. R. Ariew and D. Garber, Indianapolis: Hackett.

———. (1994) *La réforme de la dynamique: De corporum consursu (1678) et autres textes inédits*, ed. M. Fichant, Paris: J. Vrin.

———. (2001) *The Labyrinth of the Continuum: Writings on the Continuum Problem, 1672–1686*, ed. and trans. R. T. W. Arthur, New Haven: Yale University Press.

———. (2005) *Confessio philosophi. Papers Concerning the Problem of Evil, 1671–1678*, ed. and trans. R. C. Sleigh, New Haven: Yale University Press.

McDonough, J. (2009) "Leibniz on natural teleology and the laws of optics," *Philosophy and Phenomenological Research*, 78: 505–544.

———. (forthcoming) "Leibniz and optics" in ed. M. R. Antognazza (forthcoming).

Melamed Y., ed. (2016) *Eternity*, Oxford: Oxford University Press.

Murray, G., Harper, W. and Wilson, C. (2011) "Huygens, Wren, Wallis, and Newton on rules of impact and reflection" in eds. D. Jalobeanu and P. R. Anstey 2011, pp. 153–191.

Rivaud, A. (1914–1924) *Catalogue critique des manuscrits de Leibniz, fasc.2 (Mars 1672–Novembre 1676)*, Poitiers: Société Française d'Imprimerie et de Librairie.

Stan, M. (2016) "Perpetuum mobiles and eternity" in ed. Y. Melamed, *Eternity*, Oxford: Oxford University Press, pp. 173–178.

Wallis, J. (1670–1671) *Mechanica: sive, De Motu, Tractatus Geometricus*, London: William Godbid.

9 Experimental Philosophy and the Principles of Natural Religion in England, 1667–1720*

Peter R. Anstey

A new expression appeared in England in 1667. The expression was "principles of natural religion," and it was first used by Thomas Sprat in his *History of the Royal-Society of London* (Sprat 1667: 347). This chapter is about the history and meaning of this term. What were the principles of natural religion? What did the early moderns have to say about them? And why are they of philosophical interest?

Interestingly, the coining of the term coincided with two other developments in Britain in the 1660s. They were the emergence of a distinction between natural and revealed religion, which came to play an increasingly important role in the theology and philosophy of the period, and the appearance and eventual ascendancy of experimental natural philosophy. It is rather telling, therefore, that this is the context of Thomas Sprat's original use of the term. Treatments of the principles of natural religion among our protagonists cannot be understood without reference to these two developments.

The notion of principles of natural religion was deployed in at least three overlapping contexts in early modern England. One place that we find it is in the broader theological context of the dispute between Catholics and Protestants in the 1660s and 1670s over whether Protestant religion is really founded upon principles at all. For instance, Edward Worsley attacked Edward Stillingfleet in his *Protestancy without Principles* (1668) and *Reason and Religion* (1672). And, while Worsley's works are not concerned with natural religion, Stillingfleet replied to Worsley and others with interest in his *An Answer to Several Late Treatises* of 1673, making several appeals to principles of natural religion.[1]

The second context in which the notion of principles of natural religion is found is in discussions of the law of nature. Here an early source was Samuel Pufendorf's *De officio hominis et civis juxta legem naturalem* of 1673 which spoke of "propositions" of natural religion and divided them up into theoretical and practical.[2] Interestingly, Pufendorf had spelt out a theory of principles in his earlier *De jure naturae et gentium* of 1672.[3] References to principles of natural religion in the context natural law theory are not uncommon in England. We find them, for example, mentioned in passing in Samuel Parker's *A Demonstration of the Divine Authority of the Law of Nature* (1681).[4] They play a more significant role in Daniel Whitby's *A Discourse of the Necessity and Usefulness of the Christian Revelation; By Reason of the Corruptions of the Principles of Natural Religion Among Jews and Heathens* of 1705, which argues that the principles of natural religion are the moral laws of nature. On the whole, however, treatments of principles of natural religion in either the defence of Protestantism, as in Stillingfleet, or in discussions of natural law, do not attempt to treat them in the context of the theory of principles in general or to establish these principles on other foundations.[5]

By contrast, that is exactly what we find in the third context in which the principles of natural religion occur in Britain, namely, discussions of principles of natural religion in natural theology. In late seventeenth- and early eighteenth-century British natural theology, principles of natural religion are often set within the context of a theory of principles as foundations of knowledge and their epistemic status is widely discussed.

In this chapter I am concerned with the use of principles of natural religion in natural theology and, in particular, their relation to the new experimental natural philosophy. Here is a summary of the conceptual developments that, I argue, took place in discussions of, and theorising about, the principles of natural religion. From the outset principles of natural religion were regarded as a complement to the principles of revealed religion, though with one crucial difference: unlike revealed principles, the principles of natural religion were seen as analogous to the principles specific to any other systematic branch of knowledge. That is they were understood as foundational propositions from which a *scientia* could be derived. Thus, they were understood within the framework of a neo-Aristotelian theory of knowledge acquisition, which included a theory of principles. From early on it was believed that we should be able to reason about the principles of natural religion in the same way that we reason in mathematics. There were discussions of the epistemic status, modes of epistemic access, and epistemic justification of the principles of natural religion. Are they absolutely or merely morally certain? Are they innate? Are they self-evident? If not, what sorts of reasons we might have for believing them?

As the century progressed and experimental philosophy gained ascendency, this new form of natural philosophy was used to provide support for the principles of natural religion. We find general claims to this effect in works such as Robert Boyle's *The Christian Virtuoso*,[6] but most of these claims were schematic and so general as to be of little philosophical interest.

An important development occurred with the advent of Newton's natural philosophy in the *Principia*. This provided not general but specific, natural philosophical claims that were brought to bear on the principles of natural religion. The first work to combine Newton's natural philosophy with natural theology was the inaugural set of Boyle Lectures by Richard Bentley, his *The Folly and Unreasonableness of Atheism*, 1693. But Bentley's claims were rather general and amounted more to gesturing than to tight logical argument. As a theologian he was a newcomer to natural philosophy and was learning on the run, so to speak.

The decisive conceptual development came in a work that is little discussed in the literature on early modern natural theology. It is George Cheyne's *Philosophical Principles of Natural Religion* of 1705. In this work the theory of the principles of natural religion is taken in a new direction. What Cheyne argued was that *the principles of natural religion are demonstrable propositions from other principles*, namely the principles of Newtonian natural philosophy. That is, the principles of natural religion, principles such as "there exists an all powerful being," are founded on other principles, principles that Newton had discovered, such as the inverse square law of gravitational attraction.

This grafting of experimental natural philosophy, and in particular Newton's discoveries, on to the theory of the principles of natural religion initially fell on deaf ears in Britain. Few, if any, were interested in Cheyne's original arguments. It was not until Jean Le Clerc picked up the significance of Cheyne's book and published an extensive paraphrase in the *Bibliothèque ancienne et moderne* in 1715 that it received a sympathetic reading.[7] This review was soon plundered by Lambert ten Kate in his highly influential *Den Schepper en Zyn bestier te kennen in Zyne schepselen*, 1716 (the full title in English is *To Know the Creator from His Creatures, According to the Light of Reason and Mathematics, [written] to Cultivate a Respectful Religion; to Destroy the Basis of Atheism; and for an Orthodox Use of Philosophy*) which followed on the heels of Bernard Nieuwentijt's *Het regt gebruik der wereldbeschouwingen* (The Religious Philosopher: Or, the Right Use of Contemplating the Works of the Creator), 1715.[8] Recent work on Dutch Newtonianism has established that it was largely through these two works together with the pirated second edition of Newton's *Principia* published in Amsterdam in 1714 that the Dutch were introduced to Newtonianism. Soon after, the British picked up the development and it is found in the Leibniz–Clarke correspondence and reaches its high-water mark in Whiston's *Astronomical Principles of Religion, Natural and Reveal'd* which is a *tour de force* in the way in which it grafts Newtonian natural philosophy on to natural religion. This appears to be the last work that purports to give the natural philosophical principles of the principles of natural religion in England and enables us to set a *terminus a quo* for this study to around 1720.

It is worth pausing to consider the titles of Cheyne's and Whiston's books in order to capture just what they are about. Cheyne's is about the *Philosophical Principles of Natural Religion*. It is important to read the "of" here as an objective, not a subjective, genitive: it is the philosophical principles that underlie the principles of natural religion. Likewise with Whiston's *Astronomical Principles*: these are the principles, the astronomical principles, from which the principles of natural religion derive. The contrast with Descartes here brings out what is distinctive with the works of Cheyne and

Whiston. Descartes derived his laws of nature from the immutability of God whereas Cheyne and Whiston reversed the direction of explanation by arguing for the existence of God from the laws of nature. With this overview in mind, let us turn now to the widespread distinction between natural religion and revealed religion.

Natural versus Revealed Religion

Talk of a contrast class comprising natural versus revealed religion seems to have emerged in England around the same time as the notion of principles of natural religion. The earliest work to contrast natural and revealed religion was probably Richard Baxter's *Reasons of the Christian Religion,* which appeared in the same year as Sprat's *History of the Royal-Society.*[9] Before these works the term *revealed religion* is extremely rare in books published in Britain, and there is little or no evidence of a contrast between natural and revealed religion. This is in spite of the fact that Herbert of Cherbury's *De veritate,* which later came to be seen as a major source for the notion of natural religion, was published in 1633 in London (nine years after the first Paris edition).[10]

By the end of the century this distinction was a commonplace, and throughout the final three decades it was widely used by mainstream divines, such as William Sherlock, Edward Stillingfleet, and John Tillotson, as well as Christian virtuosi such as Robert Boyle. It was not until the late 1690s that natural religion, in contrast to revealed religion, began to be associated with deism and the figures of Lord Herbert and Charles Blount. Indeed, contrary to the anti-natural religion rhetoric of the opponents of deism in the early eighteenth century, natural religion was held to be an important, even necessary, complement to revealed religion throughout the latter decades of the seventeenth century. This is evident from claims that are made about the relationship between natural and revealed religion. It came to be believed that revealed religion was predicated on natural religion.

John Tillotson, for example, who was later to become archbishop of Canterbury, argued in 1679 before the King that the rules for discerning the difference between true and counterfeit doctrines and revelations include the following:

1. That Reason is the faculty whereby Revelations are to be discerned.
2. All supernatural Revelation supposeth the truth of the Principles of Natural Religion.
3. All Reasonings about Divine Revelations must necessarily be governed by the Principles of Natural Religion.
4. Nothing ought to be received as a Revelation from God which plainly contradicts the Principles of Natural Religion, or overthrows the certainty of them.[11]

Clearly for Tillotson the principles of natural religion were epistemically prior to revealed religion. This view was echoed by Robert Boyle who claimed in his *The Christian Virtuoso* that revealed religion "presupposes Natural Religion, as it's foundation."[12] Indeed, it was generally held that epistemic access to natural religion was achieved through the natural light of reason rather than through revelation. As Sherlock put it, "[i]f any thing be certain by the Light of Nature, we must acknowledge, that the Principles of Natural Religion are so."[13]

So what were the principles of natural religion? Sherlock lists them as "that there is a God, and a Providence, and a Life to come, wherein good Men shall be rewarded, and the wicked punished."[14] Boyle speaks of the principles being the existence and chief attributes of God, the immortality of the soul and that there is a future state of the soul, and that there is a divine providence.[15] Stillingfleet, for his part, claimed that the principles were "[t]hat there is a just and holy God, and a wise Providence, and a future State of Rewards and Punishments; and that God designs to bring Mankind to Happiness out of a State of Misery."[16] These were fairly standard lists, though they were often supplemented by a set of moral principles or duties of natural religion.

The Theory of Principles

The first person to develop a theory of the principles of natural religion was the divine and virtuoso John Wilkins, and I turn to his view in the next section. It is important at this point, however, that I provide a general outline of a generic theory of principles so as to evaluate just how the more specific theory of the principles of natural religion was handled.

At the beginning of the seventeenth century the predominant theory of knowledge acquisition in the sciences was a view that has its origins in Aristotle's *Posterior Analytics*. It is the view that a systematic body of knowledge or *scientia* is to be derived from a set of principles. I give a very condensed summary of the theory here, ignoring numerous qualifications and finer distinctions.

Principles in the most general sense are propositions. They are propositions that are fundamental; that is they are not derivable from other propositions. Those propositions that are principles are necessary. Principles are a genus of which there are species. There are different types of principles, such as maxims, axioms, and hypotheses. And different philosophers divide the species of principles in different ways. A common view, held, for example, by the English logician Thomas Blundeville, is that maxims, such as the whole is the sum of its parts, are self-evident principles.[17]

A number of other claims were made about principles from which one derives a science. These claims were derived from Aristotle's *Posterior Analytics*. First, principles are to be universal; that is, they were to apply to every member of a particular class or species. Second, they are to be essential; that is, they are to involve essential predication. Third, they are to be convertible; that is the subject and predicate terms can be switched without changing the truth-value of the proposition as in the following:

(1)   All humans are rational animals.
(2)   All rational animals are humans.

These three conditions are, in fact, the famous laws of Ramus and were used not merely for developing a systematic body of knowledge but also for prescribing a reticulated, bifurcating structure to that knowledge for pedagogical purposes.[18]

Over and above this general conception, principles were divided into common principles that apply in any science and proper or domain-specific principles that pertain to a particular systematic body of knowledge, such as hydrostatics, optics, or natural philosophy.[19] Once the determinate principles of a discipline were in place, the process of constructing a demonstrative science could begin. This was done by the application of the theory of demonstration, that is, through the construction of scientific syllogisms. Just what the nature of scientific syllogisms is was a vexed issue. For example, some claimed that scientific syllogisms had to be causal. This led to debates over, say, the status of mathematics: Could mathematical arguments be causal? Was there a special type of syllogism that is required for a demonstrative science?

Now it should not be thought that the process of knowledge acquisition based on principles is a kind of mechanistic procedure whereby once the correct principles are established everything else demonstratively follows. For, the process of knowledge acquisition can fail at various points. One can have poorly defined terms. One can start from false principles without knowing it. And, most important for natural religion, there are facts about human faculty psychology that can significantly inhibit our ability to reason correctly.

A pertinent example here is Pascal's theory of demonstration as spelt out in his *The Art of Persuasion*. Pascal tells us:

> This art, which I call the art of persuasion, and which is properly only the way of conducting perfect methodical proofs, consists of three principal parts: defining the terms to be used in clear definitions; proposing obvious principles or axioms to prove the point in question; and always mentally substituting in the demonstration the definitions for the things defined.
>
> (Pascal 1995: 197)

He then goes on to give us rules for definitions, rules for axioms, and rules for demonstrations. But the whole theory is prefaced by a discussion of the fact that knowledge is acquired either through the heart or the mind, that is, the will or the understanding, and that the will is persuaded by pleasure and the understanding by conviction. Pascal claims that he is only giving an account of the latter even though the former, knowledge acquired by the will, is far more important:[20] not least because it is the through the heart that we acquire knowledge of principles, even the principles of geometry.[21] And, moreover, knowledge acquired by both the mind and the heart, that is, the understanding and the will, is superior to any knowledge that is acquired by only one of those faculties.[22]

It is hardly surprising that Pascal's *The Art of Persuasion* is appended to his posthumous *On the Geometrical Spirit* in which he develops an account of geometrical demonstration. This close connection between the theory of principles and mathematical reasoning is central to our first stage in the development of the philosophy of principles of religion in Britain in the seventeenth century, as we shall see in the next section.

John Wilkins's Theory of the Principles of Natural Religion

Wilkins was the first to develop a philosophy of the principles of natural religion. It appears in his posthumous *Of the Principles and Duties of Natural Religion*, 1675. Note the reference in the title to the *duties* of natural religion. For Wilkins, the moral dimension to natural religion is on a par with the metaphysical principles. We ought to note as well that just as Wilkins was closely involved in the preparation of Sprat's *History of the Royal-Society*,[23] so his own work was seen through the press by John Tillotson and was, almost certainly, the inspiration behind Tillotson's sermon at Whitehall from which I quoted earlier.

The first point of interest in the book is the summary of chapter 2. It purports to give

> Two Schemes of Principles, relating to *Practical* things, whether *Natural* or *Moral*; proposed in the method used by Mathematicians, of *Postulata*, *Definitions* and *Axioms*.
>
> (Wilkins 1675: sig. A6r)[24]

It is clear that Wilkins's theory of principles is set within the Aristotelian theory of knowledge acquisition. Before setting out his "postulata, definitions and axioms," Wilkins provides a theory of principles which includes an account of knowledge or certainty. Knowledge "may be distinguished into three kinds," namely physical certainty, mathematical certainty, and moral certainty (Wilkins 1675: 5). For each type of certainty there is a corresponding set of propositions. These propositions are "*Self-evident* and *first Principles*." They are self-evident because there is nothing clearer than them, and they are first principles because they cannot be proved *a priori*.[25] Deductions that flow from them are as certain as the principles themselves, whether physically certain, mathematically certain or morally certain.[26] The first two types of principles are infallible, and the third is indubitable.

According to Wilkins, the principles of natural religion are only morally certain.[27] They lack the certainty that one finds in, say, geometry. Nevertheless, the deductions from them are demonstrable; that is they are adequate foundations for a science or *scientia*. He claims,

> And that to a man who is careful to preserve his mind free from prejudice, and to *consider,* they will appear *unquestionable,* and the *deductions* from them *demonstrable*: But now because that which is necessary to beget this certainty in the mind, namely, *impartial Consideration,* is in a mans power, therefore the *belief* or *disbelief* of these things is a proper subject for *Rewards* and *Punishments.*
>
> (Wilkins 1675: 31)[28]

Clearly Wilkins conceives of his principles of natural religion as akin to principles in other disciplines and that the mathematical method of reasoning will apply to them insofar as they are the foundations of a demonstrative science of natural religion. He then goes on to provide definitions of religion and natural religion:

> By *Religion,* I mean that general habit of Reverence towards the Divine nature, whereby we are inabled and inclined to worship and serve God after such a manner as we conceive most agreeable to his will, so as to procure his favour and blessing.
>
> I call that *Natural Religion,* which men might know, and should be obliged unto, by the meer principles of *Reason,* improved by Consideration and Experience, without the help of *Revelation.*
>
> (Wilkins 1675: 39)

Much of the remainder of Wilkins's book is given over to establishing the reasonableness of these morally certain principles, principles such as

1. A belief and an acknowledgment of the Divine Nature and Existence.
2. Due apprehensions of his Excellencies and Perfections.
3. Suitable Affections and Demeanour towards him.

> (Wilkins 1675: 40)

Principles of Natural Religion and Experimental Philosophy

John Wilkins was a pioneer in the 1650s in the emergence of the new experimental natural philosophy that would take shape in the early Royal Society in the following decade. However, there appears to be no connection in Wilkins's mind between the principles of natural religion and experimental philosophy. That connection had been made earlier by Thomas Sprat and

Samuel Parker,[29] though much of the polemics in the early years of experimental philosophy were defensive, justifying the new emphasis on experiment and observation and the decrying of speculation by establishing that it was not a practice that led to irreligion or even atheism. Indeed, this is precisely the context in which Sprat first introduced the expression "principles of natural religion."[30]

As the new science gained ascendancy later in the seventeenth century, however, it moved from its early defensive posture on to the front foot, so to speak, and began to use the fruits of experimental philosophy as new tools for presenting a positive case for the truth of the Christian religion. It ceased justifying itself against the charge that it led to irreligion and began to see itself as bolstering Christian apologetics in its own right. However, more often than not, this development has been construed in general terms as a success of the new science or of the triumph of Newtonianism, and not in the more specific terms as one of the achievements of experimental philosophy. Of particular importance here in the English context are the Boyle Lectures established by the Will of Robert Boyle and inaugurated in 1692.

The first Boyle Lecturer was the divine Richard Bentley. His *The Folly and Unreasonableness of Atheism*, published in 1693, was a set of eight sermons, some of which appeal to Newton's natural philosophy in order to defend theism and attack various forms of atheism. While these sermons are significant insofar as they, in an informal sense, inaugurated the trend to appeal to Newtonian natural philosophy in natural theology, they are of little significance here. This is because the lectures contain no reference at all to natural religion, let alone to the principles of natural religion. Bentley was certainly not viewing his project in as bearing any relation to the work of Wilkins. Rather, his style of natural philosophy is developed in the humanist tradition, written very much from the perspective of a classical scholar.[31] The relevant contrast here is with, say, Samuel Clarke's Boyle lectures, the second volume of which is titled *A Discourse Concerning the Unchangeable Obligations of Natural Religion*, 1706, which is unequivocally a work in the tradition of Wilkins and which is most definitely a set of lectures rather than expository sermons.[32]

Bentley's sermons are just that, sermons, that begin from a biblical text and expound the text at times using material from experimental philosophers. It is important, too, that we do not overstate the presence of Newton in Bentley's lectures. Newton really only features in sermons seven and eight. The argument of the seventh sermon is built around Newton's theory of universal gravity and argues against the formation of the world by chance. Newton's name only appears once in the entire text, in the seventh sermon, though he is clearly alluded to later in the same sermon and there are six marginal notes referring to the *Principia* in the final two sermons.[33] This is in contrast, say, to Epicurus, who is mentioned eight times, and the Epicureans, nine times.[34] Indeed, the lectures themselves suggest what Bentley's correspondence confirms, namely that Bentley had an unsure grasp of

Newton's natural philosophy and was a relative newcomer to experimental philosophy. A case in point is that the sermons contain no statement of any of Newton's laws of motion, yet these were to become a central plank of subsequent applications of experimental natural philosophy to natural religion.[35] It is also worth bearing in mind that Newton is far from the only natural philosopher whom Bentley cites. For example, in the sermon on the structure and origin of human bodies there are references to recent experiments and discoveries of Leeuwenhoek, Swammerdam, Redi, Malpighi, and van Helmont.[36]

What was needed for those wanting to defend the principles of natural religion by reference to the new experimental natural philosophy, were natural theologians who were both mathematically competent enough to understand Newton's *Principia*, and sympathetic to the sort of approach found in Wilkins. Such a natural theology would use mathematical reasoning to establish the principles of nature and then deploy the principles of nature to demonstrate the principles of natural religion.

The first competent Newtonian to move in this direction was William Whiston, who later became Newton's successor in the Lucasian Chair at Cambridge. In his *A New Theory of the Earth* of 1696, he was the first to use the new principles of Newtonian natural philosophy, that is the laws of motion and the inverse square law of gravitational attraction, to argue for the existence of God. However, this work was explicitly designed not as a defence of natural religion but of revealed religion and, in particular, the biblical account of the Deluge. Whiston claims "nothing will so much tend to the vindication and honour of reveal'd Religion, as free enquiries into, and a solid acquaintance with, (not ingenious and precarious *Hypotheses*, but) true and demonstrable principles of Philosophy" (Whiston 1696: 63), plus the history of nature and the ancient traditional texts.[37]

George Cheyne's *Philosophical Principles of Natural Religion*

The first work of natural theology explicitly to graft determinate Newtonian principles into an account of natural religion is George Cheyne's *Philosophical Principles of Natural Religion* of 1705. The title of this work alone indicates that it stands in a direct line from the work of Wilkins. Indeed, in the preface, Cheyne tells us "The End and Design of these Discourses, may be gathered from the *Title Page*, and the *Contents*" (Cheyne 1705: sig. B2r). The title sums up what the work is about: *Philosophical Principles of Natural Religion: Containing the Elements of Natural Philosophy, and the Proofs for Natural Religion, Arising from them.* So the book purports to use the elements of natural philosophy as proofs for natural religion.

Cheyne was a physician who, under the influence of Archibald Pitcairne, had attempted to develop a theory of fevers using what he believed was a kind of mathematico-Newtonian form of reasoning. This methodology is far more muted in his first foray into natural theology in the *Philosophical Principles*,[38] but the connection with the content of Newton's natural philosophy is far stronger than the earlier medical writings. For example, it is the first work of natural theology to work in discoveries from Newton's *Opticks* which had been published in the previous year.[39]

Let us examine the argument from principles. In order to set this in context we will first need to examine what Cheyne says about laws of nature. Cheyne calls his opening chapter "Of the Physical *Laws*, and the Uniform Appearances of *Nature*." He distinguishes between the laws of creation and the laws of nature. The former were used by God in the creation of the world and its creatures and the latter for the maintenance of the world.[40] He then denies that there is an *anima mundi* permeating all things.[41] Rather, God "has settled Laws, and laid down Rules, conformable to which, Natural Bodies are govern'd in their Actions upon one another, and according to which, the Changes in the material part of this *System* are brought about, which all Bodies inviolably observe" whilst God preserves them (Cheyne 1705: 4–5). Next, he tells us that he does not intend to "explain all the particular *Laws of Motion*, and of the Actions of Bodies upon one another" but that he will set down "the General *Laws of Nature*, which virtually include these others, and infer such Conclusions from 'em as I find most necessary for clearing some parts of the following Discourses" (Cheyne 1705: 6).

The sorts of conclusions that Cheyne draws from the laws are best illustrated by his use of the law of gravitational attraction. Cheyne states a version of Newton's third law:

> Repulse or Reaction is always equal to Impulse or Action, or the Action of two Bodies upon one another is always equal, but with a contrary Direction. *i.e.* The same Force with which one Body strikes upon another, is return'd upon the first by that other; but these Forces are impress'd with contrary Directions.
>
> (Cheyne 1705: 20)

He then leads the reader through a series of claims to the effect that since planets move in elliptical orbits, and since the shape of these orbits cannot be accounted for by vortices and require the action of a constant force, there must be a force acting constantly on planetary bodies. Now since matter is passive, this force, namely gravity, must have been superadded to matter and could only have been superadded by a divine being. Here is the tail end of the argument. It is stated as a corollary:

> From what has been said it appears, that the Attraction or Gravitation of Bodies toward one another, is not to be *Mechanically* accounted for, and since it has been likewise shown, that the *Planets* cannot continue their Motions in their Orbits, without the Supposition of such an Attraction or Gravitation, it is evident, that this must be a Principle annex'd to Matter by the *Creator* of the World; it is a Principle no ways

essential to Matter. . . . [Moreover] That cannot be essential to Matter which is intended or remitted, but this Property increases and diminishes reciprocally as the Squares of the Distances diminish or increase, whereas impenetrability, and the other essential Attributes of Matter are always the same. On all which Accounts, it's plain that this Universal Force of *Gravitation* is the effect of the *Divine Power* and *Virtue*, by which the Operations of Material Agents are preserv'd.

(Cheyne 1705: 47–49)

Interestingly, Whiston had earlier presented a more concise argument from the determinate principle of gravity to the existence of a divine being. Whiston's Book One, Lemma IX is the law of gravitational attraction: "The force of Attraction in several distances being reciprocally in a Duplicate Proportion" (Whiston 1696: 5). He then turns to corollaries of this lemma:

Coroll. 2. This universal force of Gravitation being so plainly above, besides, and contrary to the Nature of Matter; on the foremention'd Accounts must be the Effect of a Divine Power and Efficacy which governs the whole World, and which is absolutely necessary to its Preservation.

(Whiston 1696: 6)

He goes on to draw a number of other corollaries such as that God is omnipresent (Coroll. 3), incorporeal (Coroll. 4) and exercises his providence in the world (Coroll. 5). Then follows a general Corollary 6:

Mechanical Philosophy, which relies chiefly on the Power of Gravity, is, if rightly understood, so far from leading to Atheism, that it solely depends on, supposes, and demonstrates the Being and Providence of God; and its Study by consequence is the most serviceable to Religion of all other.

(Whiston 1696: 7)

Yet, Whiston himself quickly moves on, for this is all with a view to formulating a new theory of the formation of the Earth that is consistent with both Newton's *Principia* and the story of the Deluge, and not with natural religion.[42] It would not be for another two decades that Whiston came to recycle these arguments in a systematic defence of natural religion. Meanwhile, another Boyle Lecturer was, famously, bringing Newton's natural philosophy to bear on the question of the existence of God. This was Samuel Clarke.

Samuel Clarke and the Principles of Natural Religion

Clarke was not a practising experimental philosopher, but a mathematically competent theologian who embraced Newton's natural philosophy early and became a polemicist for Newton against Cartesian natural philosophy.

His first Boyle Lectures, delivered in 1704 and published in 1705 as *A Demonstration of the Being and Attributes of God*, were an attempt to use developments in natural philosophy, and especially Newton's celestial dynamics, to argue for the existence and nature of God. Yet the lectures themselves read as an extension of Clarke's interest in defending Newton's account of celestial dynamics against both the Cartesians and, to a lesser extent, the view of John Toland concerning an innate motive force in matter.[43] They are also concerned, like the Boyle Lectures of John Harris from 1698, to combat the atheistic consequences of the philosophy of Hobbes and Spinoza.

However, they show only minimal engagement with experimental philosophy. Indeed, Clarke's comment *en passant* that experimental philosophy might sometimes be of help to religion typifies his lack of interest. Thus, he, like Bentley before him, refers to experimental observations against spontaneous generation and claims parenthetically "[f]rom which most excellent Discovery, we may *by the by* observe the Usefulness of Natural and Experimental Philosophy, sometimes even in Matters of Religion" (Clarke 1705: 120).[44] This is a mere afterthought for Clarke: clearly experimental philosophy does not feature in the structure or content of his arguments. Like Bentley, he only uses Newton's name twice in the entire work, once to argue that matter is not a necessary being because Newton has demonstrated that gravity is a universal quality of all matter,[45] and then again in a marginal note referring to a passage in the *Principia* when arguing for the wisdom of the creator on the basis of, *inter alia*, "*exact Accommodating* densities of the Planets" (Clarke 1705: 230). Moreover, there is no sense in these lectures that the actual determinate principles of natural philosophy are playing any significant role in the arguments and certainly nothing compared to the work of Whiston or Cheyne.[46]

The terms *natural religion* and *revealed religion* do not occur at all in *A Demonstration*. It might be argued that his claim to have used a mathematical method links it to the early approach of Wilkins. For, Clarke claims in the preface, that he has "confined my self to One only Method or continued Thread of Arguing, which I have endeavoured should be as near to Mathematical as the Nature of such a Discourse would allow" (Clarke 1705: sig. A3r).[47] Yet the only indication that this is carried through is in the manner in which the theses he is defending are listed as propositions and this hardly constitutes a mathematical method. Second, the actual propositions that he defends, such as Proposition XI, "That the Supreme Cause and Author of all things, must of Necessity be Infinitely Wise" (Clarke 1705: 221), are similar to those found in Wilkins and early works of proposition-centred Christian apologetics such as Hugo Grotius's *De veritate* and the *De veritate* of Herbert of Cherbury.[48]

All of this was to change, however, in the second volume of lectures, which appeared in the following year. The title of the work signals the shift: the project is now *A Discourse Concerning the Unchangeable Obligations of Natural Religion and the Truth and Certainty of the Christian Revelation,* that is, natural and revealed religion. The term *natural religion*, totally absent from the first lectures, now appears at least nineteen times.[49] It is as if Clarke has suddenly discovered a new way of conceiving of his project in natural theology, and we can most probably put this down, in part, to his polemics with John Toland, over the latter's *Letters to Serena*, in the period between the delivery of his first and second sets of Boyle Lectures.[50] One should not rule out, however, the possible influence of Cheyne's *Philosophical Principles of Natural Religion* on Clarke's second set of Boyle Lectures. The lectures were cast as completing the twofold task of establishing both the principles of natural religion and the duties of natural religion, though this time there is no pretence to mathematical reasoning.[51]

Certainly, Clarke's second set of lectures were more concerned with the threat of deism. There is a long opening section concerned with classifying and responding to the deists, who in the view of Clarke and others, had come to be characterised as those who accept natural religion but reject revealed religion.[52] This is in spite of the fact that of the leading deist writers, neither John Toland, Anthony Collins, nor Matthew Tindal were really concerned with the project of establishing the principles of natural religion. The term *natural religion* hardly appears in any of their numerous writings. Instead, it seems that it was the opponents of deism who characterised them as proponents of natural and opponents of revealed religion.

What made matters more delicate still, was that Toland, as we see in his *Letters to Serena*, was more than capable of appropriating arguments from natural philosophy, including Newtonian natural philosophy, for his purposes, and yet several of these claims undercut the very arguments that Clarke and Cheyne were using as to establish the central principles of natural religion. For example, Toland's view that matter had an innate motive force undercut the argument from the cause of the origin of motion in matter. If such a claim was allowed to go unchecked then the project of establishing the principles of natural religion might be compromised.[53] After all, if the new natural philosophy was to provide the principles for natural religion, those principles had better be true.

Space does not permit an exploration of these issues here, but they are almost certainly the root cause of the change of the theological status of natural religion. In the seventeenth century it was endorsed and promoted by Wilkins, Stillingfleet, Tillotson, and Boyle. By the second decade of the eighteenth century, however, it was facing serious opposition and was widely seen, not as a complement to revealed religion but as an insidious alternative. This is the message of Thomas Halyburton's *Natural Religion Insufficient; and Reveal'd Necessary to Man's Happiness*, of 1714 whose subtitle claimed that it was a "Rational Enquiry into the Principles of the modern *Deists*" starting with the writings of Lord Herbert of Cherbury.

Turning back to Clarke, the shift that is so evident between his first and second sets of Boyle Lectures was to become one of the salient features of his philosophical outlook in the decades that followed, with the additional feature that he became far more explicit about the role of experimental philosophy in establishing the principles of natural religion. Thus, in the dedicatory epistle of his most widely read work, the famous correspondence with Leibniz titled *A Collection of Papers Which passed between the late Learned Mr. Leibnitz and Dr. Clarke* (1717), he says

> from the earliest Antiquity to This Day, the Foundations of Natural Religion had never been so deeply and so firmly laid, as in the Mathematical and Experimental Philosophy of That Great Man [Isaac Newton].
>
> (Clarke 1717: v)

We also find a strategic appeal to the role of natural philosophy in the defence and confirmation of natural and revealed religion:

> Natural Philosophy therefore, so far as it affects Religion, by determining Questions concerning *Liberty* and *Fate*, concerning the *Extent* of the *Powers of Matter and Motion*, and the *Proofs from Phenomena* of *God's Continual Government of the World*; is of very Great Importance.
>
> (Clarke 1717: vi–vii)

Clarke had moved from thinking that experimental philosophy was occasionally of use in matters of religion, to the claim that it was crucial and, more importantly, that natural philosophy provides proofs from phenomena of God's providence. He goes on:

> 'Tis of Singular Use, rightly to understand, and carefully to distinguish from Hypotheses or mere Suppositions, the True and Certain Consequences of Experimental and Mathematical Philosophy; Which do, with wonderful Strength and Advantage, to All Such as are capable of apprehending them, confirm, establish, and vindicate against all Objections, those *Great and Fundamental Truths of Natural Religion*, which the Wisdom of Providence has at the same time universally implanted, in some degree, in the Minds of Persons even of the Meanest Capacities.
>
> (Clarke 1717: vii)

Note the rhetoric of experimental philosophy, the decrying of hypotheses and suppositions. From then on Clarke argued that the "Mathematical Principles of Philosophy" were the key to the claim that "the State of Things . . . is such as could not arise from any thing but an *Intelligent* and *Free Cause*" (Clarke 1717: 37).

As it happened, it was Leibniz who provided the catalyst for raising the nature and role of principles in his first letter to Caroline, claiming in the opening paragraph that "*Natural Religion it self*, seems to decay in England very much," to which Clarke replies that Leibniz is correct but that this is owing to "the false Philosophy of the *Materialists* to which *the Mathematick*

Principles of Philosophy are the most directly repugnant." It is they who are "Enemies of the *Mathematical Principles of Philosophy*" (Clarke 1717: 3, 9). Leibniz replies to this in his second letter:

> *the Principles of the Materialists* do very much contribute to keep up Impiety. But I believe the Author had no reason to add, that *the Mathematical Principles of Philosophy are opposite to those of the Materialists*. On the contrary, they are the same; only with this difference, that the *Materialists*, in Imitation of *Democritus*, *Epicurus*, and *Hobbes*, confine themselves altogether to *Mathematical* Principles, and admit only *Bodies*; whereas the *Christian Mathematicians* admit also Immaterial Substances. Wherefore, not *Mathematical* Principles (according to the usual sense of that Word) but *Metaphysical* Principles ought to be opposed to those of Materialists.
>
> (Clarke 1717: 19/21)

Leibniz here introduces the distinction between mathematical and metaphysical principles. He claims that mathematics is founded upon the principle of non-contradiction and that "in order to proceed from *Mathematicks* to *Natural Philosophy*, another Principle is requisite . . . I mean, *the Principle of a sufficient Reason*" (Clarke 1717: 21). It is in response to this that Clarke clarifies his position in his second reply:

> When I said that the *Mathematical Principles of Philosophy* are opposite to those of the *Materialists*; the Meaning was, that whereas *Materialists* suppose the Frame of Nature to be such, as could have arisen from mere *Mechanical* Principles of *Matter* and *Motion*, of *Necessity* and *Fate*; the *Mathematical Principles of Philosophy* show on the contrary, that the State of Things [the *Constitution* of the *Sun and Planets*] is such as could not arise from any thing but an *Intelligent* and *Free Cause*.
>
> (Clarke 1717: 37)

Note the explicit claim here that the mathematical principles of natural philosophy show, that is, demonstrate, that the state of the world must have arisen from an intelligent and free cause, namely God. Here, is a succinct statement of the key development in the application of experimental philosophy to natural religion. As it was for Cheyne in 1705, so now it is for Clarke in 1717, the principles of natural philosophy are the principles of, or the foundations of, natural religion.

Whiston's *Astronomical Principles of Religion, Natural and Reveal'd*

The high-water mark of this view appeared in the same year as the Leibniz–Clarke correspondence. This was Whiston's *Astronomical Principles of*

Religion, Natural and Reveal'd in which he "demonstrated" that "Philosophy and Mathematicks are on the Side of both *Natural and Reveal'd Religion*" (Whiston 1717: xxx). In this work Whiston gets straight down to business, setting out in detail the "known Laws of Matter and Motion" and then giving a detailed account of the structure and arrangement of the solar system. From there he proceeds to a series of inferences from the nature, properties, and laws of the solar system. The first inference is already familiar from Cheyne:

> Since that Immechanical Power of Gravity which is constantly exercis'd in the World, is not of one even and constant Quantity, but vastly unequal, according to the Squares of the different Distances of the Bodies affected with it; it is thence also certain, that the Author of that Power must be a Being that exactly knows, and take perpetual Notice of the Distances of all those Bodies whatsoever, in all the Variety of their Parts and Magnitudes.
>
> (Whiston 1717: 87)

As for experimental philosophy, Whiston is fully cognizant of its role and its importance in establishing the principles on which the principles of natural religion are founded:

> if [the wise and examining man] spends his whole Life in the pursuit of this sort of Knowledge, he perceives new Arguments every where crowd upon him to the same purpose; which is the known Case, as to Experimental Philosophy, at this Day.
>
> (Whiston 1717: 107)

For instance, on the question of the creation versus the eternity of the world he claims,

> I have already shewed, that the present System of Things, acting according to those Laws of Motion and Nature which are now fixed in the World, cannot possibly have been *a parte ante*, and cannot possibly be *a parte post* Eternal: Much less is it possible, that one little Corner of it, such as our Earth, should have been, or should be hereafter Eternal by it self. . . . I dare appeal to the entire System of Nature, whether there appear one single Argument for, or Indication of such an Eternity, either *a parte ante*, or *a parte post*, in the whole Universe. I profess, I know none: And unless Men be so weak as to leave Fact, Nature, Experiments, and Mathematicks, for the Subtilties of Metaphysicks, and the Cobwebs of Abstract Notions, they must believe the World not to have been, nor to be, Eternal.
>
> (Whiston 1717: 108–109)

Meanwhile similar arguments were being propagated in the Netherlands as we see in the English translation of Bernard Nieuwentijt's *The Religious Philosopher*.[54]

Conclusion

I should like to make three points in conclusion. First, there has been little if any interest in the principles of natural religion in the vast secondary literature on early modern natural theology.[55] Yet, one advantage of exploring this seam within the natural theological writings of the late seventeenth and early eighteenth centuries is that it enables us to delineate some of the contours of that material more clearly and to separate out primary sources that have otherwise been lumped together.

What is important to emphasise here is that the attempt to establish the principles of natural religion on the new principles of natural philosophy was just one of a cluster of argument strategies in the dynamic field of late seventeenth- and early eighteenth-century natural theology. Other argument strategies include, first, the general attempt to bring experimental philosophy to bear on natural religion, as found in, say, Boyle; second, the use of general arguments from the structure and order of the solar system and the world, normally described as physico-theology, as found in, say, William Derham; and, third, the use of mathematical arguments as witnessed, for example, in John Arbuthnot's "An argument for divine providence" of 1710.[56]

Now, Scott Mandelbrote argued in 2007 that there were really two strands to seventeenth-century English natural theology. One represented by the Cambridge Platonists and the other by the combination of Wilkins and Boyle. The latter, Mandelbrote claims, were united in their emphasis on the law-like behaviour of the universe and its potential to point to a divine lawmaker. However, if we take account of the different argument strategies it is clear that Wilkins and Boyle's writings on natural theology lie in very different seams. For, Wilkins's project in his *Principles and Duties* is concerned with evidences for the principles of natural religion and shows no interest whatsoever in experimental philosophy. By contrast, while Boyle tacitly endorses the project of natural religion, his primary concern is to establish the positive contribution that experimental philosophy can make to Christian apologetics. Thus, what might seem like a loose pairing, when viewed through the lens of the attempt to establish the principles of natural religion, are really two quite discrete projects.

Another example of the manner in which a study of the principles of natural religion enables us to separate out writers until now have been lumped together is the triad of Bentley, Clarke, and Cheyne. One might be tempted to view Clarke's *Of the Being and Attributes of God* and Cheyne's *Philosophical Principles* as little more than extensions or elaborations of Bentley's initial grafting on of Newtonian natural philosophy to natural theology.

Yet, again with the benefit of an analysis of these texts in the light of the project to establish the principles of natural religion, these three texts come apart. Moreover, a clear development becomes discernible from Clarke's first set of Boyle Lectures to his second: it is only in the latter work that Clarke reveals an explicit commitment to the science of natural religion.

A third example, one that cannot be fully elaborated on here, is that the quest for natural philosophical principles of natural religion is really quite different from the new genre of physico-theology found in, say, the writings of Derham.[57] This is evident if we compare the contents of Whiston's *Astronomical Principles* with Derham's *Astro-Theology* of 1715. The former work sets out lemmata and attempts to argue demonstratively from them to principles of natural and revealed religion. Derham's work, by contrast, does not refer to natural religion or principles at all but, rather, surveys various cosmological theories and argues informally from the structure of the solar system to the existence of a providential God.

A second point by way of conclusion is that part of the difficulty in discerning the development and significance of the project to establish principles of natural religion in the early eighteenth century has been the complicated interplay of various agendas and groups within the community of theological writers. If the argument of this chapter is correct, the threat of deism impacted negatively on the prospect of the project for natural religion, and this was in spite of the fact that some of the seminal writings on the principles of natural religion appeared at this time. Another factor that has tended to obscure the importance of the quest to establish the principles of natural religion is the widespread acceptance of an inference from the newly discovered laws of nature, and the phenomenon of universal gravity, to the existence of a providential deity. Examples of this kind of inference are not hard to find. They appear in John Keill's lectures on mechanics in 1702, in an early autograph set of lectures on experimental philosophy given by Roger Cotes circa 1707,[58] and in Niewentijt's *The Religious Philosopher*, as well as in those works that were attempting to establish the principles of natural religion in a more formal manner. That the "natural religionists" had no monopoly on the inference has, in my view, tended to obscure the importance of this inference in their project.

Finally, there is one more loose end that needs to be tied up in our discussion of the principles of natural religion, and this has to do with the important role that experimental philosophy played in the development of the natural religion project. From the outset experimental philosophers and promoters of experimental philosophy were opposed to the use of principles, hypotheses, and maxims that were accepted without sufficient recourse to observation and experiment. There is copious material illustrating this opposition to principles amongst the experimental philosophers. Yet, I have argued here that it was the principles of the new Newtonian experimental philosophy, along with appeals to other experiments and discoveries in optics and pneumatics, that played a crucial role in the establishment of the principles of natural religion. This apparent tension, however, is quickly resolved. The case of John Locke illustrates how. Locke's *Essay concerning Human Understanding* of 1690 contains one of the most forthright critiques of principles in the later seventeenth century. And yet as Newton's achievement in the *Principia* dawned on Locke, he came to see that, in fact, there are natural philosophical principles that yield knowledge and that can form the basis or foundation of natural philosophy. These "Principles that Matter of Fact justifie" (Locke 1989: 248) were eventually accepted by Locke. The decisive development here is that the new laws of nature were grafted onto the neo-Aristotelian theory of knowledge acquisition because they were able to fulfil the same role as the necessary principles on the older view of natural philosophy. Yet these new principles were experimentally justified and as such were not speculations or hypotheses. In this way, the new natural philosophy provided continuity with the theory of principles and knowledge acquisition that had long been a mainstay of European thought.

Notes

* I should like to thank Michael Hunter, Jamie Kassler, and the participants at the colloquium on "Principles in Early Modern Thought" at the University of Sydney, August 2014 for comments and suggestions. Research for this chapter was funded by Australian Research Council, grant number FT120100282, and the University of Sydney.
1 Stillingfleet 1673: 85–86 and 354–355.
2 Pufendorf 1673: 37.
3 Pufendorf 1672: Book I, Ch. 2.
4 Parker 1681: 9 and 331.
5 Later, Jean Barbeyrac's sophisticated discussion of the science of morality makes a connection between Pufendorf and earlier English treatments of the principles of natural religion in the context of a theory of natural law. See Barbeyrac 1729. See also Barbeyrac's translation of Tillotson's sermons, Barbeyrac 1738: 69 and 91.
6 See, for example, *Christian Virtuoso I*, Boyle 1999–2000, 11: 304; *Christian Virtuoso II*, Boyle 1999–2000, 11: 304, 12: 432–433 and 481–482: "it may not be amiss to observe, that the insight of a *virtuoso* into the works of nature, may enable him to contribute much [in a general way] to them all, by solidly laying, or strongly confirming the foundation of all religion, whether natural or revealed, which is a firm belief of the existence of God, and of some of his attributes that relate to the world, and to man."
7 Le Clerc 1715.
8 See Jorink and Zuidervaart 2012, and Vermij 2003.
9 See Harrison 1990: 185, note 19.
10 The association between Herbert and natural religion was first popularized by Thomas Halyburton in Halyburton 1714. See Johnson 1994 and Serjeantson 2001 for further discussion.
11 Tillotson 1679: 6–9.
12 Boyle 1999–2000, 11: 298. See also William Popple (1690: 11) who speaks of "the first Principles upon which all religion in general is grounded."
13 Sherlock 1704: 10. See also, Nourse 1691: 170.
14 Sherlock 1704: 10. For an earlier statement, see the essay by A.W. to Charles Blount entitled "Of Natural Religion, as opposed to Divine Revelation," in Blount 1693: 195.
15 *Christian Virtuoso*, 1, Boyle 1999–2000, 11: 297–298.

16 Stillingfleet 1697: lvii.
17 Blundeville 1617: 164.
18 For further discussion, see Anstey 2015.
19 For example, for the role of principles in the discipline of mechanics, see Bertoloni Meli 2010.
20 *The Art of Persuasion* Pascal 1995: 196.
21 *Pensées*, Pascal 1995: 35.
22 *Pensées*, Pascal 1995: 195.
23 For Wilkins's involvement with the preparation of Sprat's *History*, see Hunter 1989a: 52–53.
24 See also Wilkins 1675: 12.
25 Wilkins 1675: 8.
26 Wilkins 1675: 8.
27 Wilkins 1675: 31
28 The reference to "Rewards and Punishments" here most likely refers to future divine judgement. Some promoters of natural religion argued that those who had not heard the Gospel would be judged in the light of their response to the principles and duties of natural religion. See, for example, Nourse 1691.
29 On Samuel Parker, see Levitin 2014.
30 Sprat 1667: 347–348: "With these apprehensions I come to examin the *Objections*, which I am now to satisfy: and having calmly compar'd the *Arguments* of some devout men against *Knowledge*, and chiefly that of *Experiments*; I must pronounce them both, to be altogether inoffensive. I did before affirm, that the *Royal Society* is abundantly cautious, not to intermeddle in *Spiritual things*: But that being only a general plea, and the question not lying so much on what they do at present, as upon the probable effects of their Enterprise; I will bring it to the test through the chief Parts of *Christianity*; and shew that it will be found as much avers from *Atheism*, in its issue and consequences, as it was in its original purpose.

     The public Declaration of the *Christian Religion*, is to propose to mankind, an infallible way to *Salvation*. Towards the performance of this happy end, besides the *Principles* of *Natural Religion*, which consists in the acknowledgment and Worship of a *Deity*: It has offer'd us the merits of a glorious *Saviour*: By him, and his *Apostles Ministry*, it has given us sufficient *Examples*, and *Doctrines* to acquaint us with *divine things*, and carry us to *Heven*. In every one of these, the *Experiments* of *Natural things*, do neither darken our eies, nor deceive our minds, nor deprave our hearts.

     First there can be no just reason assign'd, why an *Experimenter* should be prone to deny the essence, and properties of *God*, the universal Sovereignty of his *Dominion*, and his *Providence* over the *Creation*."
31 On the humanist approach to natural theology and experimental philosophy, see Levitin 2014 and 2015.
32 See also John Harris's Boyle Lectures (Harris 1698), which are intermediate between Bentley and Clarke insofar as they take the sermon genre and yet are concerned with establishing principles of natural religion and, like Clarke's first set of lectures, attack Spinoza and Hobbes.
33 Sermon 7 mentions Newton on pp. 8 and 31. Marginal references to the *Principia* are on pp. 8, 12, 20, 25, and 31.
34 See Bentley 1693, Sermon 1: 7, 8, 12, 27, 32; Sermon 2: 6, 7, 15, 30; Sermon 4: 8; Sermon 5: 33; Sermon 6: 6, 18; Sermon 7: 8, 20.
35 Bentley does mention the law of gravitational attraction in Sermon 7: 9.
36 Sermon 4: Redi, 28; Leuwenhoek, 31, 33; Swammerdam, 33; Malpighi, 30, 34; van Helmont, 33.
37 Another mathematician who used a similar argument form was John Keill, who in his *Introductio ad verum philosophiam* of 1702, argued from Newton's first law of motion that there must be a God. See Keill 1702: 110.

38 So Guerrini 2000: 73.
39 For Cheyne's discussion of light, see Cheyne 1705, pp. 73–98. Cheyne presented his book in person to the Royal Society on 6 June 1705. See Royal Society Journal Book 11, p. 72.
40 Cheyne 1705: 2. Here Cheyne differs from Bentley, see Sermon 6, June 1692, Bentley 1693: 9.
41 Cheyne 1705: 3.
42 The 1715 second edition of Cheyne's *Philosophical Principles* included an additional, second part on revealed religion.
43 Clarke 1705: 46–47. Clarke's knowledge of the issues arose from his Latin translation of Jacques Rohault's *Traité de Physique* which, from the second, 1702, edition contained increasing numbers of Newton-inspired critical annotations.
44 For Bentley on spontaneous generation, see Sermon 6, June 1692, Bentley 1693: 28–30.
45 Clarke 1705: 49.
46 In the preface to his *A Discourse Concerning the Unchangeable Obligations of Natural Religion and the Truth and Certainty of the Christian Revelation*, 1706, Clarke claims of his earlier *A Demonstration*, "I endeavoured, in my former Discourse, to strengthen and confirm the Arguments which prove to us the *Being and Attributes of God* . . . partly from the Discoveries (principally those that have been lately made) in Natural Philosophy," sig. A3r—v.
47 It should be pointed out that a mathematical style of reasoning was also a feature of some within the natural law tradition, including, notoriously, Richard Cumberland's *De legibus naturae*. See Cumberland 1672 and Tyrrell 1692: sigs e7v–e8r and 17.
48 Wilkins 1675: 32 cites Grotius's *De veritate*.
49 This shift in Clarke's project is reflected in the twofold critique of his Boyle Lectures by William Carroll. Carroll's reply to the first lectures (Carroll 1705) does not mention natural religion, whereas the response to Clarke's second series of lectures argues that "the very Basis both of Natural and Revealed Religion, and All genuine Morality, are fundamentally Subverted by Mr. C.," Carroll 1706: 22.
50 See Wigelsworth 2009: 82–85.
51 Clarke 1706: 17. For critical comments on Clarke's mathematical method in his book on the doctrine of the Trinity, see Roger North to Clarke, 20 February 1713, in Kassler 2014: 273.
52 Clarke 1706: 17–45.
53 See Toland 1704, Letter 5.
54 Nieuwentijt 1719: 903–906.
55 The following works on early modern natural theology or the relation between natural theology and science fail to mention Cheyne or the principles of natural religion: Buckley 1987; Calloway 2014; Gascoigne 1988; Henry 2010; Mandelbrote 2007 and 2013; Peterfreund 2012; and Topham 2010.
56 Arbuthnot 1710. For commentary, see Shoesmith 1987. See also Craig 1699.
57 Derham 1713 and 1715.
58 Bodleian Library Add. C. 272, p. 62.

Bibliography

Anstey, P. R. (2015) "Francis Bacon and the laws of Ramus," *HOPOS*, 5: 1–23.
Arbuthnot, J. (1710) "An argument for divine providence, taken from the constant regularity observ'd in the births of both sexes," *Philosophical Transactions*, 27: 186–190.
Barbeyrac, J. (1729) *An Historical and Critical Account of the Science of Morality*, London.

——. (1738) *Sermons sur diverses matieres importantes par feu Mr Tillotson, tome second*, Amsterdam.

Baxter, R. (1667) *The Reasons of the Christian Religion*, London.

Bentley, R. (1693) *The Folly and Unreasonableness of Atheism*, London.

Bertoloni Meli, D. (2010) "The axiomatic tradition in seventeenth-century mechanics" in eds. M. Domski and M. Dickson 2010, pp. 23–41.

Blount, C. (1693) *The Oracles of Reason*, London.

Blundeville, T. (1617) *The Arte of Logick*, London; 1st edn 1599.

Boyle, R. (1999–2000) *The Works of Robert Boyle*, 14 vols, eds. M. Hunter and E. B. Davis, London: Pickering and Chatto.

Buckley, M. J. (1987) *At the Origins of Modern Atheism*, New Haven: Yale University Press.

Calloway, K. (2014) *Natural Theology in the Scientific Revolution*, London: Pickering and Chatto.

Carroll, W. (1705) *Remarks upon Mr Clarke's Sermons, Preached at St Paul's Against Hobbs, Spinoza, and Other Atheists*, London.

——. (1706) *The Scepticism and Fundamental Errors Establish'd in Samuel Clark's Sermons Preach'd at St. Paul's More Fully Discovered*, London.

Cheyne, G. (1705) *Philosophical Principles of Natural Religion: Containing the Elements of Natural Philosophy, and the Proofs for Natural Religion, Arising from Them*, London.

——. (1715) *Philosophical Principles of Religion Natural and Reveal'd*, London.

Clarke, S. (1705) *A Demonstration of the Being and Attributes of God, More Particularly in Answer to Mr Hobbs, Spinoza, and Their Followers*, London.

——. (1706) *A Discourse Concerning the Unchangeable Obligations of Natural Religion, and the Truth and Certainty of the Christian Revelation*, London.

——. (1717) *A Collection of Papers which Passed between the Late Learned Mr. Leibnitz and Dr. Clarke*, London.

Craig, J. (1699) *Theologiae Christianae principia mathematica*, London.

Cumberland, R. (1672) *De legibus naturae*, London.

Derham, W. (1713) *Physico-Theology or, a Demonstration of the Being and Attributes of God from His Works of Creation*, London.

——. (1715) *Astro-Theology: or a Demonstration of the Being and Attributes of God, from a Survey of the Heavens*, London.

Domski, M. and Dickson, M., eds. (2010) *Discourse on a New Method: Reinvigorating the Marriage of History and Philosophy of Science*, Chicago and La Salle: Open Court.

Gascoigne, J. (1988) "From Bentley to the Victorians: The rise and fall of British Newtonian natural theology," *Science in Context*, 2: 219–256.

Grotius, H. (1627) *De veritate religionis Christianae*, Paris.

Guerrini, A. (2000) *Obesity and Depression in the Enlightenment: The Life and Times of George Cheyne*, Norman: University of Oklahoma Press.

Halyburton, T. (1714) *Natural Religion Insufficient: and Reveal'd Necessary to Man's Happiness*, London.

Harris, J. (1698) *The Atheistical Objections, against the Being of a God, and His Attributes, Fairly Considered, and Fully Refuted*, London.

Harrison, P. (1990) *"Religion" and the Religions in the English Enlightenment*, Cambridge: Cambridge University Press.

————, ed. (2010) *The Cambridge Companion to Science and Religion*, Cambridge: Cambridge University Press.

Henry, J. (2010) "Religion and the Scientific Revolution" in ed. P. Harrison 2010, pp. 39–58.

Herbert, Lord of Cherbury (1633) *De veritate*, London; 1st edn, Paris, 1624.

Hunter, M. (1989a) "Latitudinarianism and the 'ideology' of the early Royal Society: Thomas Sprat's *History of the Royal Society* (1667) reconsidered" in M. Hunter 1989b, pp. 45–71.

————. (1989b) *Establishing the New Science: The Experience of the Early Royal Society*, Woodbridge: Boydell Press.

Johnson, R. A. (1994) "Natural religion, common notions, and the study of religions: Lord Herbert of Cherbury (1583–1648)," *Religion*, 24: 213–224.

Jorink, E. and Maas, A., eds. (2012) *Newton in the Netherlands: How Isaac Newton Was Fashioned in the Dutch Republic*, Amsterdam: Leiden University Press.

Jorink, E. and Zuidervaart, H. (2012) " 'The Miracle of our Time': How Isaac Newton was fashioned in the Netherlands" in eds. E. Jorink and A. Maas 2012, pp. 13–65.

Kassler, J. C. (2014) *Seeking Truth: Roger North's Notes on Newton and Correspondence with Samuel Clarke c.1704–1713*, Aldershot: Ashgate.

Kate, L. ten. (1716) *Den schepper in Zyn bestier te kennen in Zyne schepselen; Volgens het Licht der Reden en Wiskonst. Tot Opbouw van Eerbiedigen Godsdienst en Vernietiging van alle Grondslag van Atheïstery, als mede tot een regtzinnig gebruyk van de philosophie*, Amsterdam.

Keill, J. (1702) *Introductio ad veram physicam*, Oxford.

Le Clerc, J. (1715) "Review of George Cheyne's *Philosophical Principles of Natural Religion*, London, 1705," *Bibliothèque ancienne et moderne*, tome 3: 41–157.

Levitin, D. (2014) "Rethinking English physico-theology: Samuel Parker's *Tentamina de Deo* (1665)," *Early Science and Medicine*, 19: 28–75.

————. (2015) *Ancient Wisdom in the Age of the New Science*, Cambridge: Cambridge University Press.

Locke, J. (1975) *An Essay concerning Human Understanding*, 4th edn, ed. P. H. Nidditch, Oxford: Clarendon Press; 1st edn 1690.

————. (1989) *Some Thoughts Concerning Education*, eds. J. W. Yolton and J. S. Yolton, Oxford: Clarendon Press; 1st edn 1693.

Mandelbrote, S. (2007) "The uses of natural theology in seventeenth-century England," *Science in Context*, 20: 451–480.

————. (2013) "Early modern natural theologies" in ed. R. Re Manning 2013, pp. 75–99.

Newton, Sir I. (1687) *Philosophiae naturalis principia mathematica*, London.

————. (1704) *Opticks*, London.

————. (1714) *Philosophiae naturalis principia mathematica*, 2nd edn, Amsterdam.

Nieuwentijt, B. (1715) *Het regt gebruik der wereldbeschouwingen, ter overtuiginge van ongodisten en ongelovigen*, Amsterdam.

————. (1719) *The Religious Philosopher*, trans. J. Chamberlayne, London.

Nourse, T. (1691) *A Discourse of Natural and Reveal'd Religion in Several Essays: Or the Light of Nature a Guide to Divine Truth*, London.

Parker, S. (1681) *A Demonstration of the Divine Authority of the Law of Nature and of the Christian Religion*, London.

Pascal, B. (1995) *Pensées and Other Writings*, trans. H. Levi, Oxford: Oxford University Press.

Peterfreund, S. (2012) *Turning Points in Natural Theology From Bacon to Darwin: The Way of the Argument from Design*, New York: Palgrave Macmillan.

Popple, W. (1690) *A Discourse of Humane Reason: With Relation to Matters of Religion*, London.

Puffendorf, S. (1672) *De jure naturae et gentium libri octo*, Lund.

———. (1673) *Samuelis Pufendorfii De officio hominis et civis juxta legem naturalem libri duo*, Lund.

Re Manning, R., ed. (2013) *The Oxford Handbook of Natural Theology*, Oxford: Oxford University Press.

Rohault, J. (1702) *Jacobi Rohaulti Physica*, trans. S. Clarke, 2nd edn, London.

Serjeantson, R. (2001) "Herbert of Cherbury before Deism: The early reception of the *De veritate*," *The Seventeenth Century*, 16: 217–238.

Sherlock, W. (1704) *A Discourse Concerning the Happiness of Good Men, and the Punishment of the Wicked, in the Next World, Part I*, London.

Shoesmith, E. (1987) "The continental controversy over Arbuthnot's argument for divine providence," *Historia Mathematica*, 14: 133–146.

Sprat, T. (1667) *The History of the Royal-Society of London*, London.

Stillingfleet, E. (1673) *An Answer to Several Late Treatises*, London.

———. (1697) *A Discourse in Vindication of the Doctrine of the Trinity*, London.

Tillotson, J. (1679) *A Sermon Preached at White-Hall, April the 4th, 1679*, London.

Toland, J. (1704) *Letters to Serena*, London.

Topham, J. R. (2010) "Natural theology and the sciences" in ed. P. Harrison 2010, pp. 59–79.

Tyrrell, J. (1692) *A Brief Disquisition of the Law of Nature*, London.

Vermij, R. (2003) "The formation of the Newtonian philosophy: The case of the Amsterdam mathematical amateurs," *British Journal for the History of Science*, 36: 183–200.

Whiston, W. (1696) *A New Theory of the Earth*, London.

———. (1717) *Astronomical Principles of Religion, Natural and Reveal'd*, London.

Whitby, D. (1705) *A Discourse of the Necessity and Usefulness of the Christian Revelation; By Reason of the Corruptions of the Principles of Natural Religion among Jews and Heathens*, London.

Wiglesworth, J. R. (2009) *Deism in Enlightenment England: Theology, Politics, and Newtonian Public Science*, Manchester: Manchester University Press.

Wilkins, J. (1675) *Of the Principles and Duties of Natural Religion*, London.

Worsley, E. (1668) *Protestancy without Principles*, Antwerp.

———. (1672) *Reason and Religion*, Antwerp.

10 A Conflict of Principles

Grotius's Justice versus Hume's Utility

Kiyoshi Shimokawa

Introduction

In this chapter I take up a conflict of principles which is found in early modern thought. It may be presented as a conflict about how to understand the relationship of justice to utility. Many students of philosophy may come to know one version or another of it by reading chapter five of John Stuart Mill's *Utilitarianism* or the opening statement of John Rawls's *A Theory of Justice*.[1] But the conflict, which manifests itself in various forms, is a very old one. Its roots may be traced back to the ancient debate between Stoics and Epicureans, or some of Plato's dialogues.[2] In the early modern period, we can find one origin of the conflict in the famous debate about justice which Grotius started in the Prolegomena to *De jure belli ac pacis* (1st edn 1625). What I should like to do is to discuss two phases of this early modern debate. I focus on two prominent participants in it, Hugo Grotius and David Hume, as representing the two phases.

Let me briefly explain the first phase of the debate. Grotius criticized the view of moral sceptics that justice is not natural but only arises from utility. He defended the natural status of justice by arguing that some Stoic principles of human nature serve as the basis of natural justice, while he also added that God annexed utility or interest to the law of nature. Thus, Grotius defended the primacy of justice over utility in the debate about the origin of justice.

John Locke and Samuel Pufendorf soon joined the debate and attacked the view that justice arises from utility.[3] But a contrary response came from Hume, who provided a full-scale defence of utility against natural justice. In fact, Hume radically challenged the Protestant tradition of natural jurisprudence which Grotius had begun, and subsequent thinkers from Pufendorf to Francis Hutcheson continued and modified. It is in opposition to Grotius and his followers that Hume developed a theory of artificial justice, that is, an interest-based, conventionalist theory of justice. This claim may sound controversial. Several recent scholars have discussed Hume's relationship to Grotius, and treated Hume as a successor to, and a friend of, the tradition of natural jurisprudence.[4] As I show in the following, this is a mistake.

Suffice it to say that Hume aligned himself with the Epicurean poet Horace; provided a refined, updated version of Hobbes's Epicurean theory of justice; and paved the way for Bentham's utilitarianism.⁵ Given these, it is appropriate to see Hume as representing the second phase of the debate where the primacy of utility was defended.

In what follows, I show what Grotius and Hume claimed about the origin of justice and try to clarify what principles they appealed to in advancing their claims. For this purpose, I not only look at their general claims about the primacy of justice or utility but also examine details of their theories of the origin of justice. The account I offer in sections 1 and 2 is designed to show not only what particular sets of principles Grotius and Hume actually adopted but also how they conflict with each other at crucial points. And in section 3, I offer some general considerations on the nature of the principles involved in the Grotius–Hume debate and try to gain a better understanding of their status. I also suggest that there is a way in which we may resolve what appears to be a very serious conflict about the relationship of justice to utility.

1. Grotius's Defence of Natural Justice

1.1 Grotius's Internal Principles: Sociability and Reason

In the Prolegomena to *De jure belli ac pacis* (hereafter DJBP),⁶ Grotius sets out to prove that there exists "Right" [*jus*] in nature. This is the task he needs to undertake because he wants to "establish" his own work "on solid foundations" (Prol. 5). Grotius presents the sceptical view of Carneades with a view to criticizing it. Carneades claims that "[l]aws were instituted by Men for the sake of Interest," and that the laws are different "not only in different Countries, according to the diversity of their Manners, but often in the same Country, according to the Times" (Prol. 5). This implies that there is no such thing as "NATURAL RIGHT" [*jus naturale*], or a universal principle of right. Carneades says that there is either no justice [*justitia*] in nature at all, or if there is any, "it is extreme Folly, because it engages us to procure the Good of others, to our own Prejudice." "Nature prompts all Men, and in general all Animals, to seek their own particular Advantage [*ad utilitates suas*]," whereas justice foolishly forbids them to do so (Prol. 5).⁷

Jus (right) and *justitia* (justice) are not distinguished from each other in this context though justice, strictly speaking, may be treated as a part of right. What Grotius wants to attack is Carneades's view that there is neither natural right nor natural justice. Another target is Horace, who was often regarded as an Epicurean because of his view that nature cannot distinguish between justice and injustice (Prol. 6).⁸ Grotius rejects their claims by presenting what he takes to be the true, Stoic conception of human nature. One of the things which is peculiar to man, says Grotius, is "his Desire of Society [*appetitus societatis*], that is, a certain Inclination to live with those of his own Kind, not in any Manner whatever, but peaceably, and in a Community regulated according to the best of his Understanding [*intellectus*]" (Prol. 6). This natural sociability, Grotius adds, is the disposition which Stoics called "oikeiosis" (Prol. 6). This is not the same as the first "oikeiosis" of animals, which Chrysippus made famous,[9] that is, the affinity to oneself which involves the impulse to self-preservation. Rather, it is the "oikeiosis" of a higher order, which is peculiar to human beings as intellectual or rational beings. Given this disposition, Grotius rejects the sceptical view that all human beings and animals by nature seek their own particular advantage.

Grotius points out that the sociability of an adult human being is accompanied by the use of speech, and by "a Faculty of knowing and acting, according to some general Principles" (Prol. 7). He stresses that this sociability with the guidance of reason, or "this Care of maintaining Society in a Manner conformable to the Light of human Understanding," is "the Fountain of Right, properly so called" (Prol. 8). By "Right, properly so called," Grotius means a particular class of rights called "perfect rights" (DJBP 1.1.5). Perfect rights are possessed by a person, but unlike imperfect ones, they can be protected by the use of force (at court or in battle).[10] So Grotius's point is that human sociability and reason[11] are the original source of those perfect rights. To this sphere of rights, he says, belong the elements of justice, that is, "the Abstaining from that which is another's [*alieni abstinentia*], and the Restitution of what we have of another's, or of the Profit we have made by it, the Obligation of fulfilling Promises, the Reparation of a Damage done through our own Default, and the Merit of Punishment among Men" (DJBP 1.1.5).

While Grotius holds that human sociability and reason are the source of perfect rights, he also refers to those human faculties as "the internal Principles of Man" [*principia homini interna*] (Prol.12). The word *principium* means "a beginning, commencement, origin," and its plural *principia* means "beginnings," "foundations," or "elements,"[12] so the internal principles of man can be taken to mean the innate powers, dispositions, or motives of man which produce external actions.[13] These inborn powers, dispositions, and motives are the basis on which Grotius's defence of natural justice ultimately rests.

1.2 God as a Great Helper

Grotius famously claims that "all we have now said would take place" even if we granted that there is no God (Prol. 11).[14] This implies that what he has said about the internal principles and their operations would be true even if God did not exist. However, Grotius believes in God and hastens to add that there is "another Original of Right," which is "the free will of God" (Prol. 12). Natural right springs from the internal principles of man, but since it was "his [i.e. God's] Pleasure that these Principles should be in us" (Prol. 12), God can also be seen as a source of what is naturally right and just. Grotius justifies this move by saying that God "by the Laws which he has given, has made these very Principles more clear and evident" and has forbidden us to give way to our passions and act contrary to the rules of reason (Prol. 13). Thus, God, the maker of the law of nature, helps to make the internal principles more evident.

Having explained the origins of natural right and justice in this manner, Grotius returns to the initial task of rebutting the sceptical view of Carneades and Horace. He rejects Horace's claim that "Interest [*utilitas*]" is the mother of what is just and equitable (Prol. 17) and suggests that the connection between justice and utility has to be explained differently. God attached interest or profit to the law of nature as an additional motivation for human beings who are to observe it: "Profit [*utilitas*] was annexed" "to the Law of Nature" because God was pleased to see that weak human beings, who lack so many things for commodious life, might eagerly seek to form a society (Prol. 17). Thus Grotius eventually brings in the notion of interest as a reward for obeying the law of nature.

1.3 Grotius's Inter-Subjective Principle: Pactum

How can human sociability and reason, with the additional help of God, determine the specific contents of natural right or natural justice? Although Grotius does not provide a clear-cut answer, he does offer a theory of justice which explains what sort of rights each human being has. This theory suggests that there are certain connections between the internal principles and the contents of natural justice. We should turn now to this substantive theory of justice.

There are two prominent features of Grotius's concept of justice which we have touched upon. First, "justice" in the proper or strict sense is defined as *alieni abstinentia*, that is, abstaining from that which is another's. Justice in this sense also requires that one ought to restore what is another's. Grotius gives this concept the special name "expletive justice." What is generally called "distributive justice" is excluded from the proper sphere of justice.[15] Second, "what is another's" (*alienum*) refers to any of the objects in which another has perfect rights. Similarly, "what is his own" (*suum*) stands for any of the objects in which he has perfect rights.

Grotius holds that every man by nature has a sphere of *suum*. It consists of his "Life," "his Body, his Limbs, his Reputation, his Honour, and his Actions" (DJBP 2.17.2.1; also, 1.2.1.[5]). In other words, every human being by nature has perfect rights in them. Grotius does not explain why he or she has those rights. But he clearly thinks that the internal principles of sociability and reason are the source of those rights. They are the source in the sense that they naturally *move* human beings to form a peaceful society where they respect, and ought to respect, one another's perfect rights.

Grotius also holds that this natural sphere of *suum* or perfect rights can be legitimately expanded by a compact (*pactum*) to incorporate external possessions. His account of how it can be expanded is offered as a narrative of the origin of property.[16] In the beginning God gave all things to mankind in common, he says, but this original community gradually disappeared because people ceased to live "in their primitive Simplicity" or "in perfect Friendship" (DJBP 2.2.2.1). They were no longer satisfied with dwelling in caves, but they "wanted to live in a more commodious and more agreeable Manner; to which End Labour and Industry was necessary" (2.2.2.4). Moreover, people were incapable of using common things because of "the distance of Places where each was settled" or "the Defect of Equity and Love" in them (2.2.2.4). What is crucial is that property arose "not from a mere internal Act of the Mind," "but it resulted from a certain Compact and Agreement, either expressly, as by a Division; or else tacitly, as by Seizure" (2.2.2.5). Why is it that people needed a certain *pactum* (translated here as "Compact and Agreement")? They needed it because "one could not possibly guess what others designed to appropriate to themselves, that he might abstain from it" (2.2.2.5), and because "several might have had a Mind to the same Thing" (2.2.2.5). Thus, Grotius takes *pactum* to be an intersubjective device which makes manifest what individuals have internally in their minds, and by which they can mutually adjust their claims.

Grotius took the notion of *pactum* from Roman law. As he remarks that "*Property* . . . as now in use, was introduced by Man's Will [*voluntas*]" (DJBP 1.1.10.4), he takes *pactum* to be an agreement of the wills of two or more persons. We suppose that the rights arising from it would count as conventional or adventitious ones, but Grotius makes them quasi-natural by joining them to their natural basis. Once some external possessions have been "admitted" as part of *suum*, the law of nature informs us that it is "a wicked Thing to take away from Any Man, against his Will, what is properly his own" (1.1.10.4). Thus, the obligation to respect property rights seems to arise from the law of nature, hence, from the internal principles and God, whereas the *pactum* itself seems to determine which particular goods are to be assigned to particular persons. This may be seen as an obscure account, or an example of the "relative" or "suppositious" natural law advocated by medieval writers and Suárez.[17] But for our purpose it is important to note that as far as the obligatory force is concerned, Grotius treats property rights as if they had originally been part of the natural sphere of *suum*. It is the natural sociability of human beings, assisted by their reason and God, that ultimately underpins the obligation of justice, or of giving to each what is his own.

1.4 Addendum: *Grotius versus Hobbes*

I should add that my interpretation of Grotius differs from some recent influential interpretations, such as Richard Tuck's and J. B. Schneewind's. Both of them rightly focus on Grotius's concern with war and controversy, but in so doing they link Grotius and Hobbes too closely.[18] Tuck, in particular, has argued that in Grotius's earlier work *De jure praedae* (which remained unpublished until 1868), the universal impulse to self-preservation rather than natural sociability was the real basis of natural law and natural rights, while also claiming that this egocentric view was still retained by Grotius in the first edition (1625) of *De jure belli ac pacis*.[19] Tuck has added that Grotius eliminated the egocentric view from the second edition (1631) in order to accommodate it to the view of "the Calvinists who surrounded the Prince of Orange."[20] This is how Tuck has assimilated Grotius's view to Hobbes's. We do not need to go into details of Tuck's interpretation. Leaving aside *De jure praedae*, I would like to point out that Grotius did refer to *appetitus societatis* in the Prolegomena to *all* editions of *De jure belli ac pacis*, including the first edition. Grotius also added a reference to *oikeiosis* in the Prolegomena to the second and later editions, so that his readers could identify his view of *appetitus societatis* with the Stoic notion of *oikeiosis* as expounded by Cicero.[21] Finally, Grotius referred to Cicero's distinction in *De finibus* between two sorts of natural principles, and stated that whereas there are some primary objects of nature (*prima naturae*) which "go before," such as the instinct of self-preservation, there are also other primary objects of nature which "come after, but ought to be the Rule of our Actions, preferably to the former" (DJBP 1.2.1.1). In short, "the Conformity of Things with Reason" ought to be "preferred to those Things, which mere natural Desire at first prompts us to" (1.2.1.2). So it is clear that in all editions of *De jure belli ac pacis*, Grotius treated natural sociability and reason, rather than self-preservation, as the basis of natural law and natural rights, no matter what he had stated in *De jure praedae*.[22]

On the other hand, Hobbes claimed that human beings by nature have little sociability. "We doe not," said Hobbes, "by nature seek Society for its own sake, but that we may receive some Honour or Profit from it; these we desire Primarily, that Secondarily" (*De cive* I.2). While Grotius took seriously the Stoic notion of natural sociability, Hobbes remained an Epicurean about justice in holding that there is no natural distinction between justice and injustice (*Leviathan* 13.13) and that justice arises solely from a preceding "covenant" of men designed for some benefits (*Leviathan* 15.2; 14.11; 14.79). In these respects, Grotius significantly differs from Hobbes. To contrast Grotius's Stoicism with Hobbes's Epicureanism, I should like to quote from the letter Grotius wrote to his brother Willem, dated 11 April 1643. In that letter Grotius stated that he had read Hobbes's *De cive*, and expressed his disagreement with Hobbes: "I cannot approve of the foundations" of Hobbes's views, above all, the foundational claim that there is "by nature a war among all human beings."[23] Unlike Hobbes, Grotius trusted natural sociability and reason, and relied on God's support while thinking about justice. Thus, we should treat Grotius as distinct from Hobbes, at least as far as the origin of justice is concerned. As we shall see in the following, it is Hume who critically responded to Grotius's Stoicism by improving upon Hobbes's Epicurean insights.

2. Hume's Defence of the Primacy of Utility

In his letter to Hutcheson dated 17 September 1739, Hume quoted the following Latin phrase from Horace: "Atque ipsa utilitas justi prope mater et aequi." The phrase can be rendered as follows: "And interest itself, you may say, the mother of justice[24] and equity." Hume clearly had in mind Grotius's criticism of Horace for Grotius had quoted the same phrase (Grotius 1667, Prol. 16). In this letter Hume endorsed Horace's view of interest as the mother of justice, suggesting that Hutcheson should reconsider his own position. Hume also claimed that "Grotius and Puffendorf [*sic*], to be consistent, must assert the same" (Hume 1969, 1: 33). As we have seen, Grotius appealed to the principle of natural sociability, while he also resorted to the idea of interest as an additional reinforcement of justice. Hume's point is, first, that this sort of inconsistency, or a combination of heavy reliance on natural sociability and a partial appeal to interest, should be removed and, second, that *interest alone* should be treated as the real mother who gave birth to justice. As Hume stressed in the *Treatise*, "the rules of justice are establish'd merely by interest" (*Treatise* 3.2.2.22).

In discussing Hume's response to Grotius, I shall take a close look at his account of the origin of justice in *Treatise* 3.2.2 and present some of the important views contained in it as exhibiting his response to Grotius.[25] I would also like to refer to *An Enquiry Concerning the Principles of Morals* whenever I find it appropriate to do so. Hume indicates the connection between justice and interest by using the word *utility*, though he uses it much more frequently in the *Enquiry* than in the *Treatise*. Utility is a tendency to bring a certain advantage or pleasure. Hume constantly keeps this sense of "utility" in mind in both works. The whole talk of the "artificial" virtue of justice that we find in the *Treatise* can be translated into the talk of our reflections on public utility, or its tendency to bring advantages to the public.

Hume's account of the origin of justice is an account of the origin of property. This is because Hume's concept of justice is restricted to the protection of property. Hume's narrow concept actually derives from Grotius. As I have argued elsewhere,[26] Hume borrowed the idea of justice as *alieni abstinentia* from Grotius, but he narrowed its scope further by limiting *what is another's* (*alienum*) to *another's possessions*. This is why Hume frequently speaks of justice as a matter of "abstaining" from "the possessions of others," or from "the property of others" if property has been created.[27] Hume limited the scope of justice by relying on a peculiar argument, one based on what he calls the "three different species of goods" (*Treatise* 3.2.2.7). This argument purports to show that our external possessions alone are the proper object of justice. We are "perfectly secure," says Hume, in the enjoyment of the first species of goods, or "the internal satisfaction of our mind." The second species or "the external advantages of our body" "may be ravish'd from us," but they "can be of no advantage to him who deprives us of them." The third species, that is our external possessions, are the only goods that need to be protected by justice because they "are both expos'd to the violence of others, and may be transferr'd without suffering any loss or alteration" (3.2.2.7). This is actually a bad argument.[28] But it is by relying on this argument that Hume transformed Grotius's notion of justice as *alieni abstinentia* into the narrower notion of justice, and excluded Grotius's personal goods (i.e. life, body, limbs, actions, reputation, and honour) from the proper sphere of justice.

2.1 Hume's Internal Principle: The Sense of Self-Interest

Given the narrow scope of justice, how does Hume explain its origin? Hume rejects Grotius's principle of natural sociability as incapable of explaining it. Hume sees that there is a "natural appetite betwixt the sexes, which unites them together, and preserves their union," which he calls "the first and original principle of human society" (*Treatise* 3.2.2.4). He also notes that a "new tye" may take place in "their concern for their common offspring," which may become "a principle of union betwixt the parents and offspring" (3.2.2.4). Nevertheless, Hume holds that nature ceases to offer a principle of union, once we go beyond a narrow circle of relatives and friends. Justice concerns one's extensive social relationships with strangers, and human beings do not have an inborn capacity to form the extensive relationships in which they can respect the strangers' possessions.

In Hume's view, people naturally tend to take the possessions of strangers. Given people's natural temper (i.e. "selfishness" and "limited generosity") and given the scarcity of external goods, "in comparison of the wants and desires of men" (*Treatise* 3.2.2.16), people are naturally motivated to take the possessions of strangers. The internal principle that operates here is *the sense of self-interest*. There are various natural passions like pity, love, envy, and revenge, but self-interest is the most pernicious passion that operates universally. This is not the narrowest self-interest of a Hobbesian kind.[29] Rather, it is the "avidity" of "acquiring goods and possessions for ourselves *and our nearest friends*" (3.2.2.12; emphasis added).

Hume offers a brilliant account of how the same sense of self-interest leads each human being to join a mutual convention for the stability of possessions and serves to establish justice and property. First, children receive training in each family, and they become "sensible of the advantages, which they may reap from society" (*Treatise* 3.2.2.4). Second, two strangers join a convention. When I meet a stranger, says Hume, I feel that "it will be for my interest to leave another in the possession of his goods, *provided* he will act in the same manner with regard to me." And "[h]e is sensible of a like interest in the regulation of his conduct. When this common sense of interest is mutually express'd, and is known to both, it produces a suitable resolution and behaviour" (3.2.2.10). Thus, a mutually advantageous scheme of actions arises easily and smoothly, at least initially. Third, my performance becomes an example, which is imitated by others (3.2.2.22). Fourth, the rule for the stability of possessions may arise slowly and gradually through "our repeated experience of the inconveniences of transgressing it" (3.2.2.10). But this experience only assures us "that the sense of interest has become common to all our fellows, and gives us a confidence of the future regularity of their conduct" (3.2.2.10). Fifth, and finally, when we have steadily regulated our conduct by the rule and acquired a stability in our possessions, "there immediately arise the ideas of justice and injustice; as also those of *property*, *right*, and *obligation*" (3.2.2.11).

By joining the convention, we learn to restrain the passion of self-interest. But Hume refuses to assign an important role to any rational faculty.[30] Reason, as he stresses in Book Two of the *Treatise*, "is, and ought only to be the slave of the passions" (*Treatise* 2.3.3.4). Hume clearly rejects Grotius's Stoic view of the role of man's reason. What is most required for the establishment of justice and property is not a rational capacity but a sense, that is, a capacity to feel the advantages of a social life. As Hume puts it, "in order to form society, 'tis requisite not only that it be advantageous, but also that men be sensible of its advantages; and 'tis impossible . . . that by study and reflection alone, they shou'd ever be able to attain this knowledge" (3.2.2.4).[31] Thus Hume assigns only a minimal role to reason, and claims that the passion of self-interest "restrains itself" "*upon the least reflection*" (3.2.2.13; emphasis added). Hume's internal principle can operate with only a little help from reason.

2.2 Hume's Inter-Subjective Principle: Convention

Though self-interest is the original motive to join a convention for the stability of possessions, it is the convention that can properly be said to be the common origin of justice and property. "The origin of justice explains that of property. The same artifice gives rise to both" (*Treatise* 3.2.2.11). This artifice is called "convention," and we now come to consider the nature and role of Hume's convention.

In one of the footnotes to Appendix 3 of the *Enquiry*, Hume states that his own theory of the origin of property, hence of justice, is "in the main, the same with that hinted at and adopted by GROTIUS" (*Enquiry*, Appendix 3.8, note 63). The point of this statement has to be explained, above all, by reference to the shared idea that property arises from a mutual agreement of men, though Hume also agrees with Grotius that justice has something to do with the defects in people's sentiments, and their preference for a refined mode of life. Hume clearly rejects the Lockean view that property arises from a unilateral act like labour and affirms that property arises from a mutual agreement of men. Grotius's *pactum*, as we have seen, is an inter-subjective agreement whereby individuals mutually express what they have in mind and adjust their claims. It is this idea of a mutual agreement that Hume seizes upon in reading Grotius's work.[32]

Although Hume's reference to Grotius exhibits the existence of a significant link between them, we should not be misled by it. Hume actually transforms Grotius's idea of *pactum*.[33] He transforms it into the idea of a mutually *advantageous* scheme. Speaking of justice, language, and money as arising from human conventions in the *Enquiry*, Hume states that "[w]hatever is advantageous to two or more persons, if all perform their part; but what loses all advantage, if only one perform, can arise from no other principle [i.e. no source other than that of convention]" (*Enquiry*, Appendix 3.8). Here Hume explains the general nature of a convention by the vocabulary of interest and advantage. That vocabulary is absent from Grotius's account of the origin of property. According to Grotius, *pactum* is an explicit or tacit agreement of the *wills* of two or more persons. The wills involved here are not specifically linked to their senses of interest, whereas for Hume, convention is a tacit agreement or convergence of *the senses of interest* of two or more persons. By making the sense of interest the central component of his concept of convention, Hume has naturalized Grotius's idea of *pactum*. Indeed, he has completed the naturalistic move that Hobbes previously made in his attempt to explain the nature of a covenant.[34]

In giving a description of the convention for the stability of possessions, Hume appears to vacillate between a sense of "common interest" and a "common sense" of interest (*Treatise* 3.2.2.10). Nevertheless, it is sufficiently clear that the "common sense" of interest arises as the convergence of one person's sense of self-interest and another's, while the "common" or "public" interest arises as the convergence gets entrenched over time. Ulpian remarked long ago that convention (*conuentio*) is the coming together of "different motions of the mind" so as to form one opinion.[35] By adapting it, we may say that Hume's convention is the coming together of senses of self-interest, so as to form one general scheme of actions that tends to public interest.[36] Within Hume's convention, each one's sense of self-interest is bound up with his or her sense of public interest. Thus he stresses that self-interest is inextricably linked to public interest: "the whole plan or scheme" of justice and property is "highly conducive" "both to the support of society, and the well-being of every individual," or "infinitely advantageous to the whole, and to every part" (*Treatise* 3.2.2.22).

The role which Hume's convention plays in his account of justice can also be contrasted with that of Grotius's *pactum* within his account. While Grotius made property rights quasi-natural, Hume thinks of property as purely conventional. He holds that there cannot be any right whatsoever prior to, or independent of, a convention. Hume even suggests that Grotius is one of those who are "guilty of a very gross fallacy" because he tried to use the word "right" or "property" prior to the convention (*Treatise* 3.2.2.11). Finally, we should note that Hume makes no reference to God in his account of justice and property. While Grotius saw God as annexing profit to the law of nature in order to help weak human beings, Hume holds that justice and property arise from the secular convention of human beings without any help from God.

2.3 Hume's Principle of Moral Approbation: Public Utility

Public interest or public utility figures prominently in Hume's account of the moral approbation of the virtue of justice, which he adds to his account of the origin of justice. Grotius offers no such account. Hume thinks that if anyone respects our property or someone else's, it produces the sense of satisfaction or pleasure in us, which sense is conveyed through sympathy to other members of society. This is a sympathy with the pleasure that arises from the virtue of justice, and this virtue tends to public interest. Moral approbation for Hume is a matter of having an agreeable feeling, so he claims that "a *sympathy* with *public* interest" is "the source of the *moral* approbation, which attends that virtue" (*Treatise* 3.2.2.24).

In the *Enquiry*, Hume is more straightforward and claims that "reflections on the beneficial consequences of this virtue [of justice] are the *sole* foundation of its merit" (*Enquiry* 3.1). "The beneficial consequences" can be rephrased as "public interest" or "public utility," so Hume is claiming that we esteem or morally approve of the virtue of justice because of its tendency to public interest. In giving this moral approbation, we are appealing to the principle of moral approbation which Hume calls "the principle of public utility" (*Enquiry* 3.47).

To understand the view Hume expresses here, we need to see it as part of his general account of moral approbation. Moral approbation, as he understands it, is a certain agreeable feeling which arises upon contemplating certain characters or qualities of mind in a cool and disinterested manner. The qualities of mind which produce this agreeable feeling are called virtues. The virtues are those qualities of mind which are *agreeable* or *useful* to the person who possesses them, or to others. Hume takes a hedonist or Epicurean position that pleasure is the only thing that is good in itself.[37] So every virtue has its merit either because it is pleasurable in itself, or because it is useful; that is, it has a tendency to bring pleasure. Given this general framework, Hume's point about the moral approbation of justice is clear: the virtue of justice is esteemed and morally approved of because it tends to bring pleasure to the public.

3. Concluding Remarks

We have seen that there is a conflict of principles relating to the question about the origin of justice and about how to relate justice to utility. Indeed, Grotius and Hume hardly share any common fundamental principles. Grotius's internal principles of sociability and reason are in conflict with Hume's internal principle, that is, the sense of self-interest. According to Grotius, God helps weak human beings by making their internal principles more evident and by annexing private interest to the law of nature. Hume, on the other hand, makes no reference to God. Instead, he takes a hedonist position and resorts to the principle of public utility in explaining the moral approbation of justice. We might say that Grotius and Hume do share the inter-subjective principle that property arises from a mutual agreement of human beings. However, Grotius's *pactum* is an agreement of wills which is influenced by the natural sociability and reason common to all human beings, whereas Hume's convention is only a convergence of people's senses of interest which artificially creates and sustains property for public interest. Thus, Grotius's principles are generally in conflict with Hume's.

Given this serious disagreement about what principles to appeal to, we may be puzzled about what to do. The existence of such a disagreement may be treated as part of the fact that in early modern thought, there was a general disagreement about which practical principles one ought to use. It may even seem necessary for us to accept the conflict of principles as part of that fact. However, I believe that we can get some positive results if we examine the debate further and add more reflections.

In this final section, I try to reflect on the debate to clarify two issues. First, I say more about the nature of principles. I want to explain my use of the word *principle*, relate it to Grotius's and Hume's usages, and go on to clarify the nature of the principles involved in the Grotius–Hume debate. This consideration will help us to connect what I have said about the particular set of principles with the more general question about what we call "principles" in a variety of situations. It may also serve to explain, at least in part, why the particular conflict we have seen seems difficult to settle.

Second, I ask what conclusion can be drawn from the Grotius–Hume debate. Since Grotius's and Hume's principles serve to support two distinct conclusions, the crucial question to ask is whether we should really consider justice to be natural and independent of utility, or rather judge it to arise from a convention and utility. To answer it, I take another look at what justice is and outline a new account of justice and utility which seeks to reconcile their conflicting accounts.

3.1 On the Nature of the Principles Involved

I begin by saying something about my use of the word *principle*. I have slightly diverged from Grotius's usage. Although he spoke of the internal principles of man, he did not refer to *pactum* as a principle in the way I did. But I believe that the description of *pactum* as an inter-subjective principle is a justifiable one. For Grotius did think of it as an inter-subjective "source" of each one's property, and "principle" in Grotius's usage primarily meant a source.

What was Hume's usage like? Just as Grotius referred to man's internal principles, Hume did speak of "the principles of human nature" (*Treatise*, Introduction para. 4; 1.3.9.12; 1.3.13.12; 1.4.2.2; 2.2.10.4; etc.). By that he meant the innate powers, dispositions, and motives of human beings, or the natural sources of human thought and action. In conformity with this usage,

Hume called a "natural appetite betwixt the sexes" "the first and original principle of human society" (3.2.2.4). One might object that Hume did not use the word *principle* for the passion of self-interest in the way I did. In one of his essays, however, Hume discussed the principles of government, and referred to "other principles" "such as *self-interest, fear, affection.*"[38] So he clearly regarded the passion of self-interest as a principle. Or again in the *Enquiry*, Hume described convention as a "principle," which we can take to mean the source or origin from which a mutually advantageous scheme arises. He cited "the principle of public utility" as one of the principles of moral approbation, and in this instance he seems to have diverged from his primary usage and meant a general law or rule of public utility rather than a source or origin of it. So we may affirm that my use of the word *principle* follows Hume's primary usage, just as it is justifiably close to Grotius's.

With that preface done, there are three points we should note about the nature of the principles involved. First, neither Grotius nor Hume appealed to what are often called "first principles." We often evoke the idea of first principles in our philosophical discussion of the status of principles, but this is not the type of principles that Grotius and Hume had in mind. "First principles," as Thomas Reid reminds us, are self-evident principles, or intuitively certain principles.[39] They are accepted intuitively, without any argument. They command immediate assent, without requiring any demonstrative reasoning or reasoning concerning matters of fact. Neither Grotius nor Hume thought it necessary to ground his theory of the origin of justice on that type of first principles. To those who classify Grotius as a rationalist, it may be surprising that his theory of justice does not begin with any self-evident principle. This may appear contrary to the requirement or expectation, associated with Pufendorf, Locke, Leibniz, and others, that a theory of morality or natural law should take the form of a deductive system based on self-evident principles.[40] But even if we allow ourselves to treat Grotius as a rationalist in some sense,[41] he clearly thinks of natural sociability and reason as the basic principles of human nature which we can confirm by experience. Indeed, in rejecting Horace's claim, Grotius stresses that "the Mother of Natural Law is human Nature" (Prol. 12). Human nature remains very important for Grotius though he disagrees with Hume about what principles of human nature are really basic.

Second, the internal principles of human beings to which Grotius and Hume appealed are *not* the principles in the propositional sense of general rules or laws. Rather, they are the principles in the ontological sense of the powers, dispositions, or motives of human actions. To put it in a different way, they are *the principles of action*, that is, those things which incite us to act,[42] or the springs or sources of action. The principles of action may be inferred from our observation of some of the external actions of human beings. As we have seen, the ontological sense of principles is primary for Grotius and Hume, while the propositional sense remains an exception. But there is a complication here. When we say that those principles or powers are in *human nature*, we suppose that they operate universally beyond any specific, cultural or political borders. When that supposition is made, we can restate the principles in the form of the general laws or propositions about human actions, so they can easily give rise to the "principles" in a more familiar, propositional sense of the word.

Third, the principles of action are distinct from the principles of knowledge or those of morality. Unlike the latter, they are very difficult to determine with precision. They are often the subject of dispute.[43] This makes indeterminate what principles of action are available, or which principle is primary and which one secondary. Besides, there is a difficult problem about how to justify the principles of action. We may want to claim that we can only justify them by appealing to our experience, but such an appeal leaves an important issue unsolved. We may justify Grotius's principles of sociability and reason by focusing on *a particular area* of experience where we find people very sociable, reasonable, and cooperative. Similarly, we may justify Hume's principle of self-interest by looking at *another area* of experience where we find people motivated to seek their own advantage and their friends'. Thus, both of these principles can be confirmed by our experience. In trying to justify the use of any principle of action, we would also have to consider whether we are primarily concerned with the effectiveness of our control over our actions, or with the ideal method of controlling them. For these reasons, the conflict of the principles of action remains difficult to settle.

3.2 On the Crucial Question

Having clarified the nature of the relevant principles, I now proceed to address the second, crucial question about what conclusion ought to be drawn from the Grotius–Hume debate. To establish a proper relationship between justice and utility, we must begin by determining what justice is. As we have seen, Hume changed Grotius's definition of justice by limiting the scope of *what is another's* to *another's possessions* and, consequently, by excluding from justice those personal goods which another has, such as his or her life, body, limbs, actions, reputation, and honour. By changing the definition, Hume actually changed the subject of the debate. Once this change is effected, their debate about "justice" becomes a debate about two different things. So in order to draw a conclusion about the proper relationship of justice to utility, we must first make clear which definition of justice we should adopt.

Which definition should we adopt, Grotius's or Hume's? If we are offered the two choices, we should definitely adopt Grotius's. We should reject Hume's narrow definition because it conceptually excludes from the sphere of justice such important personal goods as life, body, limbs, actions, reputation, and honour. This conceptual exclusion means that the treatment of our persons[44] gets divorced from the sphere of justice, and that it is left entirely to those who may arbitrarily exercise their power. In other words, no matter how our persons are treated (e.g., tortured, raped, or murdered), we cannot possibly link it to a claim about justice. This is the consequence we are bound to face if we stay with Hume's definition.[45] Of course, we may hope that within Hume's system of morals, the sense of humanity or benevolence will function to protect our persons. But that would not allow us to have a stable system of justice, one which equally protects the persons of all human beings by general rules. To achieve the equal protection of our persons, we need to expand Hume's definition so as to incorporate our persons, or adopt something like Grotius's broader definition.

What is curious about Hume's narrow definition of justice is that as long as we stick to it, we can plausibly argue that justice arises from a convention, or a convergence of our senses of interest. For our external possessions are, as a matter of fact, closely connected with our senses of interest. In particular, if we think of a commercial society, we find very close connections between external possessions and commercial value, or between their transfer and economic profit. The language of interest and advantage seems most suited to signal the existence of those connections. Once Hume's narrow definition is accepted, his strategy to link justice to people's senses of interest appears to work.

No matter how workable it may seem, however, Hume's definition does not take into account the treatment of our persons. So we must reject it. Once we adopt Grotius's definition or something like it, an appeal to interest or utility begins to look like a very inappropriate method of grounding justice. For on this definition, the objects of justice include our life, body, limbs, actions, reputation, and honour, and *the personal goods of this kind are the very elements that constitute the core of human dignity*. This is not what Grotius explicitly stated. But as I understand it, this is the most significant reason why we should include in the sphere of justice our *non*-external personal goods, or those personal goods which are *not* our external possessions. We feel their weight without considering their tendency to public interest, or independently of their connections with each one's sense of self-interest. We immediately feel injured if we are harmed in our body, actions, or reputation. And we are ready to share the sense of injury that anyone else might have if any of his physical and mental conditions were arbitrarily disturbed or destroyed. In short, we feel injured if any of us is arbitrarily detained, abused, defamed, tortured, maimed, or killed.[46] This consideration would lead us to reject Hume's claim of the sense of self-interest or public utility as *irrelevant* to the core of justice.

We are now in a position to establish a proper relationship between justice and utility. We should acknowledge, in the first place, that at least the *core* of justice, which concerns the protection of those non-external personal goods, is grounded in human nature rather than an interest-based convention. This may look like taking a Grotian line, but I do not think that the principles of natural sociability and reason can adequately explain the origin of that core. Grotius's account has little explanatory power when it comes to the protection of the core elements. It categorically affirms, without explanation, that our sociable and rational nature moves us to respect the perfect rights of another human being, or the rights another has in his or her life, body, limbs, and so on. What we really need is an account that links our vulnerable nature to the protection of those core elements. We need to see ourselves as the creatures exposed to the possibility of being attacked or harmed. As such we have an inborn sense of injury, as well as a natural desire to protect our vulnerable nature, which would get activated if violence were done to us. The sense of physical or mental injury may lead us to resentment and then to awareness of the need for justice or for a system which requires that a complaint be heard, a remedy provided, and a perpetrator punished.

An account which explains the origin of the core of justice along these lines may be combined with Grotius's account of natural sociability and reason. It may also be linked to a consequent utility, or the public interest which eventually emerges. But the public interest which emerges in this manner can be seen as presupposing the rights of individuals which are grounded in our vulnerable nature. On this new account, then, utility is not the ground of the core of justice. Rather, our vulnerable nature or our natural desire to protect that vulnerability will be seen as the ground of that core.

What I have outlined above, if duly supplemented, may explain the origin of the core of justice. The outer layer of justice, which concerns the protection of our external possessions, would require a different treatment. There are two kinds of external possessions which we need to distinguish: those possessions which are essential to our dignity and those which are not. The protection of the former type of possessions may be seen as grounded in our vulnerable nature, since we can treat them on an equal footing with those non-external personal goods which we have discussed. They are central to the protection of human dignity. But in my view, the protection of the latter type of external possessions which exceed each individual's needs to sustain his or her dignity, may be seen as being grounded in an interest-based convention. Of course, these possessions need the protection of justice, so theft and damage must be prohibited. But their status can be treated as depending on an interest-based human convention. Hume's internal principle of the sense of self-interest, as well as his brilliant conventionalist account, can be brought in at this point in order to explain the origin of the protection of those external possessions. The protection of those possessions, we should note, is subject to a change in the convention. It can change according to a change of the interest of the parties to the convention.

This is only an outline of a new account of the origin of justice, which needs further elaboration. But I think I have said enough to indicate that after dropping Hume's narrow definition of justice, there is a way in which we can reconcile Grotius's and Hume's accounts of the origin of justice, without scarifying the essential elements of human dignity.[47]

Notes

1 See Mill 1998: Ch. 5, "On the Connexion between Justice and Utility" and Rawls 1971: sec. 1, para. 1.

2 For the conflict between Stoics and Epicueans, or *honestum* and pleasure, see Cicero 1971. For an attempt to link justice to the interest of the stronger, see the debate between Socrates and Thrasymachus in Plato, *The Republic*, 338c—354b in Cooper 1997. For the contractual or conventional origin of justice, see Glaucon's exposition in *The Republic*, 358e—359b, and Callicles's argument in *Gorgias*, 482e—484c in Cooper 1997.

3 See Essay VIII in Locke 2002; and Pufendorf 1995, vols 1 and 2, 2.3.10.

4 See Buckle 1991: Preface and Ch. 5, Forbes 1975: Chs. 1 and 2, Haakonssen 1981: Chs. 1 and 2, Harris 2010.

5 I agree with James Moore and Frederick Rosen that Hume is an Epicurean about justice. For their views, see Moore 1994 and Rosen 2006: 29–57.

6 In quoting from Grotius I use Grotius 2005. When I quote Latin, I use Grotius 1667. In "A Note on the Text" in Grotius 2005, the editor Richard Tuck states that it is a reprint of "[t]he 1738 English translation of Barbeyrac's French edition," without giving any indication as to which Latin edition might have been the basis for the text. But the Latin text must have been published in 1667 or later. For there are division numbers in the Prolegomena, or "The Preliminary Discourse" in Tuck's edition, and in the main text there are not only book, chapter, and section numbers but also subsection numbers. This system of division and subsection numbers was first introduced in the 1667 Amsterdam edition. For Hume's probable connection with Grotius 1667, see note 32 in this chapter.

7 Carneades is reputed to have argued for justice as well. But this does not concern us here since we are dealing only with how Grotius represented Carneades's position.

8 For Epicurus's own view of justice, see "Principal Doctrines," XVII, XXXI, XXXIII (justice is "a kind of compact not to harm or to be harmed"), XXXIV, XXXV, XXXVI (justice is "a kind of mutual advantage in the dealings of men with one another"), XXXVII, XXXVIII, in Epicurus 1970.

9 See Laertius 2005: VII, 85–86.

10 Grotius does not offer a definition of "perfect" rights, but a brief definition offered here can be inferred from his account of just war in DJBP 1.2 and what he says about the connections between punishment and the violation of perfect rights in DJBP 2.20. See also Barbeyrac's note 21 to DJBP 1.1.4; Pufendorf 1995, vols 1 and 2, 1.7.7 and 3.1.3.

11 Here as well as elsewhere, *reason* and *understanding* are used interchangeably though the two words are not strictly identical in meaning. Grotius uses *Intellectus* (i.e. "understanding") in this context while he also evokes the idea of reason in referring to "a Faculty of knowing and acting, according to some general Principles" (quoted at the beginning of this paragraph).

12 See the entry "principium" in Lewis and Short 2002.

13 See also the entry "principle" (I. 4) in the *OED*, Simpson and Weiner 2009: "An original or native tendency or faculty; a natural or innate disposition; a fundamental quality that constitutes the source of action."

14 In English translation, the famous sentence runs as follows: "And indeed, all we have now said would take place [locum aliquem haberent], though we should even grant, what without the greatest Wickedness cannot be granted, that there is no God, or that he takes no Care of human Affairs" (Prol. 11).

15 Since distributive justice, which Grotius calls "attributive justice," concerns the fitness of things to persons, it should be linked to those virtues which are beneficial to others (DJBP 1.1.8.1).

16 I should like to acknowledge that in explaining Grotius's account of the origin of property, I have used a portion of the material found in Shimokawa 2013: 565.
17 See Tierney 1997: 328.
18 Tuck 2001: 97–102, 135. Schneewind, writing under the influence of Tuck's earlier writings, is inclined to link Hobbes to "the Grotian problematic" very closely though he is aware of Hobbes's disagreement with Grotius on the issue of sociability. See Schneewind 1998: 70–73, 82–87.
19 See Tuck 2001: 85–89, 96–99 and Grotius 2005, I: xix–xxii, xxiv, xxvi–xxvii.
20 See the editor's "Introduction" to Grotius 2005, I: xxv.
21 For Grotius's reference to *oikeiosis* and its connections with Cicero and Stoics, see Straumann 2003/04 and Brooke 2008.
22 To examine the status of self-preservation in *De jure praedae*, we should take into account the fact that a number of medieval and early modern thinkers, including Lombard, Aquinas, Vitoria, and Suárez, stressed the primacy of self-preservation and self-love over the preservation of others and charity. For this, see Tierney 1997: 322–323.
23 See Grotius's letter to his brother, dated 11 April 1643, letter no. 6166, in Nellen and Ridderikhoff 1992. Here are some original Latin sentences from the letter: "Librum de Cive vidi. Placent quae pro regibus dicit. Fundamenta tamen quibus suas sententias superstruit, probare non possum. Putat inter homines omnes a nature esse bellum et alia quaedam habet nostris non congruentia." I would like to thank Prof. Masaharu Yanagisawa, Kyushu University, for bringing this letter to my attention.
24 The word *justi* is translated here as "of justice" rather than "of right." This translation can be justified in this context since Hume himself sees Horace as treating interest as the mother of "justice."
25 In *Treatise* 3.2.1, Hume presents a famous argument to show that the virtue of justice is not natural but artificial. The argument is intended to show that there is no motive to justice in human nature, apart from the very sense of the duty to perform an act of justice. But since this argument is not specifically directed against the claim of natural sociability or reason, I shall leave it aside and treat Hume's positive account of the origin of justice as containing his critical response to Grotius.
26 See Shimokawa 2005 and Shimokawa 2011. These two papers are written in Japanese. I read an English version of the first paper at a plenary session of the 31st Hume Society conference (Keio University, 2004).
27 Hume speaks of an act of justice as that of "restoring a loan, and abstaining from the property of others" (*Treatise* 3.2.1.9). He identifies a convention for the stability of possessions with the one "concerning abstinence from the possessions of others" (3.2.2.11). The passion of self-interest changes its direction and makes each man "abstain from the possessions of others" (3.2.2.13). Speaking of the actions of "abstaining from that object [called property]" and "restoring it to the first possessor," Hume stresses that "[t]hese actions are properly what we call *justice*" (3.2.6.3). Also, he speaks of justice as "a general abstinence from the possessions of others" (*Enquiry*, Appendix 3.3).
28 For detailed comments, see my analysis of Hume's argument in Shimokawa 2005: 42–47. The quickest way to understand why Hume's argument is a bad one would be to see that "the internal satisfaction of our mind" can be disturbed easily by a threat or intimidation, while "the external advantages of our body" can also be taken away by others when they detain us, torture us, maim us, or kill us, for what they see as their advantage (e.g., money or power). Like our external possessions, our mind and our body are subject to the violence of others, and contrary to Hume's supposition, they all need the protection of justice.

29 Hume actually claims on the grounds of common experience that Hobbes's ego-istic man is like a monster that appears only in fables or romances (*Treatise* 3.2.2.5).

30 At one point, Hume speaks of "nature" as providing "a remedy in the judgment and understanding for what is irregular and incommodious in the affections" (*Treatise* 3.2.2.9). But he clearly assigns a minimal role to judgement or under-standing as I argue in this paragraph.

31 He adds that it is "our impressions and sentiments" rather than "any relation of ideas" that affect us, and give us "a concern for our own, and the public inter-est" (*Treatise* 3.2.2.20).

32 Stephen Buckle has missed this crucial point. See Buckle 1991: 269, 295–298. Hume quotes Grotius's Latin sentences from 2.2.2.4 and 2.2.2.5 of *De jure belli ac pacis*, but he stops quoting them in the middle of 2.2.2.5 as soon as Grotius's statement of "pactum" appears.

It is also worth noting that in *An Enquiry Concerning the Principles of Morals*, Appendix 3.8, note 63, Hume used subsection numbers (i.e. "art. 4 & 5") in quot-ing from *De jure belli ac pacis*, 2.2.2. Those subsection numbers were not used until the 1667 edition, and it is the very edition listed under item no. 557 in Da-vid Hume's library, which was reconstructed by Norton and Norton. See Norton and Norton 1996: 95. Though the library contains some additional volumes pur-chased by Hume's nephew, the 1667 edition is probably what Hume used himself.

33 The commentators who have tried to portray Hume as a successor to modern natural jurisprudence (see note 4 of this chapter) have failed to see his general strategic opposition to Grotius. In particular, they have not paid attention to the way Hume transformed the nature and role of Grotius's *pactum*.

34 In his Latin works, Hobbes used the term *pactum* for "covenant." Like Grotius's *pactum*, Hobbes's covenant is an agreement of wills: "to covenant, is an act of the will" (*Leviathan* 14.24). But since Hobbes defines the "will" naturalistically as "the last appetite in deliberating" (6.53), his covenant is actually a mutual agreement of different last appetites in deliberation.

35 Mommsen and Watson 1985: 2. 14. 3.

36 This is a paraphrase of Hume's own definition of convention given in the *En-quiry*, Appendix 3.7.

37 Reid rightly points out that Hume is an Epicurean in this respect and says that he rejects the view of the Stoics that there is something called *honestum* or "moral worth," which is commended in and for itself, independent of any utility, profit, or reward. See Reid 2010: 302.

38 See Hume 1985b: 34.

39 See Reid 2002: 452.

40 Pufendorf 1995, vols 1 and 2, 1.2 ("On the Certainty of the Moral Sciences"). For the thesis that morality is capable of demonstration, see Locke 1975: III. xi. 16 (p. 516); IV. xii. 8 (p. 643). Locke affirms that "the measures of right and wrong might be made out" "from self-evident Propositions, by necessary Con-sequences, as incontestable as those in Mathematicks" (Locke 1975: IV. iii. 18, p. 549). Leibniz's critical piece, known as "Opinion on the Principles of Pufen-dorf," suggests in the opening paragraph that Pufendorf should have logically deduced his conclusions from the principles he laid down. See Leibniz 2006: 65.

41 We should be careful not to impose the empiricist/rationalist distinction on Gro-tius, but he did stress the role of reason even independently of sociability, for example "NATURAL RIGHT [*ius naturale*] is the *Rule and Dictate of Right Reason, shewing the Moral Deformity or Moral Necessity there is in any Act, according to its Suitableness or Unsuitableness to a rational Nature* [natura ra-tionali]" (DJBP 1.1.10.1).

42 See the following definition given by Reid: "By *principles of action*, I understand every thing that incites us to act," Reid 2010: 74.
43 To quote from Reid again, many philosophers have propounded "very different and contradictory systems" and disagreed about how to define such key terms as "*appetite, passion, affection, interest, reason*," Reid 2010: 77–78.
44 The expression "our persons" needs to be understood here as distinct from our external possessions. It refers to our living human bodies with their modes, parts, and attributes, such as our life, limbs, actions, reputation, honour (as in Grotius), our chastity (as in Pufendorf), and our health (as in Locke).
45 Baier (2010: 86–99) has shown that Hume "enlarged" his narrow concept of justice in *The History of England*. But from my perspective, this "enlargement" reveals that there is a serious discrepancy between the concept of justice Hume *theoretically formulated* in the *Treatise*, on one hand, and the ordinary concept of justice he *actually used* for descriptive purposes in *The History of England*, on the other.
46 It is worth noting that Hume did not take into account these obvious methods of injuring people (e.g., detaining, torturing, killing, etc.) when he presented the argument from the "three different species of goods." See note 28 of this chapter.
47 I am grateful to Peter Anstey for making valuable comments on an earlier draft of this chapter.

Bibliography

Anstey, P. R., ed. (2013) *The Oxford Handbook of British Philosophy in the Seventeenth Century*, Oxford: Oxford University Press.
Baier, A. (2010) *The Cautious Jealous Virtue: Hume on Justice*, Cambridge, MA: Harvard University Press.
Brooke, C. (2008) "Grotius, Stoicism, and 'oikeiosis'," *Grotiana*, 29: 25–50.
Buckle, S. (1991) *Natural Law and the Theory of Property: Grotius to Hume*, Oxford: Clarendon Press.
Cicero, M. T. (1971) *De finibus bonorum et malorum*, trans. H. Rackham, London: William Heinemann.
Cooper, J. M., ed. (1997) *Plato: Complete Works*, Indianapolis: Hackett.
Epicurus. (1970) *Epicurus: The Extant Remains*, trans. C. Bailey, Hildesheim: Georg Olms.
Forbes, D. (1975) *Hume's Philosophical Politics*, Cambridge: Cambridge University Press.
Grotius, H. (1667) *De jure belli ac pacis libri tres, in quibus jus naturae & gentium, item juris publici praecipua explicantur,* editio nova, Amsterdam.
———. (2005) *The Rights of War and Peace*, 3 vols, ed. R. Tuck, Indianapolis: Liberty Fund.
Haakonssen, K. (1981) *The Science of a Legislator: The Natural Jurisprudence of David Hume and Adam Smith*, Cambridge: Cambridge University Press.
Harris, J. A. (2010) "Hume on the moral obligation to justice," *Hume Studies*, 36: 25–50.
Hobbes, T. (1983) *De Cive: The English Version*, ed. H. Warrender, Oxford: Clarendon Press.
———. (1996) *Leviathan*, ed. J. C. A. Gaskin, Oxford: Oxford University Press.
Hume, D. (1969) *The Letters of David Hume*, 2 vols, ed. J. Y. T. Greig, Oxford: Oxford University Press.

———. (1983–1985) *The History of England*, 6 vols, Indianapolis: Liberty Classics; 2nd edn 1762.

———. (1985a) *Essays Moral, Political, and Literary*, ed. E. F. Miller, Indianapolis: Liberty Fund.

———. (1985b) "Of the first principles of Government" in Hume 1985a: pp. 32–36.

———. (1998) *An Enquiry Concerning the Principles of Morals,* ed. T. L. Beauchamp, Oxford: Clarendon Press.

———. (2007) *A Treatise of Human Nature*, vol. 1, eds. D. F. Norton and M. J. Norton, Oxford: Clarendon Press.

Laertius, D. (2005) *Lives of Eminent Philosophers*, vol. 2, trans. R. D. Hicks, Cambridge, MA: Harvard University Press.

Leibniz, G. W. (2006) *Leibniz: Political Writings*, 2nd edn, ed. P. Riley, Cambridge: Cambridge University Press.

Lewis, C. T. and Short, C., eds. (2002) *A Latin Dictionary*, New York: Oxford University Press.

Locke, J. (1975) *An Essay concerning Human Understanding*, ed. P. H. Nidditch, Oxford: Clarendon Press; 1st edn 1690.

———. (2002) *Essays on the Law of Nature*, ed. W. von Leyden, Oxford: Clarendon Press.

Mill, J. S. (1998) *Utilitarianism*, ed. R. Crisp, Oxford: Oxford University Press.

Mommsen, T. and Watson, A., eds. (1985) *The Digest of Justinian*, 4 vols, Philadelphia: University of Pennsylvania Press.

Moore, J. (1994) "Hume and Hutcheson" in eds. M. A. Stewart and J. P. Wright 1994, pp. 23–57.

Nellen, H. J. M. and Ridderikhoff, C. M., eds. (1992) *Briefwisseling van Hugo Grotius's* Gravenhage: Instituut voor Nederlandse Geschiedenis.

Norton, D. F. and Norton, M. J., eds. (1996) *The David Hume Library*, Edinburgh: Edinburgh Bibliographical Society.

Pufendorf, S. (1995) *De jure naturae et gentium libri octo*, trans. C. H. Oldfather and W. A. Oldfather, 2 vols, Buffalo: William S. Hein.

Rawls, J. (1971) *A Theory of Justice*, Cambridge, MA: Belknap Press.

Reid, T. (2002) *Essays on the Intellectual Powers of Man*, ed. D. R. Brookes, Edinburgh: Edinburgh University Press.

———. (2010) *Essays on the Active Powers of Man*, eds. K. Haakonssen and J. A. Harris, Edinburgh: Edinburgh University Press.

Rosen, F. (2006) *Classical Utilitarianism from Hume to Mill*, London: Routledge.

Schneewind, J. B. (1998) *The Invention of Autonomy: A History of Modern Moral Philosophy*, Cambridge: Cambridge University Press.

Shimokawa, K. (2005) "Kindai Shizenhogaku no Dento to Hume no Seigi Gainen [The modern tradition of natural jurisprudence and Hume's concept of justice]," *Jinbun: Journal of the Research Institute for Humanities at Gakushuin University*, 4: 29–52.

———. (2011) "Grotius no Shizenhogaku kara Hume no Riekihogaku e [From Grotius' natural jurisprudence to Hume's interest jurisprudence]," *Shiso*, 1052: 105–126.

———. (2013) "The origin and development of property: Conventionalism, Unilateralism, and Colonialism" in ed. P. R. Anstey 2013: pp. 563–586.

Simpson, J. A. and Weiner, E. S. C., eds. (2009) *Oxford English Dictionary*, 2nd edn on CD-ROM (ver. 4.0), Oxford: Oxford University Press.

Stewart, M. A. and Wright J. P., eds. (1994) *Hume and Hume's Connexions*, Edinburgh: Edinburgh University Press.

Straumann, B. (2003/2004) "*Appetitus societatis* and *oikeiosis*," *Grotiana*, 24/25: 41–66.

Tierney, B. (1997) *The Idea of Natural Rights: Studies on Natural Rights, Natural Law, and Church Law 1150–1625*, Atlanta: Scholars Press.

Tuck, R. (2001) *The Rights of War and Peace: Political Thought and the International Order from Grotius to Kant*, Oxford: Oxford University Press.

Contributors

Peter R. Anstey is Professor of Philosophy in the School of Philosophical and Historical Inquiry at the University of Sydney.

Hon J. C. Campbell QC FAAL is an Adjunct Professor at the Sydney Law School, University of Sydney. He was formerly a judge of the NSW Supreme Court, Equity Division, and of the NSW Court of Appeal.

James Franklin is Professor of Mathematics at the University of New South Wales.

Daniel Garber is A. Watson Armour III University Professor of Philosophy at Princeton University.

Michael LeBuffe holds the Baier Chair in Early Modern Philosophy at the University of Otago.

William R. Newman is Distinguished Professor and Ruth N. Halls Professor in the Department of History and Philosophy of Science at Indiana University, Bloomington.

Sophie Roux is Professor of History and Philosophy of Science, École normale supérieure, Paris.

Kiyoshi Shimokawa is Professor of Philosophy in the Department of Philosophy at Gakushuin University.

Alberto Vanzo is an AHRC Early-Career Research Fellow of the Department of Philosophy of the University of Warwick.

Kirsten Walsh is Assistant Professor in the Department of Philosophy at the University of Nottingham.

Contributors

Index

Académie des sciences (Caen) 126
Académie royale des sciences 101–2, 126, 230
Accademia del Cimento 149–50, 152, 163n13
accident (under law) 52, 65–6, 71n33
Albertus Magnus 79–82, 92, 94n14
alchemy 36, 77–93 *passim*; *see also* *chrysopoeia*; chymical principles; high medieval alchemy
Aldrovandi, Ulisse 154
Alfred of Sareshel 78
algebra 20, 22, 202
Alsted, Johann Heinrich 105
analogical reasoning 156–7
Anaxagoras 113
Anne of Austria 104
Anstey P. R. 163n10
Anthropic Principle 32
Apian, Peter 101
appetitus societatis 272, 276–8; *see also* Grotius, Hugo; *oikeiosis*
applied mathematics 16, 20, 25–40
Arbuthnot, John 263
Archimedes 4, 16, 21, 40, 231; *Equilibrium of Planes* 29–31
Aristippus, Henricus 78
Aristotelian natural philosophy 103–9, 113, 148–9, 151, 153; *see also* Aristotle; causation; forms; four element theory
Aristotelian theory of knowledge acquisition 3–4, 9, 247, 252, 265
Aristotle 41, 72n59, 78, 82, 106, 115, 119, 150, 226; on comets 100–1, 106, 110, 113, 116–17, 120, 128; on equity 72n59; *Meteorology* 78; *Posterior Analytics* 2, 16–20, 26, 250–1; on principles 4; *Prior*

Analytics 2, 4; quantity 19; *see also* causation; forms; four element theory
arithmetic 21, 26, 35
Arnald of Villanova 81
art 7, 27
Arthur, R. 225
astrology 1, 20, 36, 128–9, 132n41; critique of in France 102–9, 111, 117, 123–6, 129; judiciary 99, 102–9, 117; natural 102, 107
astronomy 19–20, 26, 36, 88, 98–130 *passim*, 154; eccentrics 88; principles of 1, 4, 21, 202, 248, 261–3; as subordinate science 19; systems of 154; *see also* Brahe, Tycho; Copernicus, Nicolaus; Kepler, Johannes; natural philosophy; Whiston, William
Auvry, Claude 120, 137n111
Auzout, Adrien 102, 117–19; on comets 122–4, 126, 128, 130, 138n127, 139n140
Avicenna 78–80, 92, 93n5, 93n9
axioms 4–5, 7–8, 16, 18, 20–2, 38, 152, 161, 175, 177–9, 196–202, 204, 206, 211, 219n13, 227, 236, 250–2; *see also* demonstration; geometry; principles

Bacon, Francis 20, 56–7, 72n50, 74n85, 148, 158, 178, 187; Baconian experimental philosophy 16, 40, 122, 161; induction 148, 162n5; natural history 181–2, 185, 188, 191n28
Baglivi, Giorgio 156
Baier, A. 290n45
Balīnūs 78
Barbari, Giuseppe Antonio 156
Barbeyrac, Jean 265n5